5th International Conference on Advanced Materials, Mechanics and Structural Engineering (5th AMMSE 2018)

IOP Conference Series: Materials Science and Engineering
Volume 473

Seoul, South Korea
19 – 21 October 2018

ISBN: 978-1-5108-8219-5
ISSN: 1757-8981

Printed from e-media with permission by:

Curran Associates, Inc.
57 Morehouse Lane
Red Hook, NY 12571

Some format issues inherent in the e-media version may also appear in this print version.

Printed by Curran Associates, Inc. (2019)

For permission requests, please contact the Institute of Physics
at the address below.

Institute of Physics
Dirac House, Temple Back
Bristol BS1 6BE UK

Phone: 44 1 17 929 7481
Fax: 44 1 17 920 0979

techtracking@iop.org

Additional copies of this publication are available from:

Curran Associates, Inc.
57 Morehouse Lane
Red Hook, NY 12571 USA
Phone: 845-758-0400
Fax: 845-758-2633
Email: curran@proceedings.com
Web: www.proceedings.com

Table of contents

Volume 473

The 2018 5th International Conference on Advanced Materials, Mechanics and Structural Engineering

19–21 October 2018, Seoul, South Korea

Accepted papers received: 14 December 2018
Published online: 26 February 2019

Preface

Papers

IOP Conf. Series: Materials Science and Engineering **473** (2019) 011001 doi:10.1088/1757-899X/473/1/011001

Preface

The 5th AMMSE 2018 is the 2018 5th International Conference on Advanced Materials, Mechanics and Structural Engineering (5th AMMSE 2018) took place in Seoul, South Korea, on October 19-21, 2018.

The conference program covered invited, oral, and poster presentations from scientists working in similar areas to establish platforms for collaborative research projects in this field. This conference will bring together leaders from industry and academia to exchange and share their experiences, present research results, explore collaborations and to spark new ideas, with the aim of developing new projects and exploiting new technology in this field.

The committee of AMMSE expresses their sincere thanks to all authors for their high-quality research papers and careful presentations. All reviewers are also thanked for their careful comments and advices. Thanks are finally given to IOP Publication as well for producing this volume.

The Organizing Committee of AMMSE 2018

Committee Chair

Prof. Mosbeh Kaloop

Incheon Disaster Prevention Research Center

Incheon National University

IOP Conf. Series: Materials Science and Engineering **473** (2019) 011001 doi:10.1088/1757-899X/473/1/011001

Conference Photograph

IOP Conf. Series: Materials Science and Engineering **473** (2019) 011001 doi:10.1088/1757-899X/473/1/011001

Sponsors

Incheon Disaster Prevention Research Center (IDPRC)

Recently, various efforts to prevent and prepare are vitally needed for prevention of Disasters and calamities. So we understand the necessity for technology of disaster and we built up the Incheon Disaster Prevention Research Center (IDPRC) in Incheon National University (INU) in 1997.

Accordingly, Incheon Disaster Prevention Research Center (IDPRC) in Incheon National University has progressed of research on the prevention of disasters and calamities through the various seminars, conference and lectures. These research could be conducted cooperation with Incheon National University (INU) in various fields structure, soil, hydraulic and environment.

Incheon Disaster Prevention Research Center (IDPRC) will try to be a leader in the disaster of industry through the various research activities and global conference.

Incheon National University

Incheon National University established a global campus by integrating two city council funded colleges into one in March 2010 and became a national university operated by the legal entity. With the start of attracting the branch of Lawrence Berkeley Research Center that is a world famous US national policy research center for education and research, Incheon National University will attract St. Petersburg University in Russia, Polymers University in Britain, and Kent University in Belgium and other foreign universities and research centers. These efforts will make INU spring to a world-wide competitive university.

As a local development leading university, Incheon National University established INU VISON 2020 and improved its competition actively through investment in selection and concentration. In compliance with the geographic feature and areal specialty, it will concentrate on the international trade, goods distribution, applied technology convergence, BNT-based life science, urban science and the sinology- based local humanity fields.

IOP Conf. Series: Materials Science and Engineering **473** (2019) 011001 doi:10.1088/1757-899X/473/1/011001

Lists of Committees

Committee name	Organizer name	Organization
Honorary Chair	Mosbeh Kaloop	Incheon National University
Conference General Chairs	I. Mansouri	Birjand University of Technology
Honorary Chair	D.K. Kim	Dong-A University
Technical Committees	J.W. Jung	Chungbuk University, South Korea
	J.W. Hu	Incheon National University
	J.K. Ahn	Incheon National University

Peer review statement

All papers published in this volume of *IOP Conference Series: Materials Science and Engineering* have been peer reviewed through processes administered by the proceedings Editors. Reviews were conducted by expert referees to the professional and scientific standards expected of a proceedings journal published by IOP Publishing.

IOP Conf. Series: Materials Science and Engineering **473** (2019) 012001 doi:10.1088/1757-899X/473/1/012001

Lengthwise Cracks in Functionally Graded Beams Exhibiting Non-Linear Mechanical Behaviour of the Material

V Rizov[1]

[1] Professor, Department of Technical Mechanics, University of Architecture, Civil Engineering and Geodesy, 1046 - Sofia, Bulgaria

Abstract. The present paper deals with lengthwise fracture analysis of functionally graded four-point bending beams. The Ramberg-Osgood equation is used for treating the material non-linearity. A lengthwise crack located arbitrary along the beam width is studied. A solution to the strain energy release rate is derived. It is assumed that the modulus of elasticity varies exponentially along the beam width. The *J*-integral is applied for verification. The approach developed is a useful tool for assessing the influence of material non-linearity on the lengthwise fracture in functionally graded beams.

1. Introduction

Functionally graded materials play an important role in aerospace, nuclear reactors, power plants, electronics, optics and biomedicine. That is why functionally graded materials continue to attract the attention of researchers and scientists throughout the world [1-6]. Since the functionally graded materials permit spatial tailoring of the microstructure and composition, one can get maximum benefits from their inhomogeneity. A substantial condition for widening the applicability of the functionally graded materials is the study of their fracture behaviour.

The present paper is focused on the lengthwise fracture in functionally graded beams since functionally graded materials and structures can be built up layer by layer [7] which is a premise for appearance of lengthwise cracks between layers. It should be noted that lengthwise fracture in various functionally graded beams which exhibit material non-linearity has been investigated in previous works of the author mainly by using power law stress-strain relations [8-10]. A lengthwise fracture analysis of functionally graded beams has been developed assuming that the coefficient of the power-law stress-strain relation is distributed linearly along the beam height [8]. A solution to the strain energy release rate has been derived by analysing the complementary strain energy cumulated in the beam [8]. The solution has been applied for studying lengthwise fracture behaviour of functionally graded double cantilever beam configurations with considering the non-linear mechanical behaviour of the material [8]. A power law stress-strain relation has been used also in a fracture analysis of a functionally graded cantilever beam which contains a lengthwise crack [9]. A parabolic function has been adopted for describing the distribution of the coefficient of the power law stress-strain relation [9]. Fracture analysis of a functionally graded beam configuration loaded in eccentric tension has been developed with taking into account the non-linear mechanical behaviour of the material [10]. The *J*-integral approach has been applied. It has been assumed that the material is functionally graded along the thickness of the beam [10].

In contrast to previous papers, the present paper studies the lengthwise fracture in the functionally graded four-point bending beam by using the Ramberg-Osgood stress-strain relation for describing the

IOP Conf. Series: Materials Science and Engineering **473** (2019) 012001 doi:10.1088/1757-899X/473/1/012001

material non-linearity. The four-point bending beam contains a lengthwise vertical crack. The material is functionally graded along the width of the beam. A solution to the strain energy release rate is derived. It is assumed the modulus of elasticity is distributed exponentially in the width direction.

2. Analysis of the Strain Energy Release Rate

The present paper analyses the lengthwise crack in the four-point bending beam shown in figure 1. The material is functionally graded along the beam width. Besides, it is assumed that the beam exhibits non-linear mechanical behaviour of the material which is modelled by the Ramberg-Osgood stress-strain relation. A notch of depth, b_2, is introduced in the right-hand lateral surface of the beam in order to induce conditions for lengthwise fracture. A lengthwise crack of length, $2a$, is located symmetrically with respect to the mid-span (it should also be mentioned that the crack is located in beam portion, DL, which is loaded in pure bending). The crack is located arbitrary along the beam width (the widths of the left-hand and right-hand crack arms are b_1 and b_2, respectively). The right-hand crack arm is divided in two symmetric segments by the notch. The beam is loaded by two vertical forces, F, applied at the ends of the beam. It is obvious that the two segments of the right-hand crack arm are free of stresses. The beam cross-section is a rectangle of width, b, and height, h. The beam length is $2(l_1 + l_2)$. Only half of the beam, $l_1 + l_2 \leq x_3 \leq 2(l_1 + l_2)$, is considered in the analysis because of the symmetry.

The fracture is analyzed in terms of the strain energy release rate, G, by using the following formula [9]:

$$G = \frac{dU^*}{hda}, \quad (1)$$

where U^* is the complementary strain energy cumulated in half of the beam, da is an elementary increase of the crack length.

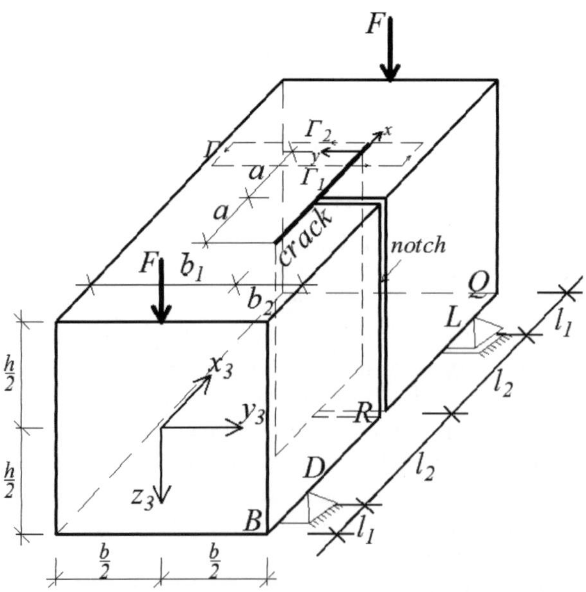

Figure 1. Functionally graded beam configuration.

Since the complementary strain energy in portions, BD and LQ, of the beam (figure 1) does not depend on a, it is enough to calculate the complementary strain energy in the portion, RL, of the beam. Therefore, U^* is written as

$$U^* = a\int_{-\frac{h}{2}}^{\frac{h}{2}}\left(\int_{-\frac{b_1}{2}}^{\frac{b_1}{2}} u_l^* dy_1\right)dz_1 + (l_2 - a)\int_{-h}^{h}\left(\int_{-\frac{b}{2}}^{\frac{b}{2}} u_u^* dy_2\right)dz_2, \quad (2)$$

where u_l^* and u_u^* are, respectively, the complementary strain energy densities in the left-hand crack arm and the un-cracked beam portion, $l_1 + l_2 + a \le x_3 \le l_1 + 2l_2$.

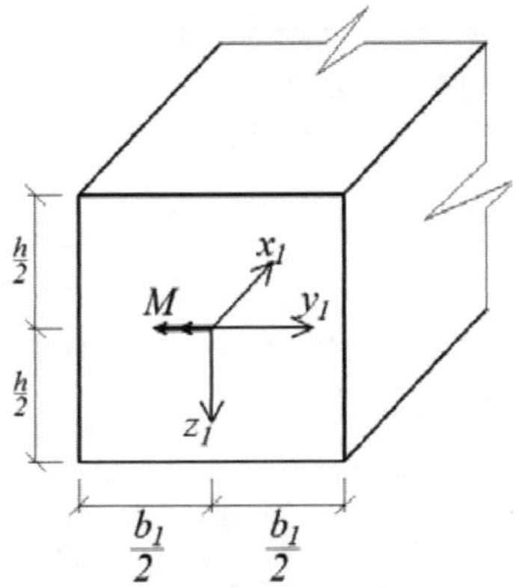

Figure 2. Cross-section of the left-hand crack arm.

In (2), y_1 and z_1 are the centroidal axes of the left-hand crack arm cross-section (figure 2), y_2 and z_2 are the centroidal axes of the cross-section of the un-cracked beam portion. The Ramberg-Osgood stress-strain relation that is applied to model the non-linear mechanical behaviour of the material is written as

$$\varepsilon = \frac{\sigma}{E} + \left(\frac{\sigma}{H}\right)^{\frac{1}{n}}, \quad (3)$$

where ε is the lengthwise strain, σ is the normal stress, E is the modulus of elasticity, H and n are material properties. The modulus of elasticity varies continuously along the beam width according to the following exponential law:

$$E = E_0 e^{\frac{q_0}{b}\left(y_3 + \frac{b}{2}\right)}, \quad (4)$$

where E_0 is the value of the modulus of elasticity at the left-hand lateral surface of the beam, q_0 is a material property which governs the material gradient, the y_3-axis is shown in figure 1. By using (4), the distribution of E in the cross-section of the left-hand crack arm is written as

$$E = E_0 e^{\frac{q_0}{b}\left(y_1 + \frac{b_1}{2}\right)}. \quad (5)$$

The complementary strain energy density in the left hand crack arm is expressed as [11]

$$u_l^* = \frac{\sigma^2}{2E} + \frac{n\sigma^{\frac{1+n}{n}}}{(1+n)H^{\frac{1}{n}}}. \quad (6)$$

By substituting of (5) in (6), one arrives at

$$u_l^* = \frac{\sigma^2}{2E_0 e^{\frac{q_0}{b}\left(y_1+\frac{b_1}{2}\right)}} + \frac{n\sigma^{\frac{1+n}{n}}}{(1+n)H^{\frac{1}{n}}}. \quad (7)$$

The span to height ratio of the beam under consideration is large. Thus, the distribution of longitudinal strains in the cross-section of the left-hand crack arm is analyzed by applying the Bernoulli's hypothesis for plane sections. Concerning the applicability of the Bernoulli's hypothesis, it should also be noted that since the beam portion, DL, is loaded in pure bending, the only non-zero strain is ε. Thus, according to the small strains compatibility equations, ε is distributed linearly in the beam cross-section:

$$\varepsilon = k_l z_1, \quad (8)$$

where the curvature of the left-hand crack arm, k_l, is determined from the following equation for equilibrium of the cross-section of the left-hand crack arm:

$$M = \int_{-\frac{h}{2}}^{\frac{h}{2}} \left(\int_{-\frac{b_1}{2}}^{\frac{b_1}{2}} \sigma z_1 dy_1 \right) dz_1, \quad (9)$$

where $M = Fl_1$ is the bending moment in beam portion, DL.

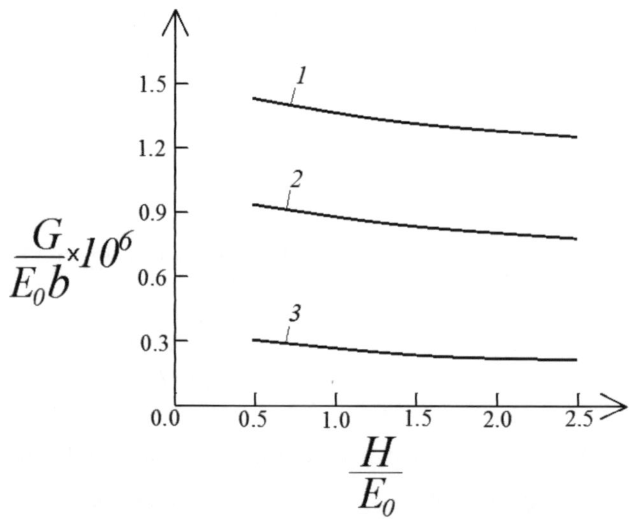

Figure 3. The strain energy release rate presented as a function of H / E_0 ratio (curve 1 – at $b_1 / b = 0.25$, curve 2 – at $b_1 / b = 0.50$ and curve 3 – at $b_1 / b = 0.75$).

Equation (9) can be used also to obtain the curvature, k_u, of the un-cracked beam portion, RL. For this purpose, b_1, σ, y_1 and z_1 have to be replaced, respectively, with b, σ_u, y_2 and z_2 (here σ_u is the normal stress in the un-cracked beam portion).

By substituting of (2) in (1), one obtains:

$$G = 2\left[\frac{1}{h}\int_{-\frac{h}{2}}^{\frac{h}{2}}\left(\int_{-\frac{b_1}{2}}^{\frac{b_1}{2}} u_l^* dy_1 \right) dz_1 - \frac{1}{h}\int_{-h}^{h}\left(\int_{-\frac{b}{2}}^{\frac{b}{2}} u_u^* dy_2 \right) dz_2 \right]. \tag{10}$$

It should be noted that the term in the square brackets in (10) is doubled in view of the symmetry (figure 1). The integration in (10) is performed by using the MatLab computer program.

The solution to the strain energy release rate (10) is verified by applying the *J*-integral approach [12]. The integration is carried-out along the integration contour, Γ, shown by a dashed line in figure 1. Since the lateral surfaces of the beam are free of stresses, the *J*-integral solution is obtained by addition of J_{Γ_1} and J_{Γ_2} (here J_{Γ_1} and J_{Γ_2} are the *J*-integral values, respectively, in segments Γ_1 and Γ_2 of the integration contour). It should be noted that Γ_1 and Γ_2 coincide with the cross-sections of the left-hand crack arm and the un-cracked beam portion, respectively.

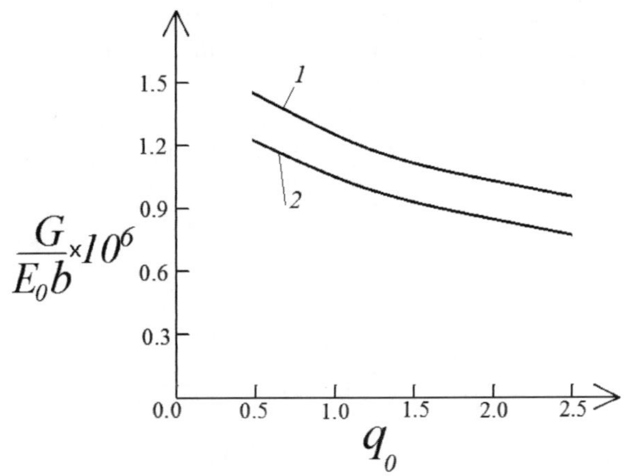

Figure 4. The strain energy release rate presented as a function of q_0 (curve 1 – at non-linear behaviour of the material, curve 2 – at linear-elastic behaviour).

The integration of the *J*-integral is performed by using the MatLab computer program. The *J*-integral values obtained match exactly the strain energy release rates. This fact is a verification of the fracture analysis developed in the present paper.

The influence of the material gradient, the lengthwise crack location along the beam width and the non-linear mechanical behaviour of the material on the fracture in the functionally graded four-point bending beam (figure 1) is investigated. For this purpose, calculations of the strain energy release rate are performed by applying (10). The results obtained are presented in non-dimensional form by using the formula $G_N = G/(E_0 b)$. It is assumed that $b = 0.025$ m, $h = 0.008$ m, $l_1 = 0.100$ m and $F = 50$ N. The crack location along the beam width is characterised by b_1/b ratio. The strain energy release rate in non-dimensional form is presented as a function of H/E_0 ratio in figure 3 at three b_1/b ratios. It can be observed in figure 3 that the strain energy release rate decreases with increasing

b_1 / b ratio. This behaviour is due to the increase of the left-hand crack arm stiffness. The curves in figure 3 indicate also that the strain energy release rate decreases with increasing of H / E_0 ratio. The effect of q_0 on the strain energy release rate is elucidated in figure 4. One can observe in figure 4 that the strain energy release rate decreases with increasing of q_0. The influence of material non-linearity on the fracture behaviour is elucidated too. For this purpose, the strain energy release rate obtained assuming linear-elastic behaviour of the functionally graded material is plotted in figure 4 for comparison with the non-linear solution. At $H \to \infty$ the Ramberg-Osgood stress-strain relation (3) transforms into the Hooke's law. Thus, the linear-elastic solution to the strain energy release rate is derived by substituting of $H \to \infty$ in (7) and (10). It can be observed in figure 4 that the material non-linearity leads to increase of the strain energy release rate.

3. Conclusions

An approach for analysis of lengthwise fracture in functionally graded four-point bending beam is developed. The Ramberg-Osgood equation is used for modelling the non-linear mechanical behaviour of the material. A lengthwise crack located arbitrary along the beam width is considered. It is assumed that the material is functionally graded in the width directions (the modulus of elasticity varies continuously along the beam width according to exponential law). The fracture is studied in terms of the strain energy release rate. The solution derived is verified by the applying the *J*-integral approach. The influence of the material gradient, the lengthwise crack location and the material non-linearity on the fracture behaviour is investigated. The approach developed in the present paper can be used to evaluate the effect of material non-linearity in design of functionally graded beam structures with considering the fracture behaviour.

Acknowledgments

The present study was carried-out with the financial support of the Research and Design Centre (CNIP) of the University of Architecture, Civil Engineering and Geodesy (UACEG), Sofia (Contract BN – 3/2018).

References

[1] Neubrand A and Rödel J 1997 *Zeit f Met* **88** 358-71.
[2] Suresh S and Mortensen A 1998 *Fundamentals of functionally graded materials* (IOM Communications Ltd, London).
[3] Hirai T and Chen L 1999 *Material Science Forum* **308-311** 509-14.
[4] Nemat-Allal M M, Ata M H, Bayoumi M R, Khair-Eldeen W 2011 *Mater. Sci. Appl.* **2** 1708-18.
[5] Uslu Uysal M and Kremzer M 2015 *Acta Phys. Polon. A* **127** 1355-7.
[6] Uslu Uysal M 2016 *Steel Comp. Struct. Int. J.* **21** 849-62.
[7] Bohidar S K, Sharma R and Mishra P R 2014 *Int. J. Res.* **1** 289-301.
[8] Rizov V 2017 *European J. Mech. A/Solids* **65** 301-12.
[9] Rizov V 2017 *Str. Fract. Compl.* **10** 11-22.
[10] Rizov V 2017 *Struct. Eng. Mech.* **64** 487-94.
[11] Rizov V 2017 *Int. J. Struct. Integr.* **8** 516-29.
[12] Broek D 1986 *Element. Eng. Fract. Mech.* (Springer).

Analysis of the Strain Energy Release Rate for a Delamination Crack in a Multilayered Beam with Material Non-Linearity

V Rizov[1]

[1]Professor, Department of Technical Mechanics, University of Architecture, Civil Engineering and Geodesy, 1046 - Sofia, Bulgaria

Abstract. The strain energy release rate for a delamination crack in a multilayered cantilever beam which exhibits material non-linearity is derived by considering the energy balance. The material non-linearity is described by the Ramberg-Osgood stress-strain relation. A cantilever beam made of adhesively bonded lengthwise vertical layers is studied. Each layer has individual widths and material properties. Besides, the delamination is located arbitrary between layers. The strain energy release rate is derived also by differentiating the complementary strain energy with respect to the crack area for verification. It is shown that the solution derived is useful for parametric investigations of delamination in multilayered beams.

1. Introduction

Multilayered materials and structures made of adhesively bonded layers of dissimilar materials have certain advantages such as high strength to weight and stiffness to weight ratios over the homogeneous materials. Therefore, multilayered materials are extensively used in many structural applications where a high performance is required. However, due to the low interlaminar strength, delamination fracture is the predominant failure mode [1-5].

The main goal of the present paper is to develop a strain energy release rate analysis for a delamination crack in a multilayered beam with considering the material non-linearity. A multilayered cantilever beam made by adhesively bonded vertical layers is analyzed. The Ramberg-Osgood stress-strain relation is applied for describing the non-linear mechanical behaviour of the material in contrast to previous works of the author in which delamination in multilayered beams of vertical layers is investigated by using power law stress-strain relations [6, 7].

2. Determination of the strain energy release rate

The delamination crack in the multilayered cantilever beam configuration shown in figure 1 is studied in the present paper. The beam is made of adhesively bonded lengthwise vertical layers of different widths and material properties. The number of layers is arbitrary. It is assumed that in each layer the material exhibits non-linear mechanical behaviour which is treated by the Ramberg-Osgood stress-strain relation. A delamination crack of length, a, is located arbitrary between layers (the widths of the right-hand and left-hand crack arms are b_1 and b_2, respectively. The external loading consists of one moment, M, applied at the free end of right-hand crack arm. Thus, the left hand crack arm is free of stresses. The cross-section of the beam is a rectangle of width, b, and height, h. The beam length is l.

The fracture is investigated in terms of the strain energy release rate, G, by considering the energy balance. By assuming a small increase, δa, of the crack length, the energy balance is written as

$$M\delta\varphi = \frac{\partial U}{\partial a}\delta a + Gh\delta a \tag{1}$$

where $\delta\varphi$ is the increase of the angle of rotation of the free end of the right-hand crack arm, U is the strain energy. From (1), G is expressed as

$$G = \frac{M}{h}\frac{\partial \varphi}{\partial a} - \frac{1}{h}\frac{\partial U}{\partial a} \tag{2}$$

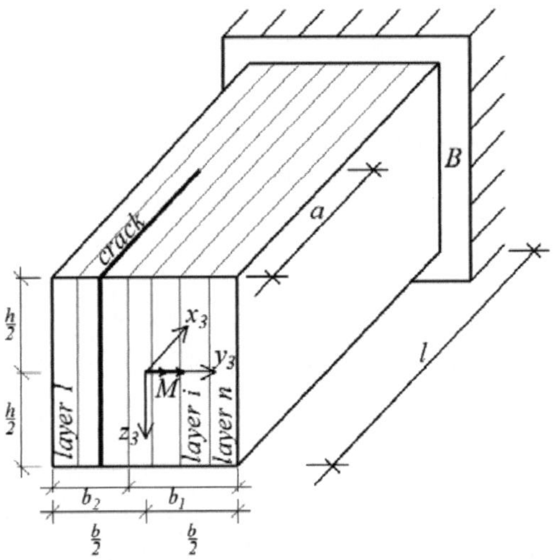

Figure 1. Geometry and loading of the multilayered cantilever beam.

The angle of rotation of the free end of the right-hand crack arm is written as

$$\varphi = a\kappa_1 + (l - a)\kappa_2 \tag{3}$$

where κ_1 and κ_2 are, respectively, the curvatures of the right-hand crack arm and the un-cracked beam portion, $a \leq x_3 \leq l$, (figure 1).

The strain energy in the beam is written as

$$U = a\sum_{i=1}^{i=n_R}\iint_{A_i} u_{0_i} dA + (l - a)\sum_{i=1}^{i=n}\iint_{A_i} u_{0R_i} dA \tag{4}$$

where n_R and n are, respectively, the numbers of layers in the right-hand crack arm and the un-cracked beam portion, u_{0_i} and u_{0R_i} are the strain energy densities in the i-th layer, respectively, of the right-hand crack arm and the un-cracked beam portion, A_i is the area of the cross-section of the i-th layer.

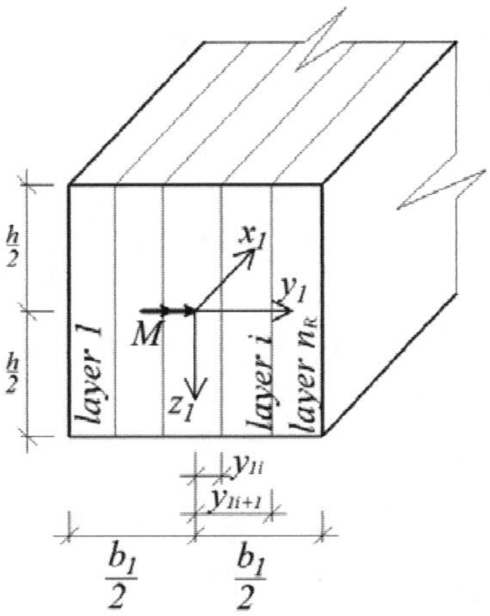

Figure 2. Cross-section of the right-hand crack arm.

The Ramberg-Osgood stress-strain relation for the material in the i-th layer is written as

$$\varepsilon = \frac{\sigma_i}{E_i} + \left(\frac{\sigma_i}{H_i}\right)^{\frac{1}{n_i}} \tag{5}$$

where ε is the lengthwise strain, σ_i is the distribution of the normal stresses in the i-th layer, E_i is the modulus of elasticity in the i-th layer, H_i and n_i are material properties in the same layer.

The distribution of lengthwise strains in the beam cross-section is analyzed by applying the Bernoulli's hypothesis for plane sections since the length to height ratio of the beam under consideration is large. Besides, since the beam is loaded in pure bending, the lengthwise strains are the only non-zero strains. Thus, according to the small strain compatibility equations, the lengthwise strains in the cross-section of the right-hand crack arm are distributed linearly

$$\varepsilon = \kappa_1 z_1 \tag{6}$$

where z_1-axis is shown in figure 2.

The curvature of the right-hand crack arm is determined from the following equation of equilibrium of the cross-section:

$$M = \sum_{i=1}^{i=n_R} \iint_{A_i} \sigma_i z_1 dA \tag{7}$$

Equation (7) can be used also to determine κ_2. For this purpose, n_R, σ_i and z_1 have to replaced, respectively, with n, σ_{R_i} and z_2 where σ_{R_i} is the distribution of the normal stresses in the i-th layer of the un-cracked beam portion, z_2 is the vertical centroidal axis of the cross-section of the un-cracked beam portion.

The strain energy density in the i-th layer of the right-hand crack arm can be obtained by the following formula [8]:

$$u_{0_i} = \frac{\sigma_i^2}{2E_i} + \frac{\sigma_i^{\frac{1+n_i}{n_i}}}{(1+n_i)H_i^{\frac{1}{n_i}}} \tag{8}$$

Formula (8) can be applied also to calculate the strain energy density in the i-th layer of the un-cracked beam portion. For this purpose, σ_i has to be replaced with σ_{R_i}.

By substituting of (3) and (4) in (2), one arrives at

$$G = \frac{M}{h}(\kappa_1 - \kappa_2) - \frac{1}{h}\sum_{i=1}^{i=n_R}\iint_{A_i} u_{0_i} dA + \frac{1}{h}\sum_{i=1}^{i=n}\iint_{A_i} u_{0R_i} dA \qquad (9)$$

The integration in (9) should be carried-out by using the MatLab program.

In order to verify (9), the strain energy release rate is determined also by using the following formula [6]:

$$G = \frac{dU^*}{hda} \qquad (10)$$

where dU^* is the change of the complementary strain energy, da is an elementary increase of the delamination crack length. The complementary strain energy is obtained by (4). For this purpose, u_{0_i} and u_{0R_i} are replaced, respectively, with $u^*_{0_i}$ and $u^*_{0R_i}$ where $u^*_{0_i}$ and $u^*_{0R_i}$ are the complementary strain energy densities in the i-th layer of the right-hand crack arm and the un-cracked beam portion, respectively. The strain energy release rate obtained by (10) is exact match of the result obtained by (9) which is a verification of the delamination fracture analysis developed in the present paper.

A parametric analysis of delamination fracture is performed by applying the solution to strain energy release rate (9).

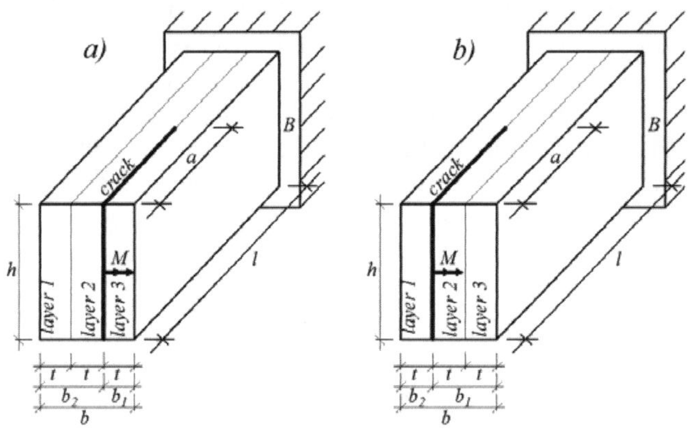

Figure 3. Two three-layered cantilever beam configurations.

The strain energy release rate is presented in non-dimensional form by using the following formula $G_N = G/(E_1 b)$. Two three-layered cantilever beam configurations are considered in order to investigate the influence if the crack location on the strain energy release rate (figure 3).

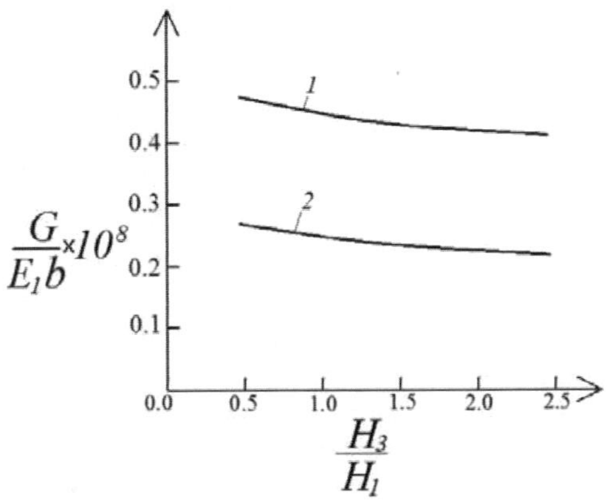

Figure 4. The strain energy release rate plotted against H_3 / H_1 ratio (curve 1 – for the beam configuration shown in figure 3a, curve 2 – for the beam configuration shown in figure 3b).

In the configuration shown in figure 3a, a delamination crack is located between layers 2 and 3. A three-layered cantilever beam configuration with a delamination crack between layers 1 and 2 is also analyzed (figure 3b). It is assumed that the width of each layer is $t = 0.005$ m. Besides, it is assumed that $h = 0.020$ m and $M = 40$ Nm. The strain energy release rate in non-dimensional form is presented as a function of H_3 / H_1 in figure 4 at $E_2 / E_1 = 0.6$, $E_3 / E_1 = 0.5$, $H_1 / E_1 = 0.7$, $H_2 / H_1 = 1.2$ and $n_1 = n_2 = n_3 = 0.75$ for the two three-layered beams shown in figure 3. The curves in figure 4 indicate that the strain energy release rate decreases with increasing of H_3 / H_1 ratio. One can observe also in figure 4 that the strain energy release rate is lower when the delamination is between layers 1 and 2. This is due to the fact that the stiffness of the right-hand crack arm is higher when the delamination is between layers 1 and 2 (figure 3a). The strain energy release rate in non-dimensional form is presented as a function of E_3 / E_1 ratio in figure 5 for the beam configuration shown in figure 3a.

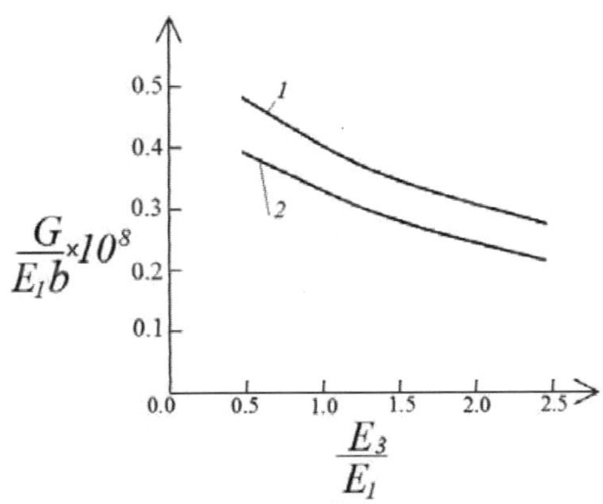

Figure 5. The strain energy release rate plotted against E_3 / E_1 ratio (curve 1 – at non-linear material behaviour, curve 2 – at linear-elastic behaviour).

It can be observed that the strain energy release rate decreases with increasing of E_3 / E_1 ratio (figure 5). This finding is attributed to the increase of the beam stiffness with increase of E_3 / E_1 ratio.

The strain energy release rate obtained assuming linear-elastic behaviour of the material in each layer is also plotted in figure 5 for comparison with the non-linear solution. It should be noted that the linear-elastic solution to the strain energy release rate is derived by substituting of $H_i \to \infty$ in (9) because at $H_i \to \infty$ the non-linear stress-strain relation (5) transforms into the Hooke's law. One can observe that the material non-linearity leads to increase of the strain energy release rate (figure 5). This finding indicates that neglecting the material non-linearity causes underestimation of the strain energy release rate. Therefore, the material non-linearity has to be taken into account in delamination fracture analyses of multilayered structural members and components.

3. Conclusions

Delamination fracture behaviour of a multilayered cantilever beam configuration is analyzed in terms of the strain energy release rate. Non-linear mechanical behaviour of the material is assumed in each layer. The beam is made of adhesively bonded lengthwise vertical layers. The beam is loaded in bending by one external moment applied at the free end of the right-hand crack arm. The Ramberg-Osgood stress-strain relation is applied to describe the non-linear mechanical behaviour of the material. A solution to the strain energy release rate is derived by considering the energy balance. The solution holds for a cantilever beam made of an arbitrary number of lengthwise vertical layers which have individual widths and material properties. Besides, the delamination crack is located arbitrary between layers (thus, the two crack arms have different widths). The strain energy release rate is obtained also by considering the complementary strain energy for verification. The solution derived is very useful for parametric analyses of delamination fracture in multilayered beams exhibiting non-linear mechanical behaviour of the material. It is found that the strain energy release rate decreases with increasing of H_3 / H_1 and E_3 / E_1 ratios. The investigation reveals also that the strain energy release rate decreases with increasing of the width of the right-hand crack arm cross-section. Concerning the influence of the material non-linearity on the delamination behaviour, the analysis shows that the material non-linearity leads to increase of the strain energy release rate. The results obtained in the present paper can by used in fracture mechanics based structural design of multilayered beams.

Acknowledgments

The present study was carried-out with the financial support of the Research and Design Centre (CNIP) of the University of Architecture, Civil Engineering and Geodesy (UACEG), Sofia (Contract BN – 3/2018).

References

[1] Dolgov N A 2005 *Str. Mater.* **37** 422-31.
[2] Dolgov NA 2016 *Str. Mater.* **48** 658-67.
[3] Szekrenyes A 2016 *European J. Mech. A/Sol.* **55** 73-99.
[4] Szekrenyes A 2016 *Appl. Math. Model.* **40** 1344–1371
[5] Rizov V 2017 *Arch. Appl. Mech.* **87** 1037-48.
[6] Rizov V 2017 Multidiscipl. Model. Mater. Struct. **13** 434-47.
[7] Rizov V 2017 *Int. J. Struct. Stabil. Dyn.* doi.org/10.1142/S0219455418500517
[8] Rizov V 2017 *Int. J. Struct. Integr.* **8** 516-29.

Reliability Assessment and Optimization of Double Random Vibration Systems based on PDEM

X J Wang[1, 2], S Z Yang[1, 2], X Wang[1], M J Xiang[1], S Q Long[1], Y Q Dong[1] and X F Zhou[1]

[1] Yunnan University, Kunming 650500, China
[2] Guangxi Key Laboratory of Disaster Prevention and Engineering Safety, Guangxi University, Nanning 530004, China

Abstract. The dynamic responses analysis method of structure with random parameters under stochastic seismic loads is proposed. The generalized probability density evolution equation for compound stochastic process is solved by the finite difference method. Then, for the first passage criterion, the dynamic reliability is analysed by establishing the one-dimensional integral in safe domain and the absorbing boundary condition. A new optimization method is coming up combining improved genetic algorithm and PDEM. Taking the parameters optimization of a three-story floor-shear frame structure with TMD under random seismic loads as an example, the correctness of the proposed method is verified by comparing interlayer displacement response and dynamic reliability.

1. Introduction

The tuned mass damper (TMD) absorbs the vibration energy by using the whiplash effect of small mass spring system attached to the permanent structure. Obviously, the reasonable selection of TMD parameters has a great influence on the effect of shock absorption. In practical engineering, random excitation such as earthquake always controls the performance of structure with uncertain parameters. Therefore, it is necessary to consider the randomness of seismic loads and structural parameters when optimizing the parameters of TMD system.

In addition, the methods of dynamic reliability analysis of engineering structure based on the two failure laws are proposed, namely, the first passage criterion and accumulated damage evaluation criterion [1]. In the analysis of dynamic reliability for the first passage criterion, the methods including leapfrog process theory, diffusion process theory and PDEM are developed. The leapfrog process analysis method [2] is widely applied at present, but it is usually difficult to grasp the probability information of structure response comprehensively. The diffusion process theory method based on Fokker-Planck-Kolmogorov (FPK) equation [3] has not been widely used because the solution of FPK equation is very difficult. In recent years, J Li and J B Chen have developed reliability analysis method of the first passage criterion based on the absorbing boundary condition [4] and PDEM [5], which considering the influence of double random factors. Dynamic reliability solved is successfully applied to the solving general multi-degree-of-freedom structural system.

In this paper, by considering the randomness of external loads and structure parameters, establishes the PDEE of double random nonlinear response characteristics and solves the system reliability in safe domain. The mass optimization design method is developed by establishing optimization model of a multi-degree-of-freedom floor-shear frame structure with TMD based on improving genetic algorithm.

The successful application of the reliability analysis and optimization design of double random vibration system with TMD is realized.

2. The Probability Density Evolution Method

Generally, the motion equation of n-dimensional double random vibration systems can be expressed as

$$M(\Theta)\ddot{X} + C(\Theta)\dot{X} + G(\Theta, X) = \Gamma F(\Theta, t) \tag{1}$$

in which $\ddot{X}$, $\dot{X}$, X is n-dimensional acceleration, velocity and displacement vector, respectively; $M(\Theta)$, $C(\Theta)$ is $(n \times n)$ mass and damping matrix; $G(\Theta, X)$ is the n-dimensional linear or nonlinear restoring force vector; Γ is $(n \times r)$ excitation influence matrix; $F(\Theta, t)$ is r-dimensional excitation vector; $\Theta = (\Theta_1, \Theta_2, \cdots, \Theta_s)$ is the random vector including structural parameters and the external excitation process, where s represents the number of all random variables in a double random system.

The physical quantity of interest is recorded as $Z = (Z_1, Z_2, \cdots Z_m)$. For well-posed dynamical system (the physical solution exists, unique and continuously depends on the basic parameters), there must be

$$Z = H_Z(\Theta, t), \dot{Z} = h_Z(\Theta, t) \tag{2}$$

in which $H_Z = (H_{Z,1}, H_{Z,2}, \cdots, H_{Z,m})^T$, $h_Z = (h_{Z,1}, h_{Z,2}, \cdots, h_{Z,m})$. Obviously, formula (2) is the solution of equation (1) if the quantity in question is displacement.

The joint probability density function $p_{Z\Theta}(z, \theta, t)$ of a conservative stochastic system (Z, Θ). According to the random event description of probability conservation principal, there is

$$\frac{D}{Dt}\int_{\Omega_t \times \Omega_\Theta} p_{Z\Theta}(z, \theta, t) dz d\theta = 0 \tag{3}$$

where $\frac{D}{Dt}$ is the total derivative; $\Omega_t \times \Omega_\Theta$ is the soluble region for any original space at time t. A series of mathematical derivations of equation (3) can obtain

$$\frac{\partial p_{Z\Theta}(z, \theta, t)}{\partial t} + \sum_{i=1}^{m} \dot{Z}(\theta, t) \frac{\partial p_{Z\Theta}(z, \theta, t)}{\partial z_i} = 0 \tag{4}$$

In particular, the equation (4) can be simplified to a first-order partial differential equation when only one physical quantity is considered.

$$\frac{\partial p_{Z\Theta}(z, \theta, t)}{\partial t} + \dot{Z}(\theta, t) \frac{\partial p_{Z\Theta}(z, \theta, t)}{\partial z} = 0 \tag{5}$$

Equation (4) and (5) are the generalized probability density evolution equation (PDEE). Obviously, the dimension of equation (4) depends only on the number of physical quantities examined, and is independent of the degree-of-freedom (DOF) of the original physical system.

J Li and J B Chen established a numerical analysis method of PDEE using finite difference method [6]. In order to fully consider the precision deviation caused by dissipation and dispersion in the calculation process, this paper uses the TVD scheme with flux limiter to solve the numerical [7].

3. Dynamic Reliability

According to the failure laws of structure, considering the response of structural interlayer displacement, the system reliability function of the first passage criterion is generally defined as

$$R(t) = \Pr\{X(\tau) \in \Omega_s, 0 \le \tau \le t\} \tag{6}$$

where $X(\tau)$ is the physical quantity that leads to structural failure; Ω_s is safe domain; t is the structure life or time period that needs to be inspected.

Mandatory absorbing boundary conditions is

$$\breve{p}_{X\Theta}(x,\theta,t)=0, x\notin\Omega_s \tag{7}$$

The residual probability density function is obtained by solving the PDEE

$$\breve{p}_X(x,t)=\int_{\Omega_\Theta}\breve{p}_{X\Theta}(x,\theta,t)d\theta \tag{8}$$

Then the dynamic reliability of structure is solved by equation (9).

$$R(t)=\int_{-\infty}^{\infty}\breve{p}_X(x,t)dx \tag{9}$$

4. Improved Genetic Algorithm

Improved genetic algorithm [8] is a stochastic search method to simulate the biological evolution mechanism. By establishing the information exchange between individuals and the strategy of group search, the problem is converted to the optimal state. Because the improved genetic algorithm does not depend on the gradient information of the objective function, taking the structural displacement is the constraint condition for solving the optimal mass of TMD in the nonlinear analysis of the multi-degree-of-freedom double-random vibration system [9]. Inter layer displacement response at top of structure $f(x)$ is the objective function when using genetic algorithm to solve the optimal mass m_d of TMD, the model can be defined as

$$\min f(x), x=x_1,x_2,\cdots,x_n \tag{10}$$

where $x=x_1,x_2,\cdots,x_n$ is the mass that TMD may achieve; $\min f(x)$ is the optimal solution of objective function. In order to facilitate the application of improved genetic algorithm, the above constrain problem is transformed into unconstrained problem, which can be expressed as

$$\varphi(x)=f(x)(1+PC) \tag{11}$$

where P is the effect of default on the objective function. P=1 represents a default, which means that the individual is more likely to be eliminated. When i individual constraint is satisfied there is $C_j=0$, otherwise $C_j=0.2$.

Fitness function represents individual adaptation to population. During the iteration process, individual fitness of each generation population needs to be determined, and the individual fitness of each generation population is

$$\mu_j=1-\frac{\varphi_j(x)}{\varphi(x)_{\min}+\varphi(x)_{\max}} \tag{12}$$

where u_j is the fitness of the j individuals in this population.

5. Numerical Example

In order to verify the correctness of study method in this paper, the solution of dynamic reliability and the parameter optimization of TMD of three-story floor-shear frame structure with random stiffness parameters under seismic excitation as an example. The mass of the structure from the bottom to the top is $2762kg$, $2760kg$ and $2300kg$. The lateral stiffness of each layer is $1.243\times10^5 kN/m$, $0.961\times10^5 kN/m$ and $0.761\times10^5 kN/m$. In the rigidity parameter, all three random variables are subject to normal distribution, and the coefficient of variation is 0.2. In the case, the damping model is based on Rayleigh damping, and the damping matrix is determined by mass matrix and rigidity matrix, namely $C=aM+bK$. Because K is a random matrix, the damping matrix is also a random matrix. Basement input acceleration time history

$$\ddot{x}_g = \xi_1 \ddot{x}_{g1} + \xi_2 \ddot{x}_{g2} \tag{13}$$

where $\ddot{x}_{g1}$ and $\ddot{x}_{g2}$ are the East-West and North-South components of the seismic acceleration record of the El Centro respectively, their peaks are all $2m/s^2$. ξ_1 and ξ_2 are the random variables of identically independent and normal distribution, the means value is all 0.5, and the coefficient of variation is all 0.2.

In addition, the value ranges of the initial design variables of TMD are $m_d \in (0.03, 10.52)t$, $c_d = 0.9 \times c_{11}$ and $k_d = 0$, where c_{11} is the first data of the damping matrix in double random vibration systems with TMD.

Optimization the mass of TMD using improved genetic algorithm, where selection probability is 0.3, crossover probability is 0.2 and mutation probability is 0.2. The population sizes are 20, 30 and 40 respectively. The number of iterations is 40 times and the calculation results are shown in table 1. The analysis of table 1 shows that the structure of TMD has a good effect on reducing the displacement response when TMD mass is 8.79t, and the maximum displacement response after optimization is 49.93% than before optimization. And the calculation results tend to be stable with the increase of population size.

The displacement response of the top layer of structure is x_l. Table 2 shows the reliability of absolute value of displacement response of top layer not exceed the given limit. It shows that there is great variability in the reliability of structural displacement when displacement limit value is small. However, TMD has a good effect on reducing the displacement response and increasing the reliability when displacement limit value is large.

Table 1. Results for the structural optimization.

Population size	The optimal mass of TMD (t)	Optimization ratio
20	8.64	48.94%
30	8.79	49.93%
40	8.79	49.88%

Table 2. The top layer displacement historical reliability.

x_l / m	Before optimization	20 populations	30 populations	40 populations
[-0.02,0.02]	0.6682	0.4689	0.4674	0.4925
[-0.04,0.04]	0.8812	0.9023	0.9024	0.8902
[-0.06,0.06]	0.9486	1.0000	1.0000	1.0000
[-0.08,0.08]	1.0000	1.0000	1.0000	1.0000

Figure 1 shows the mean and standard deviation of top layer displacement of structure before and after optimization. As can be seen from figure (a), TMD has better control effect on structure displacement response when it is larger, while the control effect is not obvious when it is smaller. In general, TMD has a good effect on reducing the displacement response of structure, and the maximum displacement response after optimization is 49.93% than before optimization. Figure (b) shows that the discreteness of displacement response of TMD structure is smaller than that of uncontrolled structure when the randomness of structural parameters and external excitation is considered. Therefore, TMD can effectively reduce the discreteness of structural response.

IOP Conf. Series: Materials Science and Engineering **473** (2019) 012003 doi:10.1088/1757-899X/473/1/012003

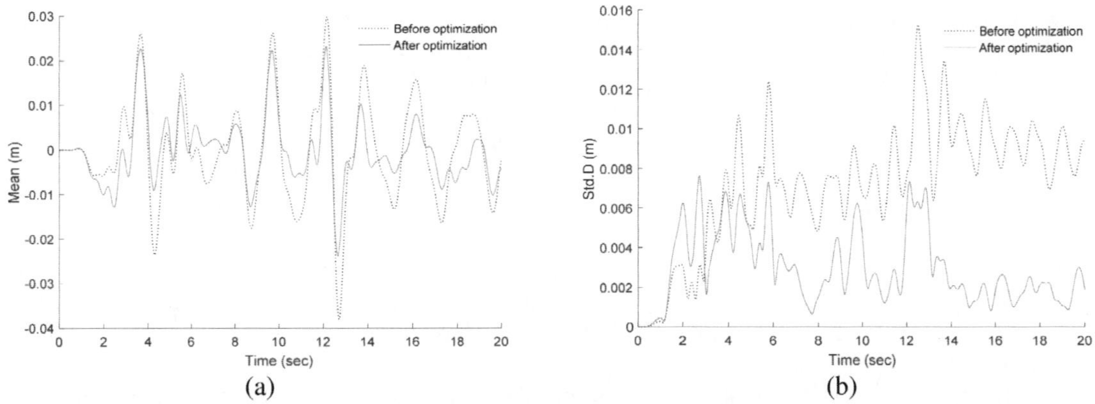

Figure 1. The mean and standard deviation (std.D) of top layer displacement of structure before and after optimization.

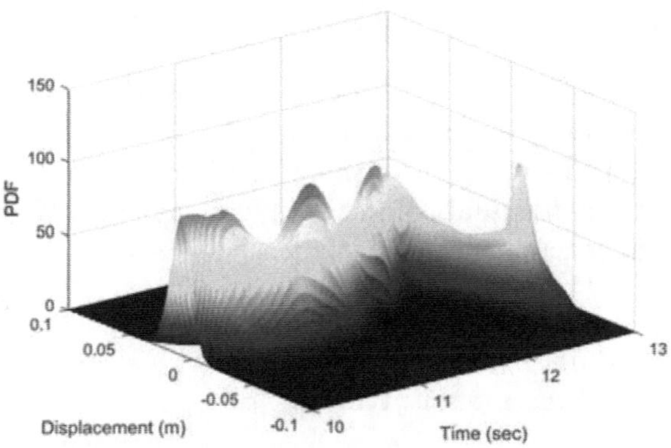

Figure 2. The displacement PDF surface of the top floor.

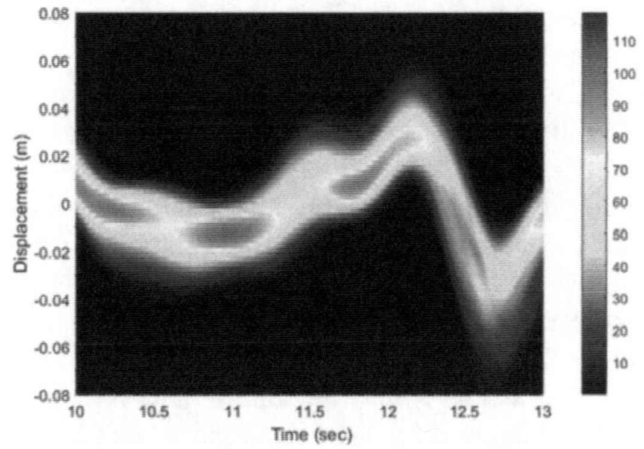

Figure 3. The contour of the PDF surface of the top floor.

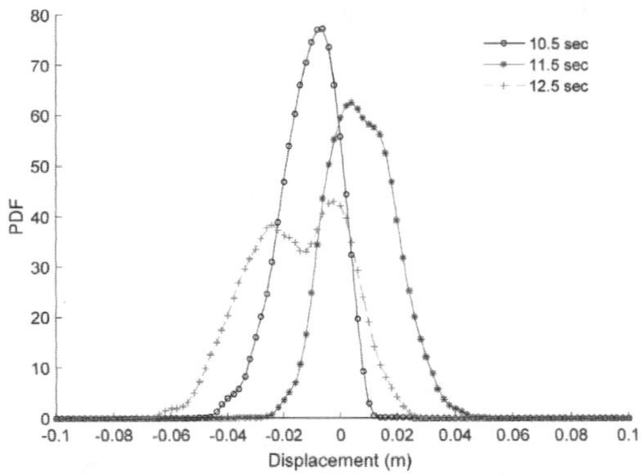

Figure 4. The PDF at certain time instants.

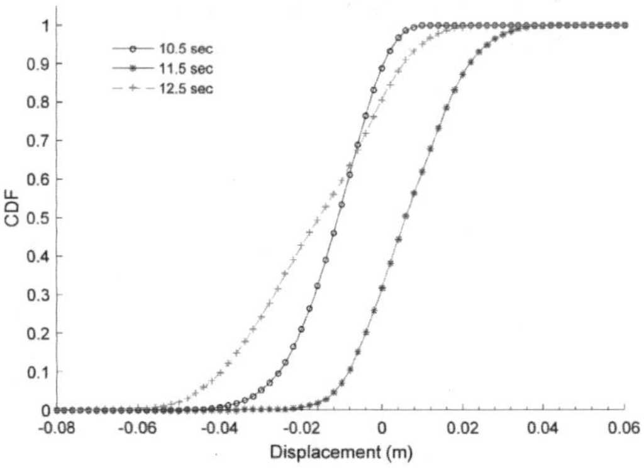

Figure 5. The CDF at certain time instants.

Combined with the displacement probability density function (PDF) surface of the top floor of structure in figure 2 and the corresponding contour of the PDF surface in figure 3, it can be seen that the PDF change with time, which reflect the result of the probability flowing in the state space. Obviously, the probability distribution of structural response is the irregular surface. Its evolution process likes a continuous mountain range, and its contour of PDF surface liking water flowing in a river. So, this is a non-stationary random process and the different widths at different times.

Figure 4 shows that the PDF at certain time instants and figure 5 shows that the cumulated probability density function (CDF) at certain time instants. Where, CDF is the integral of PDF. For the response analysis of double random vibration systems, the PDF is not the normal distribution or other special distribution, even assuming that the randomness of structural parameters and external incentives is the normal distribution. Moreover, it is a multi-peak curve at many times.

6. Conclusions

In this paper, the effect of TMD on the response of the double random vibration system is considered, and then the improved genetic algorithm and PDEM are adopted to optimize the parameters of TMD. The conclusions are as follows:

(1) TMD can effectively reduce the maximum peak and discreteness of structural displacement response. In the case, the maximum peak of structural displacement is reduced by 49.93% after optimization, which greatly improved the structural reliability. There is great variability in the reliability of structural displacement when displacement limit value is small. However, TMD has a good effect on reducing the displacement response and increasing the reliability when displacement limit value is large.

(2) The probability density distribution of the structural response is a multi-peak curve and varies with time. Therefore, it is necessary to describe the whole process with probability.

The ideas and methods presented in this paper provide a new way to solve the reliability of double random vibration system, and provide technical support for structural optimization design using TMD for shock absorption control, which has good engineering practicability.

References

[1] Zhang Z H and Yang W J 2012 *Spatial Struct.* **18** 64-75.
[2] Kawano K and Venkataramana K 1999 *Comput. Method. Appl. M.* **168** 255-72.
[3] Langley R S 1998 *J. Sound. Vib.* **123** 213-27.
[4] Li J and Chen J B 2004 *Soil. Dyn. Earthq. Eng.* **2004** 39-44.
[5] Yang J Y, Chen J B and Li J 2015 *J. Southwest Jiaotong Univ.* **50** 1047-54.
[6] Chen J B and Li J 2008 *Chinese Quart. Mech.* 21-8.
[7] Li and Chen JB 2009 *Stochastic Dyn. Struct.* (John Wiley & Sons Pte Ltd).
[8] Greco A, Cannizzaro F and Pluchino A 2016 *Eng. Struct.* 143 (2017): 152-168.
[9] Pourzeynali S, Salimi S and Kalesar H E 2013 *Sci. Iran.* **20** 207-21.

The economical design of a hand-gesture and bluetooth controlled wheel-chair by integrating indigenous components: mobility aid for the disabled

H A Khan[1], R M S U Islam[1], A W Attari[1], S I Mirza[1], M Ahmed[1]

[1]Electrical Engineering Department, Foundation University Islamabad, Pakistan

hana.amin.khan@gmail.com

Abstract. Millions of people around the world are handicapped and most of the amputees belong to a war struck third world regions. Any amputee who is unable to walk needs prosthetics which are costly or wheel chairs for movement. Manual wheel chairs are not easy to use. Most of the people cannot afford high quality motorized expensive wheel chairs. Keeping all this in mind, a very economical hand gesture-controlled wheel chair design is proposed. This wheel chair reads the signal sent from control glove and can also be switched to mobile application control where needed. The overall implementation design has simplicity and robustness which are dire features of wheelchair applications.

1. Introduction

The percentage of disabled people has increased in the world now-a-days. The disability could be by birth or due to some medical or accidental reason. The aim of this paper is to provide economical design of hand-gesture controlled wheel chair using Flex as sensor. It will help the physically disabled people in moving from one place to another just by giving direction from the hand. Specifically, in third world countries, many people are suffering from disability, there are people whose lower half of the body is paralyzed. The proposed design will add on to the comfort and make the life of people bit easier. In this era of fast growing technology and healthcare, there are still considerable amounts of physically challenged and elderly who find it difficult to move around in their house. The disabled people always find difficulties in moving from one room to another and even to do that the handicapped person was dependent on someone else who will push the wheelchair manually and take the handicapped person from one place to another. With the Hand Gesture Controlled Wheelchair the handicapped person is independent and he does not have to ask for help from any other person to move his wheelchair. A person will become independent by using a gesture-controlled mobility aid. This paper gives an overview of the work done in this field up till now, followed by the implemented method, results and conclusion.

2. Related Work

The traditional manual force driven wheel chairs require a lot of energy for movement. Also, they are bulky and cause the patient to be dependent on another person for the mobility. A lot of progress is being made in integrating technology and electronics to make designs of wheel chair more user friendly. Microelectro-Mechanical systems (MEMs) controlled sensors have been explored to see its applicability in gesture-based technology [1]. Gayatri et. al. employed MEMs based control system in their robot prototype for basic pick and place operations [2]. Ugale and Chandwaker worked on the Leap motion sensor for search and rescue robots which will be useful aid in natural disaster aid [3]. A gesture-controlled system for wheel chair and other robotic applications is the future of control schemes [4, 5]. The control of mobility through angular head movement captured by electronics is done by Babu and Anusha [6]. There are visual based techniques which do not require any external hardware like glove or other controller for sending signals for drive. A robot control is designed which implies visual recognition as a control signal. Wang et. al. developed adaptive algorithms to recognize facial feature as a control of robot [7]. Kaura et. al. also explored image processing using finger count and direction of palm to control robot [8]. Now all these techniques and methods have their own pros and cons. Most of these techniques require a complicated design and are quite expensive to implement. The design proposed in this paper is not only advanced but also very economical as compared to remote controlled wheel chairs available in market.

3. Methodology

The proposed prototype design is shown in figure 1. The wheel chair can be controlled be giving specific hand gestures or through mobile application. For hand gesture control flex sensors are implemented with wireless communication for which ZigBee module was used. Flex sensors are ideal for repetitive bending and deflection applications. They are robust, precise and ergonomic. Refer figure 2.

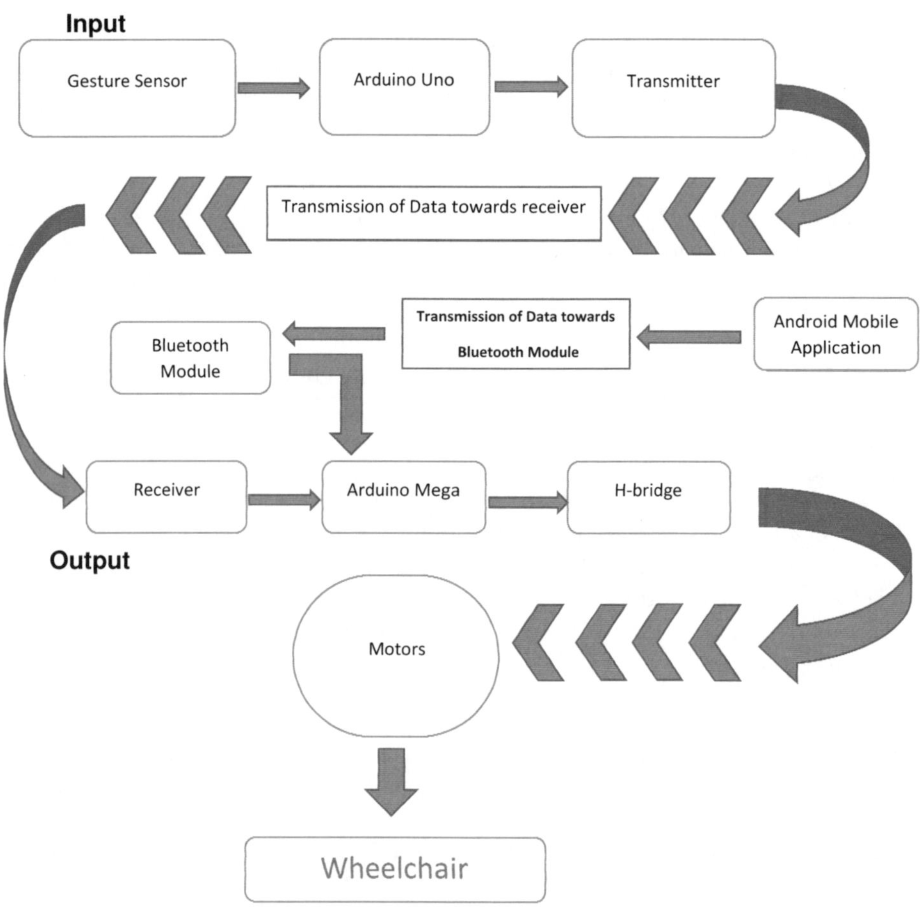

Figure 1. Flowchart of proposed wheel chair control system.

IOP Conf. Series: Materials Science and Engineering **473** (2019) 012004 doi:10.1088/1757-899X/473/1/012004

The transmitting end consists of flex sensors from which deflecting value of resistance is obtained and after processed by Arduino, required instructions are ready to transmit from zigbee x2c series module.

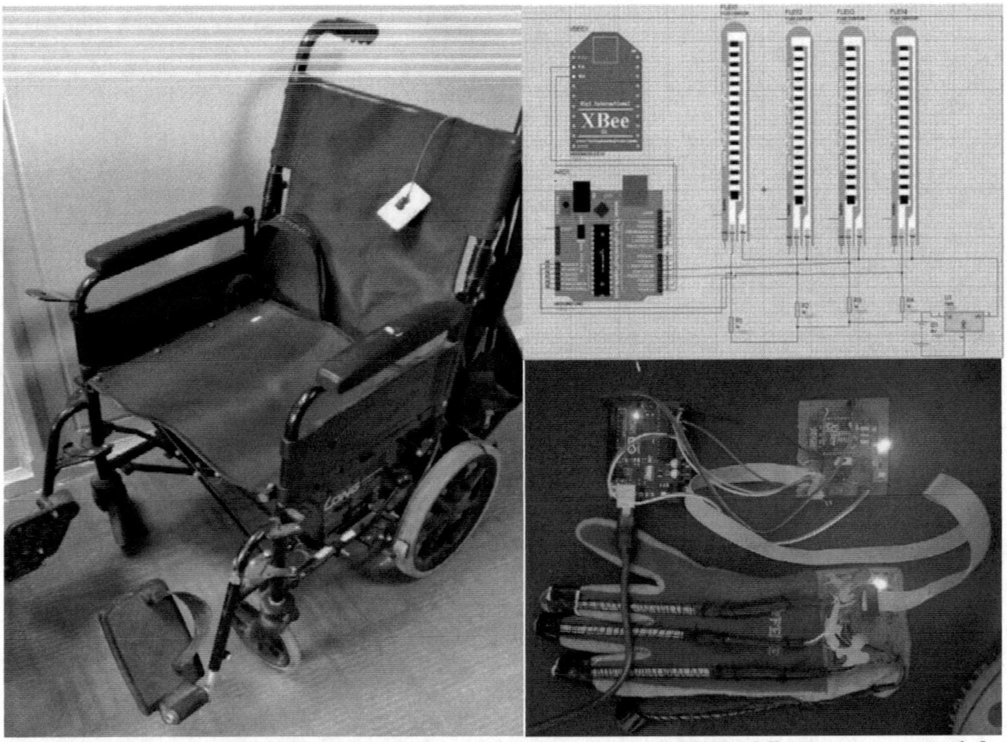

Figure 2. Indigenous wheelchair prototype (Left), Schematic design of flex sensor control & the hardware of flex sensor-based control glove (right).

Receiving end has major components which are Arduino Mega, 2x Ibt2 motor driver, Bluetooth module, ZigBee S2C module. Instructions from Transmitter are received by ZigBee on receiving end and send towards Arduino Mega using serial pins than according to these instructions H-Bridges controls wheelchair through motors. Controlling wheelchair using android Application is an

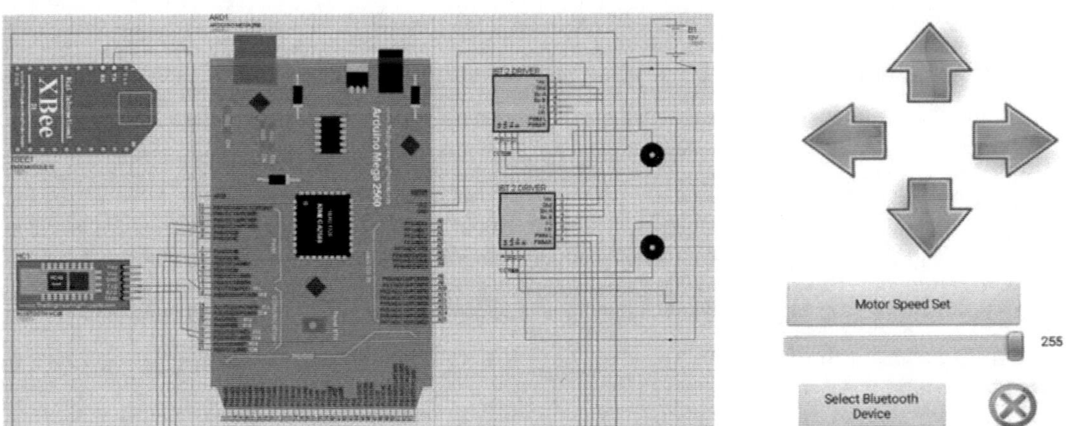

Figure 3. Complete schematic design of proposed technology (left); Interface of Bluetooth control of wheel chair (right).

additive feature of this prototype. For this purpose, Bluetooth module was used. After connecting its serial pins with Arduino Mega it was ready to communicate with Mobile, which has this app for controlling it from mobile. See figure 3.

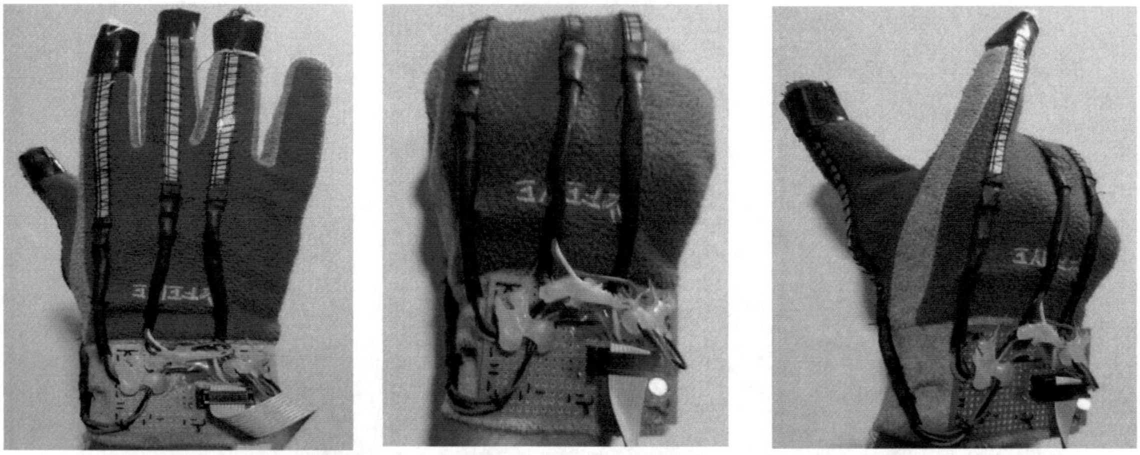

Figure 4. Specified gestures for Stopping; moving Forward; turn Left (left to right).

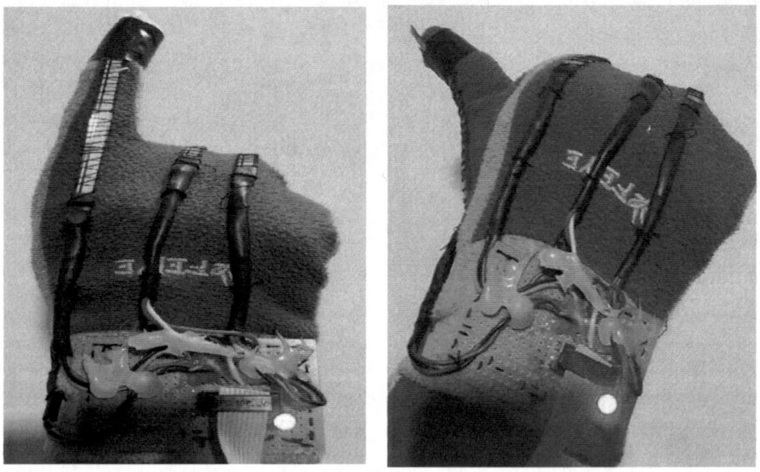

Figure 5. Specified gestures for Right turn; Reverse (left to right).

The gesture control of wheel chair for moving forward, backward, left and right is shown in the figure 4 & 5.

4. Conclusion

The economical prototype of gesture-controlled wheel chair was successfully tested for results. This wheel chair design provides force free mobility by implementing flex sensors-based control glove with and additional option of using smart mobile application for driving the wheel chair. Its features are its simple and easy to implement design which is also robust. This system can be operated both indoors and outdoors. For future work, one can make this wheelchair water and weather resistance. And improve its mechanical design for moving on stairs. Voice or head motion control for this wheel chair may also be explored.

5. References

[1] Jha P and Khurana P 2016 *Int. J. Sci. Tech.* **9(41)** 243-9.
[2] Gayatri Devi V, Vijetha T 2016 Int. J. Sci. Dev. Res. **1(6)** 82-5.
[3] Ugale A, Chandwadkar D M 2016 Int. J. Comput. Appl. **135(1)** 29-31.
[4] Goyal D and. Saini S P S 2013 Int. J. Emerg. Tech. **4(2)** 15-20.
[5] David N, Udengwu C and Onyia O 2015 Int. J. Scientific Eng. Res. **6(11)** 397-405.
[6] Sharath Babu Rao D, Anusha T 2015 "GESTURE CONTROLLED WHEELCHAIR", Int. J. Sci. Eng. Tech. Res. **4(8)**
[7] Hu C, Wang X, Mandal N K, Meng M and Li D 2003 *Canadian Conf. Electrical and Comput. Eng. Toward a Caring and Humane Tech. (Cat. No.03CH37436)* **3** 1757-62.
[8] Kaura H K, Honrao V, Patil S and Shetty P 2013 Int. J. Adv. Res. Artif. Intell. **2(5)** 69-77.

IOP Conf. Series: Materials Science and Engineering **473** (2019) 012005 doi:10.1088/1757-899X/473/1/012005

Brazing Diamond Grits Onto Stainless Steel using Active Filler Metal and Porous Nickel as an Interlayer: Analysis of the Porous Nickel/Stainless Steel Interface

Tuan Zaharinie[1, 2*], Amirul Aliff[2], Mohd Hamdi[1, 2] and Tadashi Ariga[3]

[1]Centre of Advanced Manufacturing and Material Processing, University of Malaya, 50603 Kuala Lumpur, Malaysia
[2]Department of Mechanical Engineering, Faculty of Engineering, University of Malaya, 50603 Kuala Lumpur, Malaysia
[3]Department of Materials Science, School of Engineering, Tokai University, 1117 Kitakaname, Hiratsuka-shi, Kanagawa, 259-1292 Japan

Abstract. In this research, diamond grits were brazed to stainless steel successfully using active filler metal and porous nickel as an interlayer. Brazing was accomplished at three (3) brazing temperatures (880ºC, 920ºC and 960ºC) for 10 minutes. SEM-EDS and XRD analyses were adopted to facilitate the research process. However, the analyses were specifically focused on the porous nickel (Ni)/stainless steel interface. The interface near the bonding diamond did not seem strong enough for potential grinding tool applications. It is concluded in this research that 960ºC brazing temperature is the best for brazing because it causes less and discontinuous formation of the $AgTi_2$ reaction layer in the brazed area. Otherwise, a continuously reacting layer would increase the small void formation in the area near the reaction layer, which could deteriorate the bonding joint.

1. Introduction

Demand for industrial diamonds is increasing due to the properties of diamond, which is the hardest material in nature. Diamond also has super abrasive ability for stock removal from various kinds of material surfaces. Therefore, several methods of diamond grit bonding have been investigated and patented [1, 2]. Common bonding methods used to attach diamond grits to tool matrix surfaces are sintering and electroplating. However, because of the mechanical burying of the diamond grits, it is difficult to achieve high bonding strength due to the uneven segregation of the diamond grits on the tool matrix surface [1]. The bonding process of diamond grits is important for enhancing tool performance. Hence, the technology of diamond brazing was introduced in the diamond tool industry in the 1990s to improve the bonding strength of diamond grits [3].

Significant brazing has been attained successfully using filler metals like Ni-Cr, Ag-Cu-Ti, etc. [4, 5]. Brazing is a promising method for manufacturing diamond abrasive tools. In general, diamond brazing can be performed at high temperatures (in the range of 930 - 1100ºC) but is also dependent on the type of brazing filler metal used [6]. It is interesting to note that it is difficult to braze diamond at brazing temperatures above 760ºC. This is due to diamond graphitization and oxidation that tend to occur from this temperature onwards. Brazing of diamond can also be difficult to achieve on account of the high interfacial energy between the diamond and most metal alloy materials [7]. Therefore, it is suggested that more attempts should be made to improve brazing methods to achieve better bonding performance.

In this research, a preliminary attempt was made using porous Ni and commercial active filler metal (Ag-Cu-Ti) to join diamond grits with stainless steel AISI 304. Porous nickel (Ni) was inserted as an interlayer to increase the disperse-ability of the diamond grits onto stainless steel during brazing. Active filler metal was used due to its good wettability with the diamond surface. Due to the used of interlayer, there are two interfaces were important to analyze: (1) diamond/porous metal and (2) porous metal/metal. As part of the analysis, the aim was to investigate and determine only the brazing interface of the porous metal/metal side.

2. Methodology

In this research, industrial diamond grits acquired from Japan with average size of 0.4156 mm and coated with silver (Ag) were employed. Austenitic Cr-Ni stainless steel substrate (AISI 304) with dimensions of 20 mm x 20 mm x 3 mm served as the base metal. The brazing filler metal used in this work was a commercial active filler metal, Ag-Cu-Ti. This was manufactured by Wesgo Metals, USA, consists of 63Ag-35.25Cu-1.75Ti alloy (wt.%) and has solidus and liquidus temperatures of 775ºC and 795ºC, respectively. The filler metal with a thickness of 50 μm was then cut into 15 mm x 15 mm dimensions. The 50ppi porous nickel (Ni) with an average pore size of 0.2106 mm was prepared in 15 mm x 15 mm x 1 mm dimensions. The sample was arranged in a sandwich configuration on a special clamp with stainless steel as the base, followed by filler metal, porous Ni, filler metal and diamond grits on top. The diamond grits were 0.2g per sample.

Three brazing temperatures of 880ºC, 920ºC and 960ºC and 10 min brazing time were selected as the brazing parameters. The samples were brazed in a furnace with controlled atmosphere and argon gas at a flow rate of 5 L/min. The samples were heated and cooled at a rate of 10ºC/min, respectively. After brazing, all the samples were carefully ground using SiC grit paper and polished with diamond suspension for characterization.

The characterization machine used for analysis was a Scanning Electron Microscope (SEM) model Phenom Pro X Desktop equipped with an Electron Dispersive X-ray Spectroscope (EDS). An X-Ray Diffractometer (XRD) model PANalytical Empyrean was employed for testing, while analysis was done using HighScore software to identify the elements and compounds/phases.

3. Results and Discussion

Brazing diamond grits to stainless steel was successfully achieved with the addition of porous Ni as an interlayer and Ag-Cu-Ti active filler metal. However, the bonding of the diamond with filler metal and porous Ni (the first interface) appeared to not be strong enough for potential grinding tool applications. Therefore, the brazing method needs to be improved accordingly. Further analysis was conducted on the brazed porous Ni and stainless steel (Ni/stainless steel), the second interface.

Figure 1 depicts the interface microstructure at three different brazing temperatures, at which various phases formed on the brazed interface. It is important to note that a piece of brazing filler metal was inserted between the porous Ni and stainless steel before brazing. After brazing, the brazing filler metal fully melted and diffused into the porous Ni and stainless steel. It is interesting to highlight that two colors were present in the porous Ni area, namely grey and white. Point analysis was done using EDS for all samples. Referring to Table 1, the grey area was dominated by Ni element and copper (Cu) that diffused from the brazing filler metal. It can be said that a Cu-Ni-Fe phase formed at point 2. Cu-Ni and $CuNi_2$ phases formed at points 5 and 7, while the white area indicates an Ag-rich region (point 6). At point 8, the Ag-rich area was also influenced by iron (Fe), Cu and a low concentration of Ni. The dark grey area in the brazing filler metal shows a highest percentage of titanium (Ti).

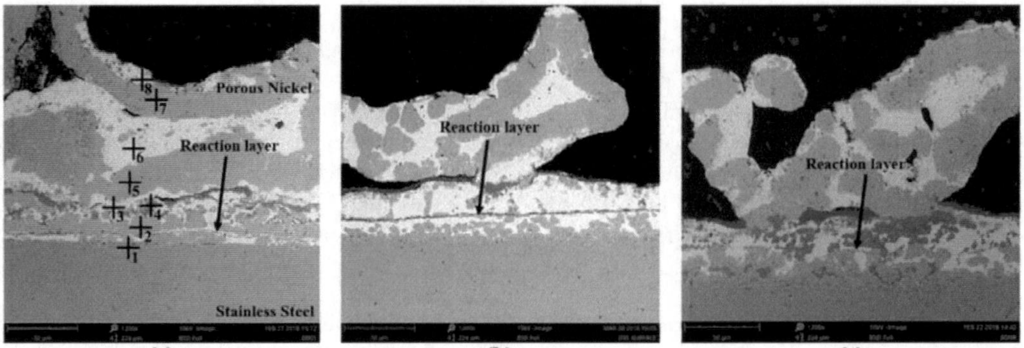

Figure 1. Microstructure of samples brazed for 10 minutes at different brazing temperatures: (a) 880ºC, (b) 920ºC and (c) 960ºC.

As seen in Figure 1 (a) and point 4 (Table 1), this area shows a higher concentration of Ti than other areas with slightly higher concentrations of Fe, manganese (Mn) and chromium (Cr) that were influenced by the diffusion from stainless steel. The possible phases in this area were Fe-Ti/Fe-Cr/Fe-Mn. The possible phases of each point marked in Figure 1 (a) are listed in Table 1.

Table 1. EDS analysis (at.%) and possible phases according to SEM micrographs of the porous Ni/stainless steel interface marked in Figure 1 (a).

Point	1	2	3	4	5	6	7	8
Copper	53.27	73.69	18.50	-	31.46	6.53	33.07	15.81
Nickel	20.55	14.35	6.68	-	60.47	-	60.39	6.28
Iron	15.78	7.41	7.58	32.59	4.85	-	3.10	13.01
Silver	3.67	4.55	67.24	13.83	3.22	93.47	3.43	64.90
Sulfur	2.37	-	-	4.47	-	-	-	-
Titanium	-	-	-	26.28	-	-	-	-
Chromium	4.36	-	-	9.36	-	-	-	-
Carbon	-	-	-	-	-	-	-	-
Manganese	-	-	-	13.45	-	-	-	-
Possible phase	Stainless steel + Cu-Ni	Cu-Ni-Fe	Ag-Cu	Fe-Ti/ Fe-Cr/ Fe-Mn	Cu-Ni	Ag	Cu-Ni	Ag-Cu-Fe + Ni

Figure 2 presents the XRD analysis of the entire brazed surface of the sample. The XRD analysis clearly shows that all elements in the sample were detected. A few compounds/phases were also evidently present, such as Cu-Ni, Cu-Ni-Fe, Fe-Ti, Fe-Cr and Dideuterium Sulfide (D_2S_1). It is noted that the peak with the highest intensity was dominated by almost all elements and compounds/phases. This analysis confirms that most of the possible phases listed in Table 1 were present.

IOP Conf. Series: Materials Science and Engineering **473** (2019) 012005 doi:10.1088/1757-899X/473/1/012005

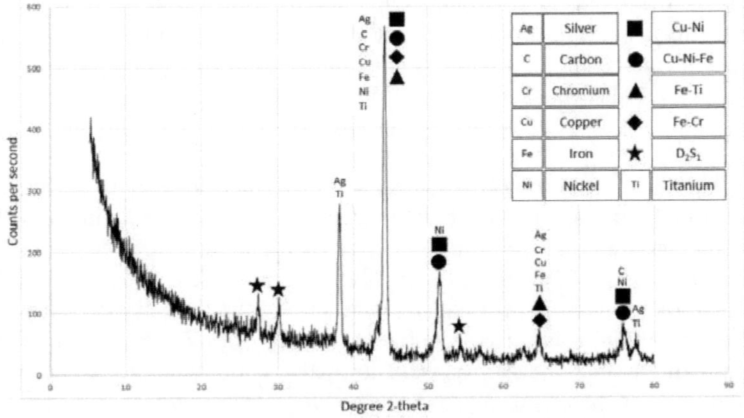

Figure 2. XRD analysis on the brazed joint.

Figure 1 indicates that the brazing layer grew with increasing brazing temperature due to the diffusion of the brazing filler metal into the stainless steel to form good bonding. During brazing, the Ag-rich liquid flowed easily and copper was depleted at high temperature. In this research, the brazing temperature of 960ºC nearly reached the melting point of Ag (961.8ºC). Figures 1 (a) and (b) show a continuous reaction layer crossing along the brazed layer. The reaction layer is obvious in the sample brazed at 920ºC for 10 minutes (Figure 1 (b)). However, there was also a potential crack. In Figure 1 (c), the reaction layer shows discontinuity with lower thickness compared to the sample brazed at 920ºC.

Therefore, the reaction layer was analyzed to identify the possible phases. Figure 3 displays a highly magnified micrograph of the sample brazed at 920ºC with EDS analysis (at. %) and the list of possible phases.

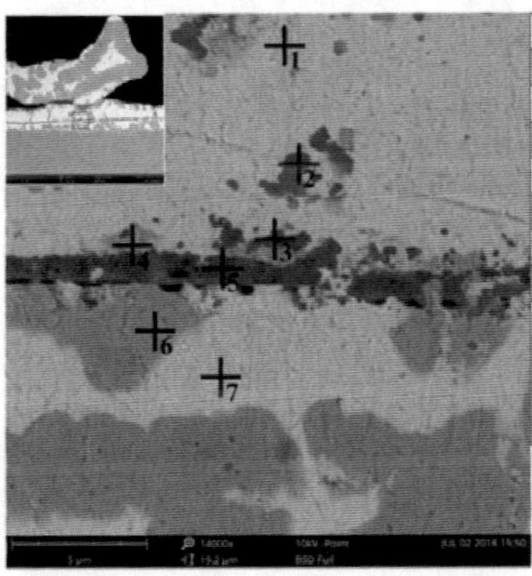

Point	1	2	3	4	5	6	7
Copper	7.85	-	-	63.47	-	78.56	10.36
Nickel	-	-	4.60	-	-	7.68	-
Iron	-	-	9.02	-	5.38	-	10.00
Silver	92.15	-	43.96	23.14	14.38	13.76	79.63
Sulfur	-	23.68	-	-	-	-	-
Titanium	-	19.10	42.42	13.39	32.32	-	-
Chromium	-	-	-	-	8.78	-	-
Carbon	-	57.23	-	-	39.14	-	-
Possible phase	Ag	β-Ti + C + S	Ag-Ti	Ag-Cu-Ti	Ag-Ti/ Fe-Cr, +C	Cu-Ag	Ag

Figure 3. SEM micrograph of the sample brazed at 920ºC with EDS analysis and list of possible phases.

The reaction layer is marked at point 5. A possible phase in the reaction layer was Ag-Ti ($AgTi_2$) or Fe-Cr with carbon content. However, the EDS value for Fe-Cr was lower and less accurate compared with the EDS value for Ag-Ti. It is also important to highlight that the phase was not TiC in accordance to the peaks in the XRD that show no match for this phase as stated in literature [8]. It can also be said that almost the same phase was detected at point 3 (AgTi) with different stoichiometric elements. In line with literature, the Ag-Ti phase was not expected to be very brittle nor cause joint strength deterioration [9]. However, small voids were observed in the Cu and Ag-rich phases near the reaction layer. Void formation is not good for bonding strength performance during servicing as crack failure could occur at this point. Hence, according to this analysis, the sample brazed at the highest temperature exhibited a good bonding interface due to the minimal and discontinuous $AgTi_2$ phase formation.

4. Conclusion

Good bonding was observed at the porous Ni/stainless steel interface. The results reveal that the filler metal and stainless steel diffused after brazing and a reaction layer of $AgTi_2$ formed and crossed along the brazing layer. The reaction layer appeared to form continuously at a low temperature (880°C). However, the reaction layer increased in thickness at 920°C brazing temperature and a crack was observed at low magnification due to the small void formation in the area near the reaction layer. At higher temperature (960°C), the reaction layer formed discontinuously due to the increasing Ag liquidity because the brazing temperature was close to the Ag melting point. The Ag-Ti phase formed was not expected to be very brittle nor cause joint strength deterioration. However, to improve the bonding performance, it is recommended to reduce the formation of this reaction layer. In this research, brazing at 960°C was more suitable for good bonding performance.

Acknowledgement

The authors are grateful for the support provided by the Ministry of Higher Education, Malaysia through Fundamental Research Grant Scheme (FRGS) project number FP060-2015A. The authors also gratefully acknowledge the support for this work by the AMMP Centre staff and support staff at the Department and Faculty of Engineering, University of Malaya.

References

[1] Klotz U E, Khalid F A and Elsener H R 2006 *Diam. Relat. Mater.* **15** 1520-4.
[2] Shiue R K, Buljan S T and Eagar T W 1997 *Sci. Technol. Weld Joining.* **2** 71-8.
[3] Huang G Q et al 2009 *Adv. Mat. Res.* **76-78** 125-30.
[4] Chen Y et al 2014 *Int. J. Refract. Metals Hard Mater.* **42** 23-9.
[5] Sun B-J and Xiao B 2018 *Diam. Relat. Mater.* **85** 98-103.
[6] Liao X et al 2017 *Int. J. Refract. Metals Hard Mater.* **66** 252-7.
[7] Long F, He P and Sekulic D P 2018 *Metals* **8** 315.
[8] Klotz U E *et al* 2008 *Mater. Sci. Eng.* A **495** 265-70.
[9] Laik A et al 2013 *Metall. Mater. Trans.* A. **44** 2212-25.

IOP Conf. Series: Materials Science and Engineering **473** (2019) 012006 doi:10.1088/1757-899X/473/1/012006

Bending Response of Doubly Curved Laminated Composite Shells using Hybrid Refined Models

J Monge[1], J Mantari[1], J Yarasca[1] and R Arciniega[2]

[1] Faculty of Mechanical Engineering, Universidad de Ingenieria y Tecnologia (UTEC), Barranco, Lima, Peru

[2] Department of Civil Engineering, Universidad Peruana de Ciencias Aplicadas (UPC), Surco, Lima, Peru

Abstract. This study presents a static analysis of laminated composite doubly curved shells using a refined kinematic model with polynomial and non-polynomial functions. In particular Maclaurin, trigonometric, exponential and zig-zag functions are employed. Refined models are based on the Equivalent Single Layer theories and obtained by using Carrera Unified formulation. The shell model is subjected to different mechanical loading such as bi-sinusoidal, uniform and point load. The governing equations are derived from the principle of virtual displacement and solved via Navier-Type closed form solutions. The results are compared with Layer-wise and higher-order shear deformation solutions available in the literature. It is shown that refined models with non-polynomial terms are capable of accurately predicting the through-the-thickness displacements and stress distributions with a low computational effort.

1. Introduction

Multilayered shell structures are utilized in many engineering fields such as aerospace, naval and automotive due to their superior characteristics. Shells are curved structures which have excellent load-carrying capacity in comparison to plates. Moreover, composite structures provide high strength, high stiffness-weight ratio and remarkable fatigue resistance compared to metal structures. Nevertheless, composite structures show local problems such as delamination that needs to be investigated with accurate and cost-efficient simulations. Consequently, reliable and efficient computational models for multilayered structures are still an important research topic.

Shell theories have been formulated for over 70 years. The First order shear deformation theory (FSDT) was proposed by Hildebrand et al. [1] in order to include the effect of transverse shear stress. The FSDT depends on a shear correction factor which is difficult to estimate for composite shells. Consequently, Higher Order Shear Deformation theories (HSDTs) [2, 3] were reported on the literature. HSDTs include additional terms on the kinematic model in order to accurately predict laminated composites displacements and stress distributions.

A generalization of HSDTs is the so-called Unified Formulations (UFs) [4]. UFs permits to write governing equations of different models within a unique computer program. The fact that in some HSDTs the number and type of shear strain functions are considered constant is a clear disadvantage in comparison to UFs. An excellent review on plates and shells theories for laminated and sandwich structures comparing HSDTs and UFs are presented by Caliri et al. [5].

Shell multilayered models can be implemented in two different approaches: Equivalent Single Layer (ESL) and Layer-wise (LW) [4]. An ESL model considers multilayered structures as a single equivalent layer. A LW model maintains each lamina separately; therefore, accurate displacements and stress distributions are obtained in comparison to ESL models. However, the high computational cost of LW models makes it difficult to implement in engineering applications. Consequently, robust ESL models with adequate precision are preferred.

The developed formulation is based on Carrera Unified formulation (CUF) with the inclusion of non-polynomial expansion function to study for the first time shell panels. According to CUF, the displacement field of the shell can be defined as an arbitrary expansion of the thickness coordinates. The governing equations are the so-called fundamental nuclei whose form does not depend on either the expansion order or the choices made for the shear strain shape functions. The highly coupled differential equations are solved via Navier-Type solutions and several static problem of shells are analyzed under different loading.

A significant number of investigations have been presented regarding shell behavior. Tornabene et al. [6] presented a stress recovery method with CUF and the differential geometry tool to obtain quasi-3D results for doubly curved anisotropic shells and panels. Dozio [7] used a space-state approach in conjunction to Levy's method for solving free vibration problem of spherical and cylindrical panels by applying a hierarchical formulation. Brischetto [8] proposed a three-dimensional elasticity solution for the static and free vibration analyses of simply supported spherical shells based on the exponential matrix method. Viola et al. [9] presented a new HSDT for free vibration and static analysis of shells. Tornabene et al. [10] applied CUF for the free vibration of doubly curved laminated shells and panels, the highly coupled differential equation were solved numerically applying radial basis functions. Cinefra and Valvano [11] combined CUF and mixed interpolation of tensorial component for solving different problems of anisotropic doubly curved shell.

The paper proposes different refined hybrid models (HRM) including non-polynomial functions for the static analysis of simply supported laminated composite doubly curved shells. The formulation includes polynomial, trigonometric, exponential and zig-zag functions. The kinematic models are adapted from the HRMs presented by Yarasca et al. [12, 13]. This paper aims to study the capability of this formulation to model shell structures.

2. Shell formulation and close form solution

Let $\{\alpha,\beta\}$ be a set of orthogonal curvilinear coordinates of a doubly curved shell. The thickness is denoted as "h", the length as "a" and the width as "b". The radii of curvature "R_α" and "R_β" are considered as constant along the midsurface domain Ω_k. The integer k denotes the layer number from the shell-bottom.

The presented paper studies the influence of different trigonometrical, exponential and zig-zag shear strain shape functions in ESL models. The displacement field contains 54 terms: 15 Maclaurin, 3 zig-zag terms, 24 trigonometrical terms and 12 exponentials terms. The non-polynomial functions were selected by Filippi et al. [14] for plates and were used by Yarasca et al. [13]. For example, the expansion along β axis is denoted as:

$$u_\beta = u_{\beta 0} + z u_{\beta 1} + z^2 u_{\beta 2} + z^3 u_{\beta 3} + z^4 u_{\beta 4} + (-1)^k \zeta_k u_{\beta 5} + \sin\left(\frac{\pi z}{h}\right) u_{\beta 6} + \sin\left(\frac{2\pi z}{h}\right) u_{\beta 7}$$
$$+ \sin\left(\frac{3\pi z}{h}\right) u_{\beta 8} + \sin\left(\frac{4\pi z}{h}\right) u_{\beta 9} + \cos\left(\frac{\pi z}{h}\right) u_{\beta 10} + \cos\left(\frac{2\pi z}{h}\right) u_{\beta 11} + \cos\left(\frac{3\pi z}{h}\right) u_{\beta 12}]$$
$$+ \cos\left(\frac{4\pi z}{h}\right) u_{\beta 13} + e^{\frac{z}{h}} u_{\beta 14} + e^{\frac{2z}{h}} u_{\beta 15} + e^{\frac{3z}{h}} u_{\beta 16} + + e^{\frac{4z}{h}} u_{\beta 17} \quad (1)$$

Refined models are defined as N hybrid refined models (N-HRM), where the N notation is the number of the variables in the model. The principle of virtual displacement for multilayered doubly curved shells is given by:

$$\sum_{k=1}^{N_l} \int_{\Omega_k} \int_{A_k} \left\{\delta \varepsilon_p^{k^T} \sigma_p^k + \delta \varepsilon_n^{k^T} \sigma_n^k\right\} H_\alpha{}^k H_\beta{}^k \, d\Omega_k dz = \sum_{k=1}^{N_l} \int_{\Omega_k} \int_{A_k} \{\delta u^k p^k\} H_\alpha{}^k H_\beta{}^k d\Omega_k dz \quad (2)$$

where Ω_k and A_k represents the integration domains in plane and the thickness direction, respectively. p^k is the mechanical load applied at a certain k layer.

The numerical results were obtained via the Navier closed-form solution for simply supported orthotropic panels. The displacement expressed as a summation of harmonics:

$$u_{\alpha_s}^k = \sum_{m,n} U_{\alpha_s}^k \cos\left(\frac{\pi m}{a}\alpha\right) \sin\left(\frac{\pi n}{b}\beta\right)$$
$$u_{\beta_s}^k = \sum_{m,n} U_{\beta_s}^k \sin\left(\frac{\pi m}{a}\alpha\right) \cos\left(\frac{\pi n}{b}\beta\right) \quad (3)$$
$$u_{z_s}^k = \sum_{m,n} U_{z_s}^k \sin\left(\frac{\pi m}{a}\alpha\right) \sin\left(\frac{\pi n}{b}\beta\right)$$

where "*m*" and "*n*" are the number of waves and $U_{\alpha_s}^k$, $U_{\beta_s}^k$ and $U_{z_s}^k$ are related to the amplitudes of the Fourier series displacement.

A 2D elastic theory for shells with constant radii of curvature is solved. This paper presented a simplified model of the 3D elasticity theories with the assumptions of the displacement field shown in equation (1). The presented refined model can capture the real behavior of shell structures for displacement and in-plane stresses. However, a refinement is needed for transverse shear and normal stresses. The 3D elasticity are valid for solving problems of shells with simply supported boundary conditions. The shear stresses and their derivatives are replaced into the 3D elasticity equations in order to calculate the out-plane stresses. The 3D elasticity equations are reported by Brischetto [8] and given by:

$$H_\beta^k \frac{\partial \sigma_{\alpha\alpha}^k}{\partial \alpha} + H_\alpha^k \frac{\partial \sigma_{\alpha\beta}^k}{\partial \beta} + H_\alpha^k H_\beta^k \frac{\partial \tau_{\alpha z}^k}{\partial z} + \left(\frac{2H_\beta^k}{R_\alpha^k} + \frac{H_\alpha^k}{R_\beta^k}\right)\tau_{\alpha z}^k = 0$$
$$H_\beta^k \frac{\partial \sigma_{\alpha\beta}^k}{\partial \alpha} + H_\alpha^k \frac{\partial \sigma_{\beta z}^k}{\partial \beta} + H_\alpha^k H_\beta^k \frac{\partial \tau_{\beta z}^k}{\partial z} + \left(\frac{2H_\alpha^k}{R_\beta^k} + \frac{H_\beta^k}{R_\alpha^k}\right)\tau_{\beta z}^k = 0 \quad (4)$$
$$H_\beta^k \frac{\partial \tau_{\alpha z}^k}{\partial \alpha} + H_\alpha^k \frac{\partial \tau_{\beta z}^k}{\partial \beta} + H_\alpha^k H_\beta^k \frac{\partial \sigma_{zz}^k}{\partial z} - \frac{H_\beta^k}{R_\alpha^k}\sigma_{\alpha\alpha}^k - \frac{H_\alpha^k}{R_\beta^k}\sigma_{\beta\beta}^k + \left(\frac{H_\beta^k}{R_\alpha^k} + \frac{H_\alpha^k}{R_\beta^k}\right)\sigma_{zz}^k = 0$$

The formulation must satisfy the zero shear strain condition at the bottom and the top of the shell structure and interlaminar continuity. Equations (4) could not be solved analytically, therefore, an approximate solution using Taylor series is adopted. These equations are valid for doubly curved shells with constant radii of curvature.

3. Results and discussions

In this section, some results of the static analysis of composite doubly curved shell panels are presented. The boundary conditions are considered as simply supported. The geometry data is considered as: $a = b = 1$ and $R_\alpha = R_\beta = R$. Different side-to-thickness (*a/h*) and curvature-radius-to-side (*R/a*) ratios are considered as parameters. The first case is related to a laminated doubly-curvature panel with the stacking sequence 0°/90°/0° subjected to bi-sinusoidal loading. Some results are compared with different HSDTs [2, 6, 15, 16] available in the literature. The second case is a benchmark solution of two laminae. The original problem was presented by Demasi [4] for plates and it is used to analyze doubly curved panels in this paper.

The mechanical properties of each layer for the first case were reported by Reddy and Liu [2]:

$$E_1 = 25E_2, G_{12} = G_{13} = 0.5E_2, G_{23} = 0.2E_2, \upsilon_{12} = \nu_{13} = 0.25 \quad (5)$$

In this paper, the transverse displacement $\bar{w}$ and transverse normal stress $\bar{\sigma}_{zz}$ is evaluated at (a/2, b/2, 0). The normal stress $\sigma_{\alpha\alpha}$ is evaluated at (a/2, b/2, h/2). The transverse shear stress $\tau_{\alpha z}$ and $\tau_{\beta z}$ are computed at (0, b/2, 0) and (a/2, 0, 0), respectively.

Figure 1 shows dimensionless stresses $\bar{\sigma}_{\alpha\alpha}$, $\bar{\sigma}_{\alpha z}$, $\bar{\sigma}_{\beta z}$ and $\bar{\sigma}_{zz}$ distributions through the thickness. The results demonstrated a good accuracy overall the thickness distributions with respect to the LD4. Figure 2 shows the relation between the dimensionless $\bar{u}_z$ and the inverse of curvature-radius-to-size *(a/R)* ratio for a moderate thick shell *a/h* = 10. The displacements are compared with other formulations available in the literature such as the sinusoidal shear deformation theory proposed by Ferreira et al. [16], the third order shear deformation proposed by Reddy and Liu [2], etc.

IOP Conf. Series: Materials Science and Engineering **473** (2019) 012006 doi:10.1088/1757-899X/473/1/012006

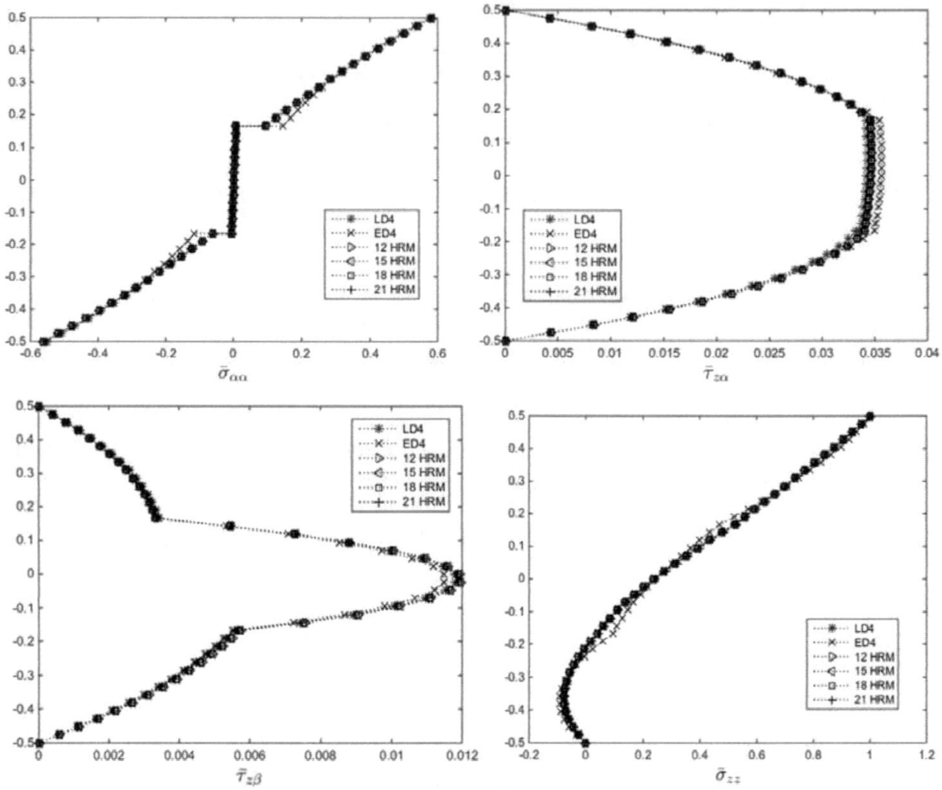

Figure 1. Dimensionless stress distributions for 0°/90°/0°, a/h = 10, R/a = 5.

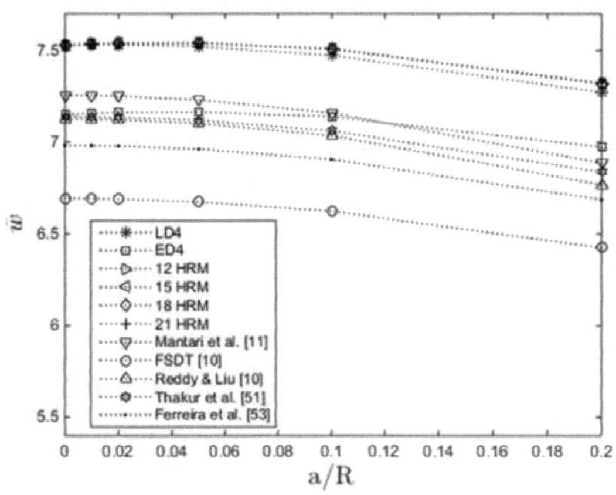

Figure 2. Dimensionless transverse displacement $\bar{u}_z$ versus a/R ratio for 0°/90°/0°, a/h = 10.

The second case problem was studied by Demasi [4] for plate configuration. In this paper, it is extended to doubly-curvature panel. The mechanical properties of layers are: $E_{11} = 25E_{22}$, $E_{11} = 25E_{33}$, $E_{11} = 50\,G_{12}$, $E_{11} = 50G_{13}$, $E_{11} = 125G_{23}$, $\varphi = 0$ for the bottom and $E_{11} = 25E_{22}$, $E_{11} = 2.5E_{33}$, $E_{11} = 50\,G_{12}$, $E_{11} = 50G_{13}$, $E_{11} = 125G_{23}$, $\varphi = 90$ for the top. The Poisson ratio is considered as: $v_{12} = v_{13} = v_{21} = 0.25$ for both layers. The thickness ratio is considered as *a/h = 4*. Different curvature-radius-to-size ratios are investigated. Stresses $\sigma_{\alpha\alpha}$, $\tau_{\alpha z}$ and $\tau_{\beta z}$ are calculated at (a/2,b/2,h/4), (0,b/2,h/4) and (a/2,0,h/4), respectively. Figure 3 shows the through thickness distributions of these the stresses $\bar{\sigma}_{\alpha\alpha}$, $\bar{\tau}_{\alpha z}$, $\bar{\tau}_{\beta z}$ and $\bar{\sigma}_{zz}$. Good agreements in the results are obtained overall the thickness distribution.

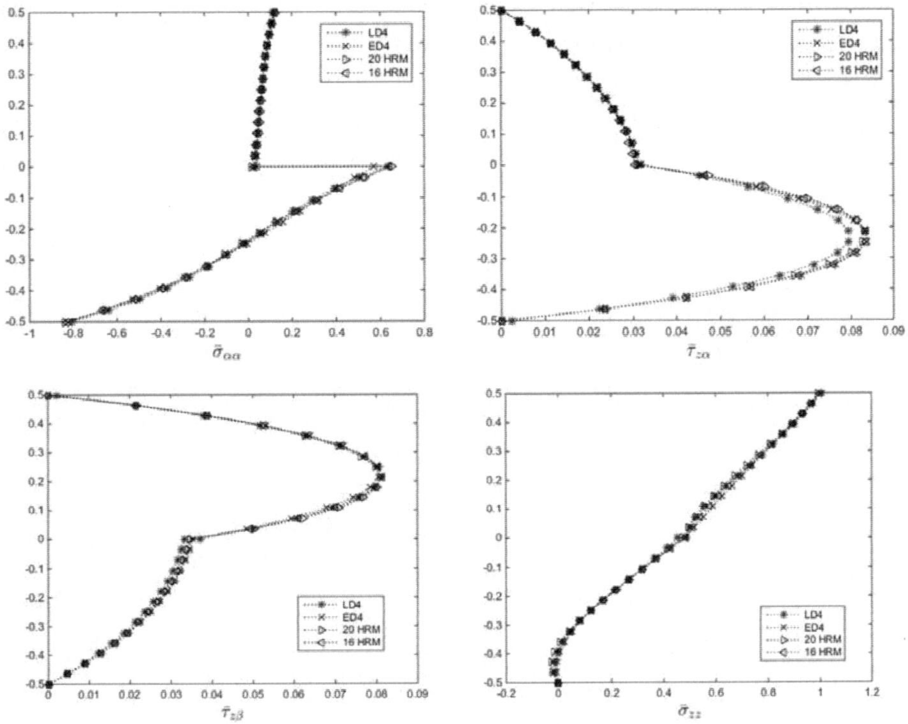

Figure 3. Distribution of dimensionless stresses for Demasi benchmark.

4. Conclusions

This paper presented an analytical solution for the static analysis of doubly curved shells with constant curvature. The shear strain shape functions were considered as hybrid with different combinations of trigonometric, exponential, zig-zag and Maclaurin polynomials thickness expansions. Simply supported shells subjected to bi-sinusoidal loading are analyzed by using the principle of virtual displacement and solved via Navier-closed form solutions. Stresses for different thickness ratios and curvature-radius-to-size are calculated. This work investigated how different types of loadings and the effect of the curvature affected the HRM for shells. The following conclusions are obtained:

(1) Trigonometric functions showed better results over exponential functions for 0/90/0 lamination.

(2) The used of exponential, trigonometric and zig-zag shear functions in the kinematic models are mandatory for the adequate simulation of complex laminations as the one proposed by Demasi.

References

[1] Hildebrand F, Reissner E and Thomas G 1949 *Notes on the foundations of theory of small displacements of orthotropic shells* NACA technical notes No. 1883.
[2] Reddy J N and Liu C 1985 *Int. J. Eng. Sci.* **23** 319-30.
[3] Mantari J and Guedes Soares C 2012 *Comp. Struct.* **94** 2640-56.
[4] Demasi L 2009 *Comp.* **88** 1-16.
[5] Caliri M *et al* 2006 *Comp. Struct.* **156** 63-77.
[6] Tornabene F *et al* 2014 *Comp. Struct.* **107** 675-97.
[7] Dozio L 2016 *Comp. Part B* **98** 97-107.
[8] Brischetto S 2017 *Comp. Part B* **119** 230-52.
[9] Viola E *et al* 2013 *Comp. Struct.* **101** 59-93.
[10] Tornabene F *et al* 2013 *Comp. Part B* **55** 642-59.
[11] Cinefra M and Valvano S 2016 *Mech. Adv. Mater. Struct.* **23**.
[12] Yarasca J *et al* 2017 *Comp. Struct.* **161** 362-83.
[13] Yarasca J *et al* 2017 *Comp. Struct.* **176** 860-76.

[14] Filii M *et al* 2016 *Comp. Struct.* **150** 103-14.
[15] Sandipan N 2017 *Acta Mech.* **228** 69-87.
[16] Ferreira A J M 2011 *Comp.: Part B* **42** 1276-84.

IOP Conf. Series: Materials Science and Engineering **473** (2019) 012007 doi:10.1088/1757-899X/473/1/012007

Fracture Analysis for Torsion Problems of a Deep Sea Spar Platform Main Body

Nan Liu[1], Jisen Liu[2], Wei Liu[3]

[1] School of transportation civil engineering, Shandong Jiaotong University, China
[2] Branch of water administration supervision, Jinan, China
[3] School of transportation civil engineering, Shandong Jiaotong University, China

Abstract. Attention should be paid to variations of crack effects on the stability of offshore structure during building and using. The body of the deep-water Spar platform is the important support of the whole structure, once produced edge cracks in certain parts, the overall structure of the platform would be severely affected. By using boundary integral equations which applied to the edge cracks of the cylinder, this paper discussed singularity of the crack tip; then used the boundary element numerical calculation method to divide the crack boundary into many units, used different interpolation functions to calculate the stress intensity factor of crack tip; Finally, calculated torsional fracture of spar platform with edge cracks under different wind loads, got the allowed maximum crack length of Spar's body with cracks under different wind scale. It was concluded that used the boundary element calculation method was practical in ocean engineering.

1. Introduction

Deep-sea Spar platform has been served in the Marine environment for a long time, various complex environmental loads (such as earthquakes, etc.) may cause the torsional damage of the platform, which may be one of the most important factors in the whole structure damage [1], and the wind load may cause the upper structures of the platform to have a torsion effect on its main part, the main reason is that the wind surface of the whole platform is asymmetric [2]. This article analyzed torsional fracture of the main structure of Spar platform with edge cracks, used the boundary element numerical method to divide the boundary into many units, calculated the stress intensity factor at the crack tip by using different interpolation functions, calculated torsional rigidity of a cylinder with a straight edge crack by programming, and calculated the maximum allowable crack length of the cracked body under different wind conditions at last, which proves that boundary element method is practical in marine engineering.

2. Basic structure of the Spar platform

Main body plays an important role in the whole spar structure. Figure 1 shows a cross-section of the main part of Truss Spar platform which has been put into use at present. From top to bottom, it is divided into hard cabin, middle section and soft cabin. As we can see, the body of the Spar platform is not a simple cylinder structure, but a structure made by many different materials from the inside out. Its internal also contains a lot of facilities and equipment, and its structural type is quite complex, so it's very difficult to analysis directly. We should simplify the main structure appropriately during torsional failure analysis under various marine environmental loads [3]. Each component of the main body as if it is made of one material, as shown in figure 2, the section is a circular structure, the boundary between the outermost and the outer layer is denoted as S_1, the boundary between the inner layer and hollow part is denoted as S_2, thus the circular structure is divided into two parts: the outside is denoted as Ω_1, the edge contains curve crack $\Gamma_k (k = 1,2,...,n)$, and the inside is denoted as Ω_2, which without cracks. The whole main structure is subjected to torsion, the torque is T, shear modulus of materials and poisson ratio are G and ν respectively.

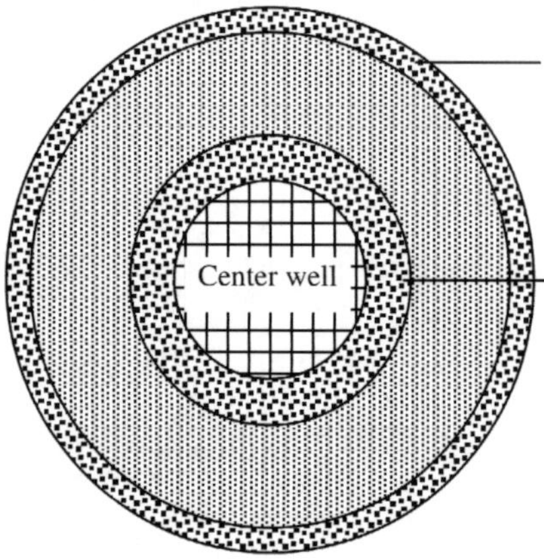

Figure. 1 Major structures of spar platform.

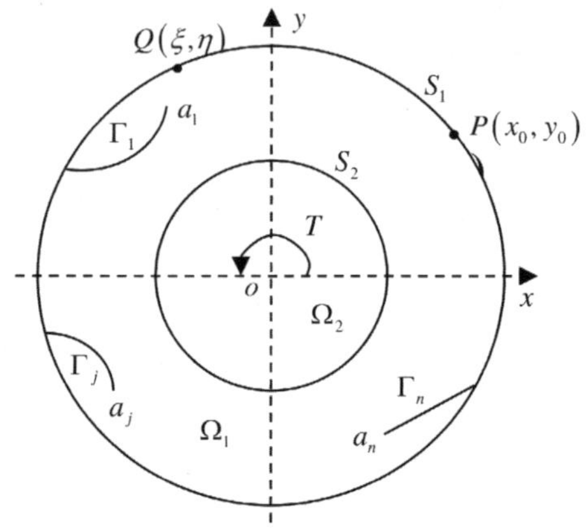

Figure. 2 Section geometry of a cracked cylinder.

3. Crack tip singularity for torsion of a cylinder with edge crack

In this section, the singularity of crack tip is considered. Take the local area with cracks on the boundary as an example, as shown in figure 3.

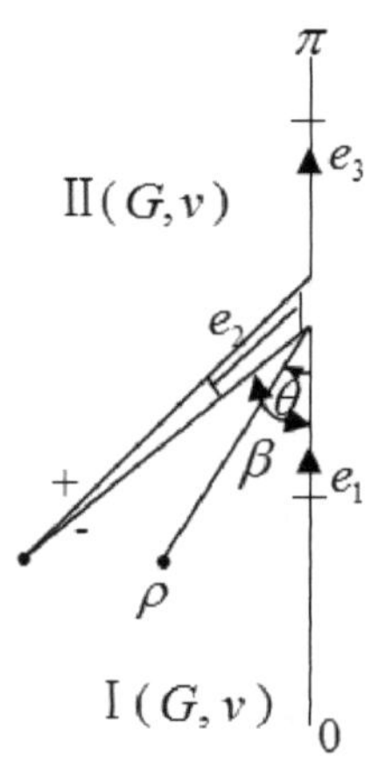

Figure 3. Areas and units division.

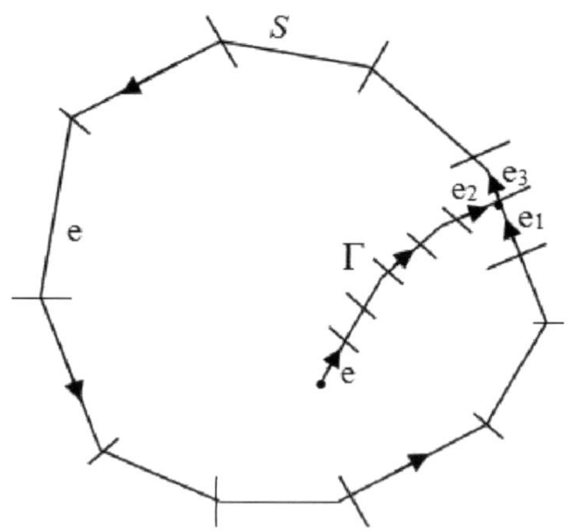

Figure 4. Division on boundary and crack.

(1) For area I ($0 < \theta < \beta$), (G, ν):

$$w_1 = \rho^{\lambda_1} A_1 \cos(\lambda_1 \theta)$$

$$\tau_{\theta z1} = -G\lambda_1 \rho^{\lambda_1 - 1} A_1 \sin(\lambda_1 \theta)$$

$$\gamma_{\rho z1} = \lambda_1 \rho^{\lambda_1 - 1} A_1 \cos(\lambda_1 \theta)$$

Where, $\lambda_1 = \pi / \beta$ ($A_1 \neq 0$, $\lambda_1 > 1$).

(2) For area II ($\beta < \theta < \pi$), (G, ν):

$$w_2 = \rho^{\lambda_2} C_2 \cos[\lambda_2 (\theta - \beta)]$$

$$\tau_{\theta z2} = -G\lambda_2\rho^{\lambda_2-1}C_2\sin[\lambda_2(\theta-\beta)]$$

$$\gamma_{\rho z2} = \lambda_2\rho^{\lambda_2-1}C_2\cos[\lambda_2(\theta-\beta)]$$

Where, $\lambda_2 = \pi/(\pi-\beta)$ ($A_2, B_2 \neq 0, 0<\beta<\frac{\pi}{2}$, $\lambda_2>1$), $C_2 = \frac{A_2}{\cos(\lambda_2\beta)}$.

We should note that when $\beta = \frac{\pi}{2}$, $\lambda_1 = \lambda_2 = 2$, this will cause the interpolation function on the different units repeated. In order to make the following interpolation function meaningful, when $\beta = \frac{\pi}{2}$, we choose $\lambda_1 = \lambda_2 = 4$, this will correspond to the characteristic solution $k = 2$.

4. Boundary element numerical calculation

4.1. Selection of interpolation functions

The resulting boundary integral equation satisfies the condition of displacement singleness, but it is very difficult to solve [4]. Therefore, other methods are considered instead of direct solution. Much practice proves that the boundary element method is a feasible numerical method. The boundary and cracks are divided into M plus N linear units, as shown in figure 3 and 4, that is $S+\Gamma = \sum_e L_e$, $S = \bigcup_{j=1}^{M} L_j$, $T = \bigcup_{j=1+M}^{M+N} L_j$, each linear unit L_e is mapped into the interval $-1 \le t \le 1$, the point Q on the unit L_e is denoted as

$$Q = Q_1^e N_1(t) + Q_2^e N_2(t), (|t|<1, Q \in L_e) \quad (1)$$

Where, Q_1^e and Q_2^e are the two terminal nodes of unit e, take shape functions as follows:

$$N_1(t) = \frac{1}{2}(1-t),\quad N_2(t) = \frac{1}{2}(1+t)$$

The interpolation function on the boundary S and crack Γ is defined as follows. The function of the interpolation function is to close the linear equations after the boundary integral equation is discretized and to satisfy the displacement single value condition. First, the linear interpolation function of the boundary S is defined as follow:

$$F(Q) = F_1^e N_1(t) + F_2^e N_2(t) \quad (|t| \le 1,\ Q \in L_e \subset S) \quad (2)$$

The unit on the crack Γ can be divided into general linear unit and crack tip unit, where the general linear unit does not contain crack tip, and its interpolation function is

$$F(Q) = F_1^e N_1(t) + F_2^e N_2(t) \quad (|t| \le 1,\ Q \in L_e \subset \Gamma) \quad (3)$$

If the initial node $(t=-1)$ of the unit is a crack tip point, $F(Q)$ is denoted as

$$F(Q) = \frac{F_1^e}{\sqrt{N_2(t)}} + (F_2^e - F_1^e)N_2(t) \quad (|t| \le 1,\ Q \in L_e \subset \Gamma) \quad (4)$$

The unit e_2 which is adjacent to unit e_1 on the boundary S, as shown in figure 4, whose interpolation function is

$$F(Q)=(F_1^{e_1}-F_2^{e_1})N_1(t)+\frac{F_2^{e_1}}{[N_1(t)]^{1-\lambda_1}} \quad (|t|\le 1, Q\in L_{e_1}\subset S) \tag{5}$$

The unit e_2 which is adjacent to unit e_3 on the boundary S, the unknown function $F(Q)$ is denoted as

$$F(Q)=\frac{F_1^{e_3}}{[N_2(t)]^{1-\lambda_2}}+(F_2^{e_3}-F_1^{e_3})N_2(t) \quad (|t|\le 1, Q\in L_{e_3}\subset S) \tag{6}$$

For the unit e_2 where the crack Γ intersects with the boundary, the following interpolation function is introduced [5]

$$F(Q)=\frac{F_{2c}^{e_2}}{[N_1(t)]^{1-\lambda_2}}+(F_{1c}^{e_2}-F_{2c}^{e_2}-Q_{2c}^{e_2})[N_1(t)]+\frac{Q_{2c}^{e_2}}{[N_1(t)]^{1-\lambda_1}} \quad (|t|\le 1, Q\in L_{e_2}\subset \Gamma) \tag{7}$$

Thus, on the unit e_1,

$$F(Q)=\frac{\partial\varphi(Q)}{\partial s(Q)}=-\left.\frac{\partial\varphi(Q)}{\partial\rho}\right|_{\theta=0}=-\frac{1}{\alpha}\frac{\partial w_1(\rho,0)}{\partial\rho}=-\frac{1}{\alpha}\lambda_1 A_1\rho^{\lambda_1-1}$$

$$F(Q)=(F_1^{e_1}-F_2^{e_1})N_1(t)+\frac{F_2^{e_1}}{[N_1(t)]^{1-\lambda_1}}, \quad \rho=l_{e_1}N_1(t)$$

Therefore,

$$\lim_{\rho\to 0}[(\rho)^{1-\lambda_1}F(Q)]=-\frac{1}{\alpha}\lambda_1 A_1=(l_{e_1})^{1-\lambda_1}F_2^{e_1}$$

Consequently,

$$F_2^{e_1}=-\frac{1}{\alpha}\lambda_1 A_1(l_{e_1})^{\lambda_1-1} \tag{8}$$

In the same way, on unit e_2,

$$F_{2c}^{e_2}=-\frac{1}{\alpha}\lambda_2 C_2(l_{e_2})^{\lambda_2-1} \tag{9}$$

$$Q_{2c}^{e_2}=-\frac{1}{\alpha}\lambda_1 A_1(l_{e_2})^{\lambda_1-1} \tag{10}$$

On unit e_3,

$$F_1^{e_3}=-\frac{1}{\alpha}\lambda_2 C_2(l_{e_3})^{\lambda_2-1} \tag{11}$$

Then,

$$F_{2c}^{e_2}=(l_{e_2})^{\lambda_2-1}(l_{e_3})^{1-\lambda_2}F_1^{e_3} \tag{12}$$

$$Q_{2c}^{e_2}=(l_{e_1})^{1-\lambda_1}(l_{e_2})^{\lambda_1-1}F_2^{e_1} \tag{13}$$

4.2. Boundary element numerical calculation

The following singular integral formula can be obtained [2]:

$$\int_{-1}^{1} F(Q(t))\frac{dt}{t} = F_2^e - F_1^e, \text{ (for liner unit on the boundary } S,\ L_e \subset S\text{);}$$

$$\int_{-1}^{1} F(Q(t))\frac{dt}{t} = F_2^e - F_1^e, \text{ (for common unit on a crack that does not contain a crack tip, } L_e \subset \Gamma\text{);}$$

$$\int_{-1}^{1} F(Q(t))\frac{dt}{t} = F_1^e\sqrt{2}\ln(3-2\sqrt{2}) + (F_2^e - F_1^e), \text{ (for the initial node unit, } L_e \subset \Gamma\text{).}$$

For unit e_1, e_2, e_3, When β takes any value, λ_1 and λ_2 are not natural numbers, so the Newton's binomial theorem no longer holds, here we only calculate in the case of $\beta = \pi/2$, then, $\lambda_1 = \lambda_2 = 4$, so

$$\int_{-1}^{1} F(Q(t))\frac{dt}{t} = -F_1^{e_1} + \frac{1}{6}F_2^{e_1}, \text{ (for unit } e_1 \text{ on boundary } S,\ L_e \subset S\text{);}$$

$$\int_{-1}^{1} F(Q(t))\frac{dt}{t} = -F_{1c}^{e_2} + \frac{1}{6}F_{2c}^{e_2} + \frac{1}{6}Q_{2c}^{e_2}, \text{ (for unit } e_2 \text{ where crack intersects with the boundary, } L_e \subset \Gamma\text{);}$$

$$\int_{-1}^{1} F(Q(t))\frac{dt}{t} = -\frac{1}{6}F_1^{e_3} + F_2^{e_3}, \text{ (for unit } e_3 \text{ on boundary } S,\ L_e \subset S\text{).}$$

The algebraic equations above can solve out the unknown variables $\{F_j\}$, $Q_{2c}^{e_2}$ and λ_i of the functions $F(Q)$, then the stress intensity factor can be calculated as follow:

$$K_{\text{III}}(a_j) = -\frac{G\alpha}{2}\sqrt{2\pi d_e}F_1^e \tag{14}$$

5. An example of torsion analysis of Spar platform under wind load

In this section, the torsional fracture of the main body in the deep sea Spar platform with linear edge crack under different wind loads is calculated, as shown in figure 5. The main part of the platform is regarded as a simplified cylinder composed of high strength concrete materials containing linear edge cracks[6]. Building facilities on the deck of the platform are 16 to 26*m* above sea level, and the distance between the main center and the point of acting force of wind load is $b = 16m$, and $A = 200m^2$.

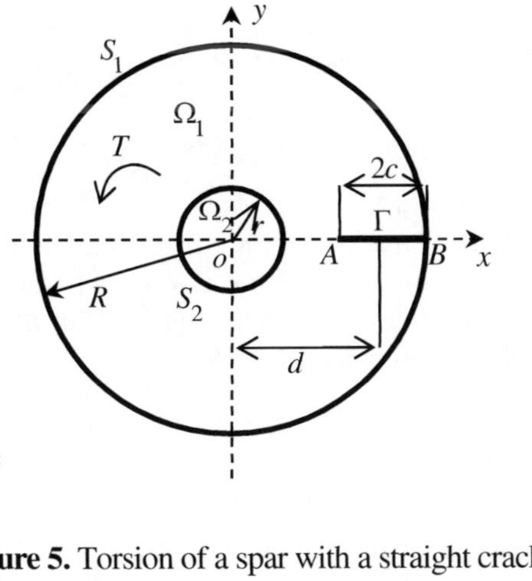

Figure 5. Torsion of a spar with a straight crack.

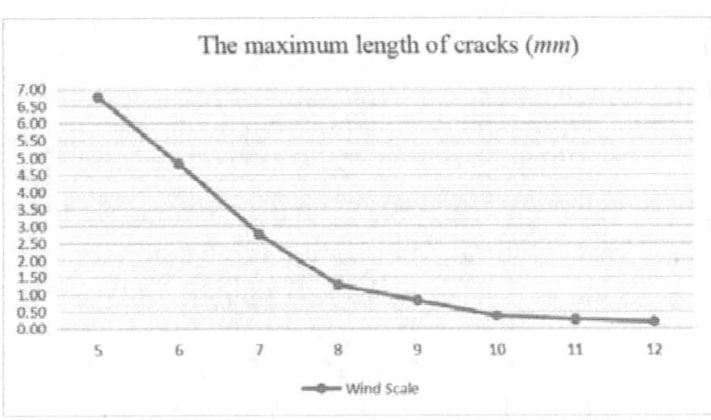

Figure. 6 The maximum length of cracks of main body

The main parameters involved in the calculation are as follows:

$k = 1.5$, $k_z = 1.1$, $\kappa = 0.613.N \cdot s^2 / m^4$, $K_{IC} = 1.105 MN / m^{3/2}$, $[\tau] = 2.22 Mpa$; $R = 16.155m$, $r / R = 0.4$, $c / R = 0.3$, $d / R = 0.7$

By using the present method and a FORTRAN program [7-10], the numerical results are obtained as

$$D^* = 0.996652,\ K^*_{\mathrm{III}}(A) = -0.597362,\ K^*_{\mathrm{III}}(B) = 0$$

Where, $D^* = D / (\frac{\pi}{2} G R^4)$, $K^*_{\mathrm{III}} = K_{\mathrm{III}} / (G \alpha R \sqrt{\pi c})$

From the strain energy density factor criterion (Li, 1998), we can obtain:

$$K_{\mathrm{III}c} = \sqrt{1 - 2\nu} K_{\mathrm{I}c} = 0.816 K_{\mathrm{I}c} = 0.90168 (MN / m^{3/2}),\ (\nu = 0.167)$$

According to the fracture criterion, the allowable maximum crack length of the Spar platform containing the main part of the linear edge crack under different wind loads can be calculated, as shown in figure 6. It is known that when there is fresh breeze, the maximum allowable crack size is 6.7577mm, but when hurricane is coming, the maximum is 0.1805mm. With the increase of wind load, the maximum crack length permitted of the main body with cracks in the platform decreases. This shows that in marine engineering, if we want to avoid the fracture failure of the structure caused by torsion, we must pay attention to the existence of cracks. The security of the platform can be guaranteed only by minimizing the length of cracks in the platform structure, which is also applicable to the design and use of various platforms on the sea [11].

6. Conclusions

The following main conclusions are obtained by analysis:

(1) Under the action of wind load and other factors, cracks in the structure cannot be completely avoided, and cracks are the most important reason for torsional fracture damage in the most critical supporting part of Spar platform.

(2) Analysis of Torsion fracture is very necessary for the safety of platform structure, so it should be carried out throughout the design, construction and using of ocean platform.

(3) The computing method of boundary element numerical is practical in Marine engineering. At the same time, if the structural fracture damage caused by torsion is to be avoided as much as possible, we must pay attention to the cracks, only by minimizing the crack length of the platform structure can we keep its safety, which is also applicable to the design and using of various platforms on the sea.

Acknowledgments

This work was supported by the PhD research start-up and College fund of Shandong Jiaotong University.

References

[1] Zhang M Z 2001 Earthquake Resistant Engineering **3** p 11-14

[2] Lu Z Z 2005 Fracture Analysis for Torsion Problem of Isotropic or Anisotropic Bar with Curvilinear Cracks Master's Dissertation of Ocean University of China chapter **4** pp 41–50

[3] Pan T W 2010 Static and Dynamic Analysis for the Structure of Gravity Platform in Shallow Sea Doctoral Dissertation of Ocean University of China chapter **5** pp 61–85

[4] Guo Z H 1973 Journal of Peking University **1** p 43

[5] Guo Z H 1979 J. of Mech. **3** p 298

[6] Guo Z H 1980 J. of Mech. **2** p 208

[7] Paker A P 1981 E.&F.N. Spon ltd p 36-38

[8] Li Q F 1998 Fracture Mechanics and Its Application in Engineering (Harbin: Harbin engineering university press) p 61-65

IOP Conf. Series: Materials Science and Engineering **473** (2019) 012007 doi:10.1088/1757-899X/473/1/012007

[9] Zhang X E 2000 Sichuan Arch. **20** 4 p 51

[10] Lu Z Z, Wang Y B and Chen L Z 2007 Eng. Mech. **24** 11 p 41-46

[11] Liu N 2013 Fracture Analysis of Offshore Platform Structures Doctoral Dissertation of Ocean University of China chapter **6** pp 76–102

Green Design of Novel Metal Matrix Composites

R Singh[1], N Singh[1], I Farina[2], I Mascolo[3], M De Piano[3], A Amendola[3], F Fraternali[3*]

[1]Department of Production Engineering, Guru Nanak Dev Engineering College, Ludhiana (India)
[2]University of Naples "Pathenope", Naples (Italy)
[3]Department of Civil Engineering, University of Salerno (Italy)

Abstract. In the present research work investigations have been made for surface hardness of Al-SiC metal matrix composite (MMC) prepared by combining fused deposition modelling (FDM), vacuum moulding(VM) and stir casting process. This combination resulted into novel method for preparation of MMC as green process in single step with significant reduction in time. The input process parameters (namely:SiC particle size, proportion of SiC, vacuum pressure and VM sand grin size) were studied at three levels using Taguchi L9 orthogonal array to find out there effect on surface hardness of casted specimens. Microstructure analysis was conducted to support hardness data.

1. Introduction

MMC is engineered combination of the metal or metal alloy and hard particle/ceramic which are reinforced to base metal to get essential properties [1]. It is used to make engine blocks which helps to reduce weight of engine and also improved fuel efficiency [2]. MMCs are functional materials and provide an alternative to conventional monolithic alloys. The use of Al based matrix as functional materials resulted into improvement of physical and mechanical properties (such as: high specific modulus, strength, thermal stability etc.) [3]. Besides, Al based matrix performs better in different conditions such as, higher speed and different load which is beneficial for different fields [4, 5]. Several methods (like: stir casting, infiltration, powder metallurgy and spray forming etc.) are used to prepare Al-SiC MMC [6, 7].The base matrix phase for the present study selected was Al6063, and particles of SiC. In this work, Acrylonitrile-Butadiene-Styrene (ABS) material based pattern was prepared on FDM. This pattern was used for VM assisted stir casting to produce of Al-SiC MMC. FDM is one of the commercial additive manufacturing processes which reduce cycle times, material costs etc. [8-11]. In FDM process, parts are fabricated by extruding a molten ABS material through a heated nozzle in a prescribed pattern onto a platform where material is deposited; it cools, solidifies and bonds with the adjoining material [12, 13]. The previous studies highlights that VM process gives better mechanical properties [14-16]. It was due to the solidification time which can be controlled by controlling the permeability in the mould. Increasing the permeability decreases the solidification time which in turn results in less hardness. In order to increase the hardness, the permeability in the mould has to be less, due to which the heat transfer rate from the cavity-mould interface to the outer of the mould reduces. This could be achieved by increasing the vacuum pressure and by using the sand of fine grit size. High vacuum pressure and finer sand grains increases the mould strength because of tightly packing of sand grains [17, 18]. Tightening the mould could bring more sand into the box and thus increases the time to solidify. This means that increasing the quantity of sand in the mould, reduces the heat transfer rate from the mould. Increasing the sand amount can also be done by providing correct degree of vibration which helps in uniform distribution of sand and completely fills all the corners of the mould. The various steps of VM process shows in figure 1. For this research work, standard geometry of circular disc was taken 50mm diameter and thickness was 10mm. Small holes were made in the FDM pattern for the suction of vacuum to provide proper sticking of plastic film. In stir casting stage, electrical stirrer was used to mix SiC abrasive particles in molten Al. The slurry was heated to a liquid state and then automatic electrically mixing was carried out for 10 min at a normal stirring rate of 500 rpm. The furnace temperature was controlled up to $700^{\circ} \pm 20^{\circ}C$. VM process specimens were casted by Al-6063 alloy.

Figure 1. Steps in VM process.

2. Experimentation

Based upon pilot experimentation, table 1 shows different parameters and their levels for final experimentation. Further table 2 shows control log for experimentation as per Taguchi L9 OA. Based upon table 2, table 3 shows observations for surface hardness. It has been observed that keeping the vacuum pressure below 300mm of Hg results in distorted mould cavity on pouring molten metal during casting. However, when the pressure was above 400 mm of Hg it results in tearing of plastic film which drops the pressure of the mould well below 200mm of Hg hence results in distorted cavity. So the 3 levels for vacuum pressure were selected between 300-400mm of Hg pressure. Three particle size of SiC i.e. single particle size (SPS) 100μm, dual particle size (DPS) 100μm+120μm and triple particle size (TPS) 100μm+120μm+150μm are used as reinforcement.

Table 1. Input parameters.

(**A**) Particle size		(**B**) Vacuum pressure (mm of Hg)	(**C**) Sand grain size AFS No	(**D**) Percentage Of SiC
1	SPS	300	50	5%
2	DPS	350	60	7.5%
3	TPS	400	70	10%

The hardness tests on vacuum molded components were performed using Mitutoyo micro hardness tester (Vickers hardness). The surface obtained after casting was cleaned Based on control log of experimentation (see table 2), three sets of experiments were made as R1, R2 and R3 for measurement of hardness as shown in table 3.

Table 2. Control log for experimentation.

S. no.	Particle size	Vacuum pressure (mm of Hg)	grain size AFS No	Percentage of SiC
1	SPS	300	50	5%
2	SPS	350	60	7.5%
3	SPS	400	70	10%
4	DPS	300	60	10%
5	DPS	350	70	5%
6	DPS	400	50	7.5%
7	TPS	300	70	7.5%
8	TPS	350	50	10%
9	TPS	400	60	5%

Table 3. Observations of hardness.

S. no.	Hardness(HV)		
	R_1	R_2	R_3
1	47.6	49.4	50.7
2	54.6	52.6	53.6
3	59.2	65.5	62.4
4	72.1	68.3	70.2
5	60.5	54.2	54.2
6	64.0	65.0	64.5
7	57.1	58.3	60.4
8	54.6	53.4	54.0
9	61.3	58.9	63.7

The best results for hardness was obtained at DPS particle size of SiC with 10% of SiC and 400mm of Hg vacuum pressure with 60 grit size of sand particle. Particle size of SiC-DPS: It has been observed that hardness of the MMC was best when MMC was prepared using DPS. TPS was found to be the intermediate size among the three particle sizes. Due to this, the amount of DPS added during casting was more than SPS or TPS for the same percentage of addition. This high amount of addition results in more hardness. Percentage of SiC-10%: SiC found to be harder than Al because of which, when added to the Al during casting, hardness of the AMC increases. Which shows that the hardness of the AMC increases with an increase in the percentage of SiC content?

The microstructure images of 9 experiments were taken at 100X as shown in figure2, which shows SiC abrasive particles present in AMC.

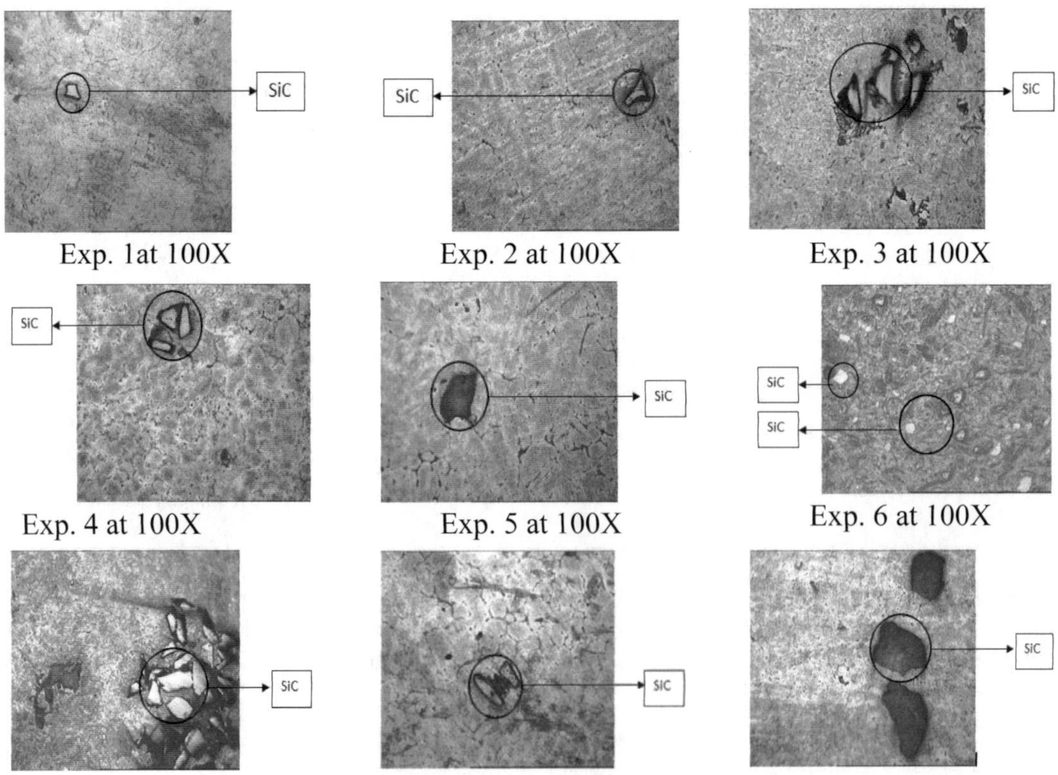

Exp. 1at 100X Exp. 2 at 100X Exp. 3 at 100X

Exp. 4 at 100X Exp. 5 at 100X Exp. 6 at 100X

Exp. 7 at 100X Exp. 8 at 100X Exp. 9 at 100X

Figure 2. Microstructure images of 9 experiments.

Vacuum pressure–400mm of Hg vacuum pressure increases the compaction ratio of the mould increases and gives tight mould which in turn allows extra sand to come into the moulding box and thus reduces the heat transfer rate. This increases the time required for solidification and gives more hardness and grain size–60: As the mesh number increases, the grain becomes finer. So 60 grain size was the intermediate size among the selected sizes and got closely packed on application of pressure. This reduces the permeability in the sand and also the time required for solidification. If finer sand particles are use then it goes in tiny holes of mould box which block the path of vacuum suction. At optimized parametric conditions (i.e. DPS particle size of SiC with 10% of SiC and 400mm of Hg vacuum pressure with 60 grit size of sand particle) the dimensional measurements were made in order to understand the capability of combined process. It has been observed that the dimensional accuracy/ deviation (expressed as difference between nominal dimension and measured dimension) 'Δd' of the castings prepared by combined process are in the range of 0.35-0.39mm with IT grade of castings in range of IT12-13, acceptable as per UNIEN 20286-1 (ISO standard) with process capability index 'Cp' of 1.38.Based upon table 3, table 4 shows percentage contribution of input parameters for hardness. The confirmatory experiments were conducted at suggested settings and the percentage improvement for hardness is 14.71%, which was calculated using the modelling $(F.V - I.V)/I.V *100$. These results are valid for 95% confidence level and are in line with the observations made by other investigators.

Table 4. Percentage contribution of input parameters for hardness.

Parameter	Particle size	Vacuum pressure	% SiC	Sand size
Hardness	34.04%	31.04%	18.84%	15.56%

3. Conclusions

Improvement in hardness has been observed by combining of VM process, FDM and stir casting. The results suggest that particle size of SiC, vacuum pressure, sand grain size and percentage of SiC significantly affects the hardness. The improvement in hardness is found to be 14.71%. The best parameter setting for hardness is: particle size of Sic-DPS, vacuum pressure-400mm/Hg, sand grain size-60 and percentage of SiC-10% respectively. The hardness was also justified by comparing microstructure images. This study may be further carried out with more number of input parameters and their levels. We address studies on the employment of the analyzed materials for the fabrication and testing of innovative materials and structures at different scales to future work [19-59].

Acknowledgements

The authors are thankful to manufacturing research lab, Department of Production Engineering, GNDEC Ludhiana (India) for technical support. IF acknowledges financial support from the Department of Engineering, University of Naples Parthenope. AA, IM and FF acknowledge financial support from the Italian Ministry of Education, University and Research (MIUR) under the 'Departments of Excellence' grant L.232/2016.

References

[1] Singla M, Dwivedi, D D, Singh, L. and Chawla V 2009. *J. Min. Mater. Charact. Eng.* **8(06)** 455.
[2] Shibata K and Ushio H 1994 *Trib. Int.* **27(1)** 39-44.
[3] Nair F and Karamis M B 2010 *Trib. Int.* **43(1-2)** 347-55.
[4] Kiourtsidis G E and Skolianos S M 2002 *Wear* **253(9-10)** 946-56.
[5] Lim S C, Gupta M, Ren L and Kwok J K M 1999 *J. Mater. Proc. Tech.* **89** 591-6.
[6] Eslamian M, Rak J and Ashgriz N 2008 *Powd. Tech.* **184(1)** 11-20.
[7] Singh K and Singh R 2013 Appl. Mech. Mater. **91**.
[8] Lee C W, Chua C K, Cheah C M, Tan L H and Feng C 2004 *Int. J. Adv. Manuf. Tech.* **23(1-2)** 93-101.
[9] Anitha R, Arunachalam S and Radhakrishnan P 2001 *J. Mater. Proc. Tech.* **118(1-3)** 385-8.
[10] Peng A, Xiao X and Yue R 2014 *Int. J. Adv. Manuf. Tech.* **73(1-4)** 87-100.

[11] Chang D Y and Huang B H 2011 *Int. J. Adv. Manuf. Tech.* **53** 1027-37.
[12] Noriega A, Blanco D, Alvarez B J and Garcia A 2013 *Int. J. Adv. Manuf. Tech.* **69(9-12)** 2301-13.
[13] Li Y and Zhang J 2013 *Int. J. Adv. Manuf. Tech.* **69(5-8)** 1819-31.
[14] Bakhtiyarov S I, Overfelt R A, Black M G and Weiss D J 2005 *Trans. American Foundrymen's Soc.* **113** 879.
[15] Bishop D and Bose S 1983 *Trans. American Foundrymen's Soc.* **91** 441-6
[16] Jain C K and Gaindhar J L 1986 *Trans. American Foundrymen's Soc.* **94** 657-62.
[17] Jain C K and Gaindhar J L 1986 *Trans. American Foundrymen's Soc.* **94** 657-62.
[18] Singh R, Singh J and Singh J 2012 *J. Institution Eng. (India): Series C* **93(4)** 325-30.
[19] Ascione L and Fraternali F 1992 *Comput. Struct.* **45(5-6)** 985-99.
[20] Fraternali F and Reddy J N 1993 *Int. J. Solids Struct.* **30(24)** 3337-55.
[21] Fraternali F and Bilotti G 1997 *Comput. Struct.* **62(5)** 837-59.
[22] El Sayed T, Mota A, Fraternali F and Ortiz M 2008 *Comput. Methods Appl. Mech. Eng.* **197(51-52)** 4692-701.
[23] El Sayed T, Mock W, Mota A, Fraternali F and Ortiz M 2009 *Comput. Mech.* **43(4)** 525-34.
[24] Ngo D, Fraternali F and Daraio C 2012 *Phys. Rev. E* **85(3)** 036602.
[25] Raney J R, Fraternali F, Amendola A and Daraio C 2011 *Compos. Struct.* **93(11)** 3013-18.
[26] Fraternali F 2011 *Mech. Res. Commun.* **38(3)** 176-80.
[27] Colangelo F, Cioffi R, Liguori B and Iucolano F 2016 *Compos. Part B: Eng.* **106** 234-41.
[28] Fraternali F, Marino A, Sayed T E and Cioppa A D 2011 *Mech. Adv. Mater. Struct.* **18(4)** 225-43.
[29] Fraternali F, Lorenz C D and Marcelli G 2012 *J. Comput. Phys.* **231(2)** 528-40.
[30] Schmidt B and Fraternali F 2012 *J. Mech. Phys. Solids* **60(1)** 172-80.
[31] Fraternali F, Senatore L and Daraio C 2012 *J. Mech. Phys. Solids* **60(6)** 1137-44.
[32] Blesgen T, Fraternali F, Raney J R, Amendola A and Daraio C 2012 *Mech. Res. Commun.* **45** 58-63.
[33] Amendola A, Carpentieri G, De Oliveira M, Skelton R E and Fraternali F 2014 *Compos. Struct.* **117** 234-43.
[34] Fraternali F, Carpentieri G, Modano M, Fabbrocino F and Skelton R E 2015 *Compos. Struct.* **134** 247-54.
[35] Fabbrocino F, Farina I, Berardi V P, Ferreira A J M and Fraternali F 2015 *Compos. Part B: Eng.* **83** 297-305.
[36] Fraternali F, De Chiara E and Skelton R E 2015 *Smart Mater. Struct.* **24(10)** 105032.
[37] Farina I, Fabbrocino F, Carpentieri G, Modano M, Amendola A, Goodall R, Feo L and Fraternali F 2016 *Compos. Part B: Eng.* **90** 76-85.
[38] Amendola A, Smith C J, Goodall R, Auricchio F, Feo L, Benzoni G and Fraternali F 2016 *Compos. Struct.* **142** 254-62.
[39] Singh R, Singh S and Fraternali F 2016 *Compos. Part B: Eng.* **98** 244-9.
[40] Farina I, Fabbrocino F, Colangelo F, Feo L and Fraternali F 2016 *Compos. Part B: Eng.* **99** 305-11.
[41] Singh R, Kumar R, Feo L and Fraternali F 2016 *Compos. Part B: Eng.* **101** 77-86.
[42] Carpentieri G, Skelton R E and Fraternali F 2017 *Struct. Multidiscipl. Optim.* **55(2)** 449-58.
[43] Fabbrocino F, Farina I and Modano M 2017 *Compos. Part B: Eng.* **115** 376-83.
[44] Fabbrocino F, Farina I, Amendola A, Feo L and Fraternali F 2016 June. Optimal design and additive manufacturing of novel reinforcing elements for composite materials. In *Proceedings of the 7th European Congress on Computational Methods in Applied Sciences and Engineering, Crete, Greece* pp. 5-10.
[45] Carpentieri G, Modano M, Fabbrocino F, Feo L and Fraternali F 2017 *Meccanica* **52(7)** 1561-76.
[46] Singh R, Singh N, Fabbrocino F, Fraternali F and Ahuja I P S 2016 *Compos. Part B: Eng.* **105** 23-9.
[47] Singh R, Bedi P, Fraternali F and Ahuja I P S 2016 *Compos. Part B: Eng.* **106** 20-7.
[48] Boparai K S, Singh R, Fabbrocino F and Fraternali F 2016 *Compos. Part B: Eng.* **106** 42-7.

[49] Fabbrocino F, Modano M, Farina I, Carpentieri G and Fraternali F 2017 *Compos. Struct.* **169** 167-72.
[50] Cimmino M C, Miranda R, Sicignano E, Ferreira A J M, Skelton R E and Fraternali F 2017 *Compos. Part B: Eng.* **115** 275-81.
[51] Colangelo F, Messina F, Di Palma L and Cioffi R 2017 *Compos. Part B: Eng.* **116** 46-52.
[52] Fraternali F and Amendola A 2017 *J. Mech. Phys. Solids* **99** 259-71.
[53] Amendola A, Benzoni G and Fraternali F 2017 *Compos. Part B: Eng.* **115** 117-23.
[54] Orefice A, Mancusi G, Feo L and Fraternali F 2017 *Compos. Struct.* **160** 1126-35.
[55] Singh R, Kumar R, Ranjan N, Penna R and Fraternali F 2018 *Compos. Struct.* **184** 704-13.
[56] Modano M, Mascolo I, Fraternali F and Bieniek Z 2018 *Front. Mater.* **5** 5.
[57] Amendola A, Fraternali F 2018 *Compos. Struct.* **191** 145–53.
[58] Mascolo I, Amendola A, Zuccaro G, Feo L, Fraternali F 2018 *Front. Mater. (Mech. Mater.)* **5(16)**.
[59] Messina F, Ferone C, Colangelo F, Roviello G and Cioffi R 2018 *Comp. Part B: Eng.* **132** 161

IOP Conf. Series: Materials Science and Engineering **473** (2019) 012009 doi:10.1088/1757-899X/473/1/012009

Mathematical Modeling of Surface Roughness in the Forming of Innovative Materials

R Singh[1], N Ranjan[1], I Farina[2], M. De Piano[3], A Amendola[3] and F Fraternali[3*]

[1]Department of Production Engineering, Guru Nanak Dev Engineering College, Ludhiana (India)
[2]University of Naples "Pathenope", Naples (Italy)
[3]Department of Civil Engineering, University of Salerno (Italy)

Abstract. Vacuum moulding (VM) process has many potential engineering applications. Not much work hitherto has been reported for modeling the surface roughness (SR) in VM process. In the present study, outcome of Taguchi model has been used for developing a mathematical model for SR; using Buckingham's π-theorem. Three input parameters namely type of metal/ pouring temperature; shape factor and vacuum pressure were selected to give output in form of SR. This study provides main effect of these variables on SR and shed light on the mechanism of SR in VM process.

1. Introduction

VM process has the ability to obtain high dimensional accuracy and surface finish as compared to other casting processes [1]. Today, this process is used for the casting of pressure pipe fittings, thin-sectioned curtain walls, pulleys, bathtubs, railroad bolsters, machine tools, and engine parts [2]. The size of the castings is ranging from a few ounces up to 10 metric tons. Aluminum alloys, copper alloys, cast iron, carbon steel, manganese and stainless steels are some of common metals, which are currently used in VM process [3-5]. This is one of the casting processes, which is distinctly different from other sand casting processes as this process requires no binders for holding the sand grains together in the mould [6, 7]. The vacuum inside the mould results in a net pressure pushing in, holding the sand rigidly in the shape of the pattern, even after the pattern is removed. The process uses a specially designed, strong, highly flexible polymer film to seal the open ends of the sand mould and form the mould cavity [8]. The literature review reveals that lot of work has been reported on optimization of VM process [9-12]. Various process parameters like vacuum imposed, vibration frequency, pouring temperature and plastic film thickness for the sound casting produced by VM process have been reported. But hitherto no work has been reported for modeling the SR in VM. So, the present investigation has been focused to develop mathematical model for SR in VM. For VM process, an approach to model SR was proposed and applied by some researchers [13]. This model was an attempt for predicting the SR as macro model in VM and is based upon robust design concept of Taguchi technique. The model was mechanistic in sense that parameters can be observed experimentally from a few experiments for a particular material and then used in prediction of SR over a wide range of process parameters. This was demonstrated for VM, where very good predictions were obtained using an estimate of multi parameters at a time. In that study, effects of three process parameters (type of metal/ pouring temperature; vacuum pressure; and shape factor/ geometrical shape) were revealed. Table 1 shows various input and output parameters used in experimental study.

Table 1. Various input and output parameters.

Input Process parameters	Levels			Output process parameter
	1	2	3	
Type of metal/pouring temperature	Pb/375°C	Zn/510°C	Al/700°C	SR (μm)
Vacuum pressure (KPa)	43	48	53	
Geometrical shape	Cylinder	Hemi-sphere	Cuboid	

The relationships were studied by considering interaction between these variables. Table 2 shows control log of experimentation (based upon Taguchi L9 OA) and experimental observations for SR.

Table 2. Control log of experimentation and experimental observations for SR.

Control log of Experimentation				Observations		
S. No.	Material's pouring Temp. (℃)	Geometrical Shape	Vacuum Pressure (KPa)	SR (μm) R1	R2	R3
1	375 (Pb)	Cylinder	43	3.71	3.73	3.76
2	375 (Pb)	Hemisphere	48	3.72	3.75	3.79
3	375 (Pb)	Cuboid	53	3.39	3.41	3.43
4	510 (Zn)	Cylinder	48	3.18	3.19	3.22
5	510 (Zn)	Hemisphere	53	3.31	3.32	3.34
6	510 (Zn)	Cuboid	43	3.42	3.43	3.44
7	700 (Al)	Cylinder	53	3.09	3.12	3.14
8	700 (Al)	Hemisphere	43	3.19	3.21	3.24
9	700 (Al)	Cuboid	48	3.14	3.15	3.16

Note: R1, R2 and R3 are the three repetitions of the experiments to reduce the experimental error.

Based on this model, Pandher and Singh (2011) studied the relationships between SR and controllable process parameters [13]. These relationships agree well with the trends observed by experimental observations made otherwise [3-7].

1.1.Description of the VM process

Figure 1 shows the setup of VM process.VM process can be limited to as few as five items (namely: vacuum system, Film heater, vibrating table, pattern carrier and flasks) for automation or for higher production [9].

Figure 1. VM process [13].

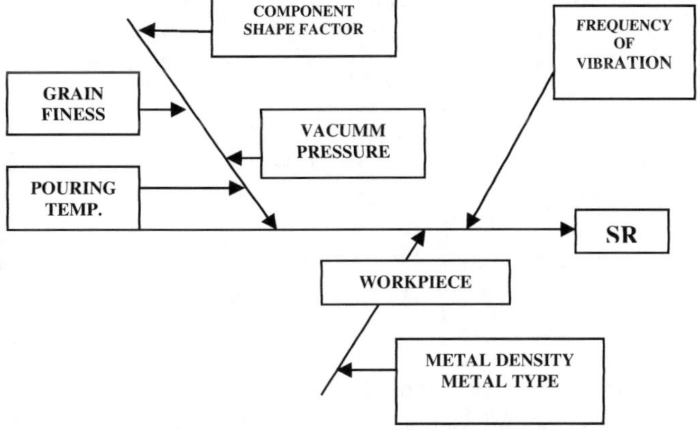

Figure 2. Cause and effect diagram of SR in VM.

There is no need for heavy, noisy jolt squeeze equipment, ramming of slingers. Any shape or size can be produced in a VM from thin walls to thick sections, or from castings weighing ounces to several tons. Fine surface finish and excellent dimensional accuracy, no moisture related defects, no cost for binders, excellent sand permeability, and no toxic fumes from burning the binders are key advantages of VM. The major VM process variables affecting SR are shown as cause and effect diagram (see figure 2).

The study presented in this paper is based on a previously published macro model based on Taguchi robust design [13]. The parameters like AFS No. 70 /grain size (0.210mm), frequency of vibration (30Hz) and casting volume (57697.5mm^3) was kept constant for present study.

2. Mathematical modelling of 'SR'

As per Taguchi design SR in VM was significantly dependent on type of metal and vacuum pressure. Table 3 and 4 respectively shows percentage contribution of input parameters and geometric model for SR [13]. The case study under consideration deals primarily with obtaining optimum system configuration in terms of response parameters with minimum expenditure of experimental resources. The best settings of control factors have been determined through experiments.

Table 3. Percentage contribution for SR.

Parameters	Sum of square	Percentage contribution
Metal type/Pouring Temp.	2.235101	77.1295
Geom. Shape	0.107156	3.697764
Vacuum pressure	0.29454	10.16406
Error	0.261058	9.008678

Table 4. Geometric model of SR.

Optimized SR conditions	
Metal type/ pouring temperature	Pb/ 375°C
Vacuum pressure	43 KPa
Geometrical Shape	Hemi sphere

The Buckingham's π-theorem proves that, in a physical problem including "n" quantities in which there are "m" dimensions, the quantities can be arranged in to "n-m" independent dimensionless parameters. In this approach dimensional analysis is used for developing the relations [14, 15].

Since SR, 'Ra' depends upon type of metal; cast metal density has been selected to represent this input parameter. Therefore by selecting basic dimensions: M (mass); L (length);T (time); and θ (temperature) The dimensions of foregoing quantities would then be:-

1. Surface finish "Ra" (μm) L
2. Grain fineness number "N" (μm) L
3. Cast metal density "ρ" (kg/mm^3) ML^{-3}
4. Shape factor "F" (dimensionless) $M^0 L^0 T^0$
5. Pouring Temperature "θ" (°C) θ
6. Vacuum pressure "P" (kgf/mm^2) $M L^{-1} T^{-2}$
7. Frequency of vibration "γ" (1/sec) T^{-1}

$$\text{Now, } Ra = f(N, \rho, F, \theta, P, \gamma) \tag{1}$$

In this case n is 7 and m is 4. So, we can have (n-m = 3) π_1, π_2, and π_3 three dimensionless groups. After some lengthy calculations, we finally obtain

$$Ra = K1 \{ F. \rho. N3.\gamma 2 / P\} \tag{2}$$

Here 'K1' represents vacuum moulding correction factor.

Now by keeping F. ρ. $N^3.\gamma^2$ fixed, experiments were performed for different values of P, to find out 'Ra' and 'K1' in equation (2). The actual experimental data for three different types of metals/ metal density have been collected and plotted in figure 3.

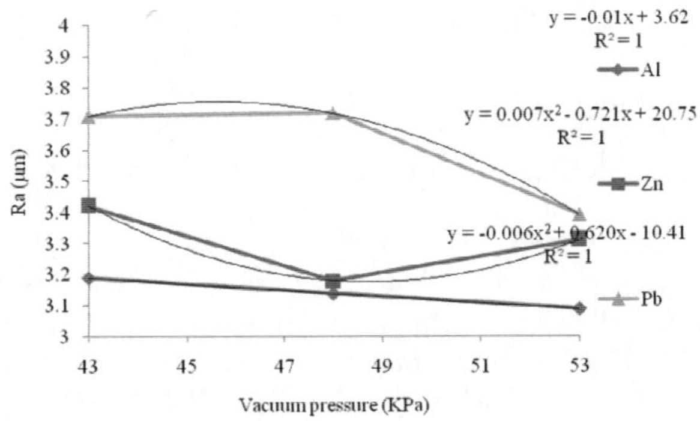

Figure 3. Ra Vs vacuum pressure for different type of metals.

The data collected has been further used for finding best fitting curve with coefficient of co-relation equals to "1". Thus equation (2) of Ra for this case becomes-Ra = K1 {F. ρ. N^3. γ^2/ P}

For metal Al (ρ=2.7g/cm³), pouring temperature 700°C

$$Ra = (-0.01P + 3.62)\ F.\ N^3.\ \gamma^2 \quad (3)$$

For metal Zn (ρ=7.14g/cm³), pouring temperature 510°C

$$Ra = (0.0074P^2 - 0.7214P + 20.758).\ F.\ N^3.\ \gamma^2 \quad (4)$$

For metal Pb (ρ=11.34g/cm³), pouring temperature 375°C

$$Ra = (-0.0068P^2 + 0.6208P - 10.411).\ F.\ N^3.\ \gamma^2 \quad (5)$$

It has been experimentally observed that dependence of SR in form of Ra is of the form given in equation 3-5. Since metal type/ pouring temperature is coming as most contributing factor, three different equations (3-5) have been separately deduced for expressing three different conditions of metal type. So, this final form of equations with any consistent system of units for the quantities may be used to get the main effect of variables on SR.

3. Conclusions

The Buckingham's π-theorem has been used for mathematical modeling of SR in VM process. The contribution of input parameters to SR of casting is: Metal type/pouring temperature 77.12%, vacuum pressure 10.1%, and geometrical shape 3.6%. The optimized settings for minimum SR is: pouring temperature 700(°C), geometrical shape (Cuboid) and vacuum pressure 53(KPa). The mathematical equation developed here sufficiently express all significant input parameters (see equation3-6) with coefficient of co-relation ≈ 1. The results of the study presented in this paper pave the way to relevant future research on the optimal fabrication of innovative engineering materials and structures [16-54], taking advantage of the high dimensional accuracy and surface finish of vacuum molding.

Acknowledgements

The authors are thankful to manufacturing research lab, Department of Production Engineering, GNDEC Ludhiana (India) for technical support. IF acknowledges financial support from the Department of

Engineering, University of Naples Parthenope. AA and FF acknowledge financial support from the Italian Ministry of Education, University and Research (MIUR) under the 'Departments of Excellence' grant L.232/2016.

References

[1] Bakhtiyarov S I, Overfelt R A, Black M G and Weiss D J 2005 *Trans. American Foundrymen's Soc.* **113** 879.
[2] Barua P B, Kumar P and Gaindhar J L 1996 *Qual. Reliab. Eng. Int.* **12(6)** 421-7.
[3] Bishop D and Bose S 1983 *Trans. American Foundrymen's Soc.* **91** 441-6.
[4] Gaindhar J L, Jain C K and Subbarathnanmaiah K 1985 *Trans. American Foundrymen's Soc.* **93** 303-6.
[5] Gouwens P R 2005 *Trans. American Foundrymen's Soc.* **98**-174.
[6] Jain C K and Gaindhar J L 1986 *Trans. American Foundrymen's Soc.* **94** 657-62.
[7] Kumar P and Creese R C 1999 *Trans. American Foundrymen's Soc.* **107** 95-8.
[8] Kumar P and Creese R C 2001 *Trans. American Foundrymen's Soc.* **109** 221-36.
[9] Kumar P and Gaindhar J L 1995 *Qual. Reliab. Eng. Int.* **11(3)** 175-81.
[10] Kumar P, Barua P B and Gaindhar J L 2000 *Qual. Reliab. Eng. Int.* **16(6)** 475-85.
[11] Kumar P and Goel P 2002 *Qual. Eng.* **15(1)** 1-8.
[12] Kumar S, Kumar P and Shan H S 2007 *Mater. Manuf. Process.* **22(7-8)** 879-86.
[13] Pandher S C and Singh R 2011 *Int. J. Adv. Mechatron. Robot.* **3** 97
[14] Singh R and Khamba J S 2009 *Int. J. Adv. Manuf. Tech.* **43(5-6)** 573-80.
[15] Singh R and Khamba J S 2008 *Int. J. Abrasive Tech.* **2(1)** 3-24.
[16] Amendola A, Smith C J, Goodall R, Auricchio F, Feo L, Benzoni G and Fraternali F 2016 *Compos. Struct.* **142** 254-62.
[17] Singh R, Singh S and Fraternali F 2016 *Compos. Part B: Eng.* **98** 244-9
[18] Farina I, Fabbrocino F, Colangelo F, Feo L and Fraternali F 2016 *Compos. Part B: Eng.* **99** 305-11.
[19] Singh R, Kumar R, Feo L and Fraternali F 2016 *Compos. Part B: Eng.* **101** 77-86.
[20] Carpentieri G, Skelton R E and Fraternali F 2017 *Struct. Multidiscipl. Optim.* **55(2)** 449-58.
[21] Carpentieri G, Modano M, Fabbrocino F, Feo L and Fraternali F 2017 *Meccanica* **52(7)** 1561-76.
[22] Singh R, Singh N, Fabbrocino F, Fraternali F and Ahuja I P S 2016 *Compos. Part B: Eng.* **105** 23-9.
[23] Singh R, Bedi P, Fraternali F and Ahuja I P S 2016 *Compos. Part B: Eng.* **106** 20-7.
[24] Boparai K S, Singh R, Fabbrocino F and Fraternali F 2016 *Compos. Part B: Eng.* **106** 42-7.
[25] Fabbrocino F, Modano M, Farina I, Carpentieri G and Fraternali F 2017 *Compos. Struct.* **169** 167-72.
[26] Cimmino M C, Miranda R, Sicignano E, Ferreira A J M, Skelton R E and Fraternali F 2017 *Compos. Part B: Eng.* **115** 275-81.
[27] Fraternali F and Amendola A 2017 *J. Mech. Phys. Solids* **99** 259-71.
[28] Farina I, Fabbrocino F, Carpentieri G, Modano M, Amendola A, Goodall R, Feo L and Fraternali F 2016 *Comp. Part B: Eng.* **90** 76.
[29] Amendola A, Benzoni G and Fraternali F 2017 *Compos. Part B: Eng.* **115** 117-23.
[30] Orefice A, Mancusi G, Feo L and Fraternali F 2017 *Compos. Struct.* **160** 1126-35.
[31] Singh R, Kumar R, Ranjan N, Penna R and Fraternali F 2018 *Compos. Struct.* **184** 704-13.
[32] Modano M, Mascolo I, Fraternali F and Bieniek Z 2018 Front. Mater. **5** 5.
[33] Amendola A, Fraternali F 2018 Compos. Struct. **191** 145–53.
[34] Mascolo I, Amendola A, Zuccaro G, Feo L and Fraternali F 2018 *Front. Mater. (Mech. Mater.)* **5(16)**.
[35] Colangelo F and Cioffi R 2017 *Comp. Part B: Eng.* **115** 43.
[36] Fabbrocino F, Farina I and Modano M 2017 *Compos. Part B: Eng.* **115** 376-83.

[37] Fabbrocino F, Farina I, Amendola A, Feo L and Fraternali F 2016 June. Optimal design and additive manufacturing of novel reinforcing elements for composite materials. In *Proceedings of the 7th European Congress on Computational Methods in Applied Sciences and Engineering, Crete, Greece* pp. 5-10.
[38] Ascione L and Fraternali F 1992 *Comput. Struct.* **45(5-6)** 985-99.
[39] Fraternali F and Reddy J N 1993 *Int. J. Solids Struct.* **30(24)** 3337-55.
[40] Fraternali F and Bilotti G 1997 *Comput. Struct.* **62(5)** 837-59.
[41] El Sayed T, Mota A, Fraternali F and Ortiz M 2008 *J. Biomech.* **41(7)** 1458-66.
[42] El Sayed T, Mock W, Mota A, Fraternali F and Ortiz M 2009 *Comput. Mech.* **43(4)** 525-34.
[43] Ngo D, Fraternali F and Daraio C 2012 *Phys. Rev. E* **85(3)** 036602.
[44] Raney J R, Fraternali F, Amendola A and Daraio C 2011 *Compos. Struct.* **93(11)** 3013-8.
[45] Fraternali F 2011 *Mech. Res. Commun.* **38(3)** 176-80.
[46] Fraternali F, Marino A, Sayed T E and Cioppa A D 2011 *Mech. Adv. Mater. Struct.* **18(4)** 225-43.
[47] Fraternali F, Lorenz C D and Marcelli G 2012 *J. Comput. Phys.* **231(2)** 528-40.
[48] Schmidt B and Fraternali F 2012 *J. Mech. Phys. Solids* **60(1)** 172-80.
[49] Fraternali F, Senatore L and Daraio C 2012 *J. Mech. Phys. Solids* **60(6)** 1137-44.
[50] Blesgen T, Fraternali F, Raney J R, Amendola A and Daraio C 2012 *Mech. Res. Commun.* **45** 58-63.
[51] Amendola A, Carpentieri G, De Oliveira M, Skelton R E and Fraternali F 2014 *Compos. Struct.* **117** 234-43.
[52] Fraternali F, Carpentieri G, Modano M, Fabbrocino F and Skelton R E 2015 *Compos. Struct.* **134** 247-54.
[53] Fabbrocino F, Farina I, Berardi V P, Ferreira A J M and Fraternali F 2015 *Compos. Part B: Eng.* **83** 297-305.
[54] Fraternali F, De Chiara E and Skelton R E 2015 *Smart Mater. Struct.* **24(10)** 105032.

Mechanical and Experimental Study on the use of Sustainable Materials for Additive Manufacturing

R Singh[1], R Kumar[1], I Farina[2], M De Piano[3], A Amendola[3], F Fraternali[3*]

[1] Dept. of Production Engineering, Guru Nanak Dev Engineering College, Ludhiana (India)
[2] Department of Engineering, University of Naples Parthenope, Naples (Italy)
[3] Department. of Civil Engineering, University of Salerno (Italy)

Abstract. In the present study multi-material 3D printing of acrylonitrile butadiene styrene (ABS), polylactic acid (PLA) and high impact polystyrene (HIPS) has been performed by using a low cost printing technology for manufacturing of innovative and sustainable composite materials. Further thermal (heat capacity at glass transition temperature, thermal conductivity) and mechanical (break elongation, break load, percentage elongation at break, break strength and young's modulus in tensile as well as pull out test) properties have been investigated for 3D printed multi-material components.

1. Introduction

3D printing- a better known as additive manufacturing (AM) forms the object by successive layers of materials for different application areas. 3D printing has the potential/impact to transform manufacturing supply chain, distribution channel and business model. There are some of the 3D printing technologies which are presently used by the society to apply as specific solutions. In recent years, AM has become the most common technique for fabricating periodic lattices and mechanical and acoustic meta materials. Several fabrication methods have been proposed in this field, with resolution ranging from the centimeter- to the nanometer-scale. Worth mentioning here are: polyjet 3-D printing technologies; multi-material fused deposition modelling (FDM), electron beam melting; x-ray lithography; deep ultraviolet lithography; deep Ultraviolet lithography; soft lithography; two-photon polymerization; atomic layer deposition; and projection micro-stereolithography, among other available methods [1-5]. FDM is a low-cost technique of AM which is used to prepare the functional prototypes of polymers/composites [6-8]. In FDM, part is build layer by layer by heating a thermoplastic filament to a semi-liquid state and extruding it through a small nozzle per 3D CAD models usually in STL format [9-11]. Manufacturing and technology environments are fuelling a new generation of engineer, scientist and designers. Just as it made at- home polymer 3D printers a commonality today, the world could eventually see a metal or ceramic 3D printer become common in an average household. To achieve this goal, future inventions in next generation structures using existing materials via AM will surely need to revolve around cost reduction, improved performance, and advanced structural design [12]. The study conducted for 3D printing of multilateral component of ABS and thermoplastic polyurethane (TPU) reveals with support of 3D imaging that interface properties are found in control with good layers connectivity [13]. Multi-material 3D printing potentials to be a milestone in rapid manufacturing, customized design and structural applications. Being compatible of functionally graded materials in a single structural form can be applied potentially in structural application to get the benefit of combined/hybridised properties. The multi-material printing provides a fast and robust structure with compact functionality of all combined materials [14]. The multi-material 3D printing applicable to fascinating application of new smart 4D structures which can provide the achievement of shape/property/functionality [15]. It was highlighted that existing AM techniques such as FDM can be modified to hybrid deposition manufacturing (HDM) with embedded component to produce more complex, integrated multi-material component than traditional technique [16].

Fewer studies have been reported in recent past to enhance the properties of thermoplastics by reinforcing with metals/ non-metals through extrusion and finally 3D printing of same by FDM. But hitherto no work has been reported on multi material printing of thermoplastics to enhance its mechanical properties. In the present study ABS, PLA and HIPS were 3D printed in multi nozzle FDM to explore the applicability of final product in structural application (repair/maintenance of heritage buildings).

2. Experimental details

Experimentation stage contains evaluation of melting and solidification characteristics, glass transition temperature determination, extrusion and multi-material 3D printing based upon an experimental design. ABS has high toughness, high degree of mould-ability and low thermal conductivity and PLA has good biodegradability/crystallinity whereas HIPS is of low cost with high impact resistance. These three thermoplastics have been selected for the fabrication/multi-material printing operation with FDM. Table 1 shows mechanical, thermal and rheological properties (average value of 3 consecutive repetitions) of feedstock material. It should be noted that ABS, PLA and HIPS are having significant differences in their melt floe index (MFI), glass transition temperature (GT), peak load (PL), peak strength (PS), peak elongation (PE), Young's modulus (YM) and yield stress (YS). The aim of the present study is to fabricate the new composite of all three polymers so that final product must be comprised of advantages of all polymers. Following the design of experiment based upon Taguchi L9 orthogonal array, the multi-material parts have been prepared on FDM and mechanical properties have been evaluated as shown in table 2.

Table 1. Mechanical, rheological and thermal properties ABS, PLA and HIPS.

Polymer	MFI (g/10 min)	GT (°C)	PL (N)	PS (MPa)	PE (mm)	%PE (%)	YM (MPa)	YS (MPa)
ABS	8.7	109.76	207.0	10.78	4.75	6	175	0.49
PLA	13.5	62.57	282.4	14.71	5.13	7	47.9	0.27
HIPS	7.5	100.41	80.8	4.21	1.90	3	112.5	3.44

Table 2. Experimental design for multi-material 3D printing on FDM and mechanical properties of 3D printed multi-material component.

Exp no.	Material Combination	Infill (%)	Speed (mm/s)	PL (N)	PS (MPa)	PE (mm)	PE (%)	YM (MPa)	YS (MPa)
1	APH	60	50	133.9	6.97	2.85	4	72.92	2.73
2	APH	50	60	179.9	9.37	2.85	4	264.58	2.68
3	APH	100	70	206.9	10.78	4.37	6	73.29	0.21
4	PHA	60	60	161.3	8.40	3.04	4	325.00	1.00
5	PHA	80	70	189.9	9.89	3.99	5	79.17	4.54
6	PHA	100	50	187.9	9.79	3.23	4	108.33	5.13
7	HAP	60	70	149.0	7.76	3.99	5	85.42	0.28
8	HAP	80	50	174.8	9.10	3.42	5	161.84	0.17
9	HAP	100	60	164.4	8.56	3.61	5	249.67	0.16

The thermal properties have been checked by differential scanning calorimeter (DSC). The endothermic reaction was carried under the heating rate of +10°C/min from 30°C to 250°C, whereas exothermic reaction was carried under -10°C/min from 250°C to 30°C.In the present case extrusion with TSE was performed under the temperature of 230°C, rotational speed of 50rpm and with applied load of 10Kg to prepare the feedstock filaments of 1.75±0.05mm. The extrusion parameters were determined by pilot experimentation based upon uniformity and dimensional accuracy. Commercial open source FDM setup (Company: Divide by Zero) configured with two nozzle head was selected for multi-material 3D printing. The static parameters for fabrication of composite parts were nozzle diameter of 0.3mm, filament diameter of 1.75±0.05mm, layer height 0.27mm, 3 perimeters (by adjusting 3 top and 3 bottom layers), rectilinear fill pattern, 30mm/sec perimeter speed, travel speed of 130mm/sec, at extrusion temperature of 250°C and bed temperature of 55°C. The two parameters were varied for fabrication (i) infill percentage of 60, 80 and 100%, (ii) under printing speed of 50, 60 and 70mm/sec. The multi-material printing was customized by total 12 layers with 4 layers of each material. The multi-material printing was configured as APH (4 layers of ABS on bottom, 4 layers of PLA in middle and 4 layers of HIPS on top), PHA (4 layers of PLA on bottom, 4 layers of HIPS in middle and 4 layers of ABS on top) and HAP (4 layers of HIPS on bottom, 4 layers of ABS in middle and 4 layers of PLA on top).

3. Results and discussion

It was observed that the extruded feedstock of ABS, PLA and HIPS resulted in the significant differences as regards to mechanical properties are concerned. The experiment average by three repeated trails outlined that ABS was having maximum YM, PLA was having maximum PL, PS, PE and minimum YM and YS, whereas HIPS resulted in minimum PL, PS, PE and minimum YM and YS (See table 1). Figure 1 shows 3D printed multi-material component and the load Vs. deflection curves of ABS, PLA and HIPS material undergone tensile failure.

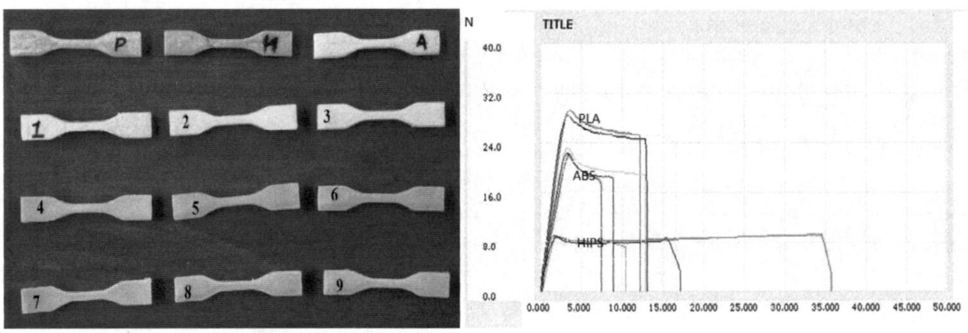

Figure 1. (a) 3D printed multi-material component,
(b) load Vs. deflection curves of ABS, PLA and HIPS.

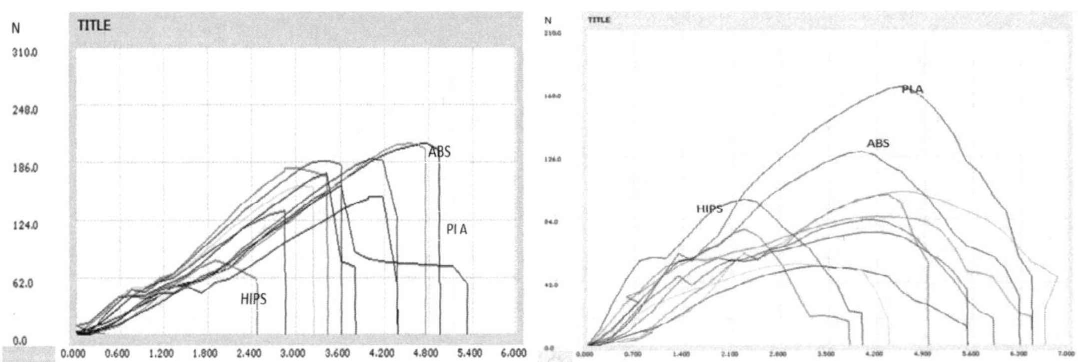

Figure 2. Load Vs. deflection curve (a) for tensile properties (b) for pull out properties.

As per ASTM D 638 type IV, the material tested through ultimate testing machine. After the fracture of each sample, data were recorded and shown in table 2. It was observed that at experiment no. 3 with APH multi-material configuration, 100% infill percentage and 70mm/sec printing speed resulted in the maximum value of PL, PS, elongation properties and minimum YM whereas at experiment no. 1 with APH, 60% infill and 50mm/sec printing speed configuration it resulted in minimum values of PL, PS and PE properties. Component at experiment no. 4 resulted in the maximum YM. It was observed from here that at best setting of input variable, the PL, PS, PE, percentage of PE was achieved greater that HIPS but lower than PLA and ABS. The most important fact was observed in case of YM where at experiment no. 2, 4 and 9 resulted in the values greater than any of the parent material. Again, the YS at experiment no. 3, 8 and 9 resulted in the values lesser than each parent material (See figure 2 (a)). Since pull out test is one of the most important consideration to justify it for the structural applications. The pull-out test was conducted on all the samples with provision of evaluating the PL, PS, PE and percentage changes of PE. It was observed that strength of multilateral component was resulted greater than HIPS but lesser than ABS and PLA as shown in figure 2(b).

4. Concluding remarks

Following conclusions have been drawn from the present study of 3D printing of multilateral component for structural applications:

- Multi-material 3D printing of ABS, PLA and HIPS polymer is feasible because these thermoplastic possess similar heat capacities (13.63mJ for ABS, 14.71mJ for PLA and 11.71mJfor HIPS).
- From tensile properties investigations, it was observed that the peak strength of HIPS (4.21 MPa) was minimum to other material (PLA and ABS). But 3D printing of multi-material resulted in improved properties to a significant level (10.78 MPa) under controlled input conditions.
- It was observed that strength of multilateral component was observed greater than HIPS but lesser than ABS and PLA in pull out test.

The use of multi-material 3D printing techniques for the rapid prototyping of next-generation meta materials and structures, whose mechanical response is largely derived from the geometry of the microstructure will form the subject of future studies [16]-오류! 참조 원본을 찾을 수 없습니다..

Acknowledgements

The authors are thankful to manufacturing research lab, Department of Production Engineering, GNDEC Ludhiana (India) for technical support. IF acknowledges financial support from the Department of Engineering, University of Naples Parthenope. AA, LF and FF acknowledge financial support from the Italian Ministry of Education, University and Research (MIUR) under the 'Departments of Excellence' grant L.232/2016.

Reference

[1] Truby R L, Lewis J A 2016 *Nature* **540** 371-8.
[2] Amendola A, Carpentieri G, De Oliveira M, Skelton R E and Fraternali F 2014 *Compos. Struct.* **117** 234-43.
[3] Amendola A, Nava E H, Goodall R, Todd I, Skelton R E and Fraternali F 2015 *Compos. Struct.* **131** 66-71.
[4] Fraternali F, Carpentieri G and Amendola A 2015 *J. Mech. Phys. Solids* **74** 136-57.
[5] Amendola A, Smith C J, Goodall R, Auricchio F, Feo L, Benzoni G and Fraternali F 2016 *Compos. Struct.* **142** 254-62.
[6] Kumar R, Singh R, Ahuja I P S, Amendola A and Penna R 2018 *Compos. Part B: Eng.* ***132*** 244-57.
[7] Kumar R, Singh R, Ahuja I P S, Penna R and Feo L 2017a *Compos. Part B: Eng.* **137** 1-15.
[8] Kumar R, Singh R, Hui D, Feo L and Fraternali F 2017b *Compos. Part B: Eng.* **134** 193-206.
[9] Singh R, Kumar R, Feo L and Fraternali F 2016 *Compos. Part B: Eng.* **101** 77-86.
[10] Singh R, Kumar R, Mascolo I and Modano M 2018a *Compos. Part B: Eng.* **143** 132-40.
[11] Singh R, Kumar R, Ranjan N, Penna R and Fraternali F 2018b *Compos. Struct.* **184** 704-13.
[12] Bandyopadhyay A and Heer B 2018 *Mater. Sci. Eng. R: Reports* **129** 1-16.

[13] Guessasma S, Nouri H and Roger F 2017 *Polym.* **9(8)** 372.
[14] Muguruza A et al 2017 *Procedia Manuf.* **13** 746-53.
[15] Momeni F, Liu X and Ni J 2017 *Mater. Des.* **122** 42-79.
[16] Ma R R, Belter J T and Dollar A M 2015 *J. Mech. Robot.* **7(2)** 021002.
[17] Ascione L and Fraternali F 1992 *Comput. Struct.* **45(5-6)** 985-99.
[18] Fraternali F and Reddy J N 1993 *Int. J. Solids Struct.* **30(24)** 3337-55.
[19] Fraternali F and Bilotti G 1997 *Comput. Struct.* **62(5)** 837-59.
[20] El Sayed T, Mota A, Fraternali F and Ortiz M 2008 *Comput. Methods Appl. Mech. Eng.* **197(51-52)** 4692-701.
[21] El Sayed T, Mock W, Mota A, Fraternali F and Ortiz M 2009 *Comput. Mech.* **43(4)** 525-34.
[22] Ngo D, Fraternali F and Daraio C 2012 *Phys. Rev. E* **85(3)** 036602.
[23] Raney J R, Fraternali F, Amendola A and Daraio C 2011 *Compos. Struct.* **93(11)** 3013-18.
[24] Fraternali F 2011 *Mech. Res. Commun.* **38(3)** 176-80.
[25] Colangelo F, Cioffi R, Liguori B and Iucolano F 2016 *Compos. Part B: Eng.* **106** 234-41.
[26] Fraternali F, Marino A, Sayed T E and Cioppa A D 2011 *Mech. Adv. Mater. Struct.* **18(4)** 225-43.
[27] Fraternali F, Lorenz C D and Marcelli G 2012 *J. Comput. Phys.* **231(2)** 528-40.
[28] Schmidt B and Fraternali F 2012 *J. Mech. Phys. Solids* **60(1)** 172-80.
[29] Fraternali F, Senatore L and Daraio C 2012 *J. Mech. Phys. Solids* **60(6)** 1137-44.
[30] Blesgen T, Fraternali F, Raney J R, Amendola A and Daraio C 2012 *Mech. Res. Commun.* **45** 58-63.
[31] Amendola A, Carpentieri G, De Oliveira M, Skelton R E and Fraternali F 2014 *Compos. Struct.* **117** 234-43.
[32] Fraternali F, Carpentieri G, Modano M, Fabbrocino F and Skelton R E 2015 *Compos. Struct.* **134** 247-54.
[33] Fabbrocino F, Farina I, Berardi V P, Ferreira A J M and Fraternali F 2015 *Compos. Part B: Eng.* **83** 297-305.
[34] Fraternali F, De Chiara E and Skelton R E 2015 *Smart Mater. Struct.* **24(10)** 105032.
[35] Farina I, Fabbrocino F, Carpentieri G, Modano M, Amendola A, Goodall R, Feo L and Fraternali F 2016 *Compos. Part B: Eng.* **90** 76-85.
[36] Amendola A, Smith C J, Goodall R, Auricchio F, Feo L, Benzoni G and Fraternali F 2016 *Compos. Struct.* **142** 254-62.
[37] Singh R, Singh S and Fraternali F 2016 *Compos. Part B: Eng.* **98** 244-9.
[38] Farina I, Fabbrocino F, Colangelo F, Feo L and Fraternali F 2016 *Compos. Part B: Eng.* **99** 305-11.
[39] Singh R, Kumar R, Feo L and Fraternali F 2016 *Compos. Part B: Eng.* **101** 77-86.
[40] Carpentieri G, Skelton R E and Fraternali F 2017 *Struct. Multidiscipl. Optim.* **55(2)** 449-58.
[41] Fabbrocino F, Farina I and Modano M 2017 *Compos. Part B: Eng.* **115** 376-83.
[42] Fabbrocino F, Farina I, Amendola A, Feo L and Fraternali F 2016 June. Optimal design and additive manufacturing of novel reinforcing elements for composite materials. In *Proceedings of the 7th European Congress on Computational Methods in Applied Sciences and Engineering, Crete, Greece* pp. 5-10.
[43] Carpentieri G, Modano M, Fabbrocino F, Feo L and Fraternali F 2017 *Meccanica* **52(7)** 1561-76.
[44] Singh R, Singh N, Fabbrocino F, Fraternali F and Ahuja I P S 2016 *Compos. Part B: Eng.* **105** 23-9.
[45] Singh R, Bedi P, Fraternali F and Ahuja I P S 2016 *Compos. Part B: Eng.* **106** 20-7.
[46] Boparai K S, Singh R, Fabbrocino F and Fraternali F 2016 *Compos. Part B: Eng.* **106** 42-7.
[47] Fabbrocino F, Modano M, Farina I, Carpentieri G and Fraternali F 2017 *Compos. Struct.* **169** 167-72.
[48] Cimmino M C, Miranda R, Sicignano E, Ferreira A J M, Skelton R E and Fraternali F 2017 *Compos. Part B: Eng.* **115** 275-81.
[49] Colangelo F, Messina F, Di Palma L and Cioffi R 2017 *Compos. Part B: Eng.* ***116*** 46-52.

[50] Fraternali F and Amendola A 2017 *J. Mech. Phys. Solids* **99** 259-71.
[51] Amendola A, Benzoni G and Fraternali F 2017 *Compos. Part B: Eng.* **115** 117-23.
[52] Orefice A, Mancusi G, Feo L and Fraternali F 2017 *Compos. Struct.* **160** 1126-35.
[53] Singh R, Kumar R, Ranjan N, Penna R and Fraternali F 2018 *Compos. Struct.* **184** 704-13.
[54] Modano M, Mascolo I, Fraternali F and Bieniek Z 2018 *Front. Mater.* **5** 5.
[55] Amendola A, Fraternali F 2018 Compos. Struct. **191** 145–53.
[56] Mascolo I, Amendola A, Zuccaro G, Feo L and Fraternali F 2018 *Front. Mater. (Mech. Mater.)* **5(16)**.

IOP Conf. Series: Materials Science and Engineering **473** (2019) 012011 doi:10.1088/1757-899X/473/1/012011

Lateral-Torsional Buckling of C-Beams with Varying Inertia

I Mascolo[1], M Modano[2], A Amendola[1] and F Fraternali[1]

[1] Department of Civil Engineering, University of Salerno, Italy
[2] Department of Structures for Engineering and Architecture, University of Naples Federico II, Italy

Abstract. The present work deals with a numerical study on the flexural-torsional buckling of a thin-walled beam with uniformly varying C-section. Approximate solutions to the examined problem are obtained through the Dirichlet variational method and the Rayleigh-Ritz discretization procedure. The presented results prove to be asymptotically convergent for increasing numbers of test functions. This convergence property highlights the technical utility of the proposed approach, on considering the difficulty of obtaining analytic solutions for buckling problems of thin-walled beams with non-uniform cross-section.

1. Introduction

Thin-walled open-section members with continuous variation of the cross-sectional areas are increasingly employed in structural, mechanical and aeronautical engineering applications, since they allow for optimising strength, stiffness and stability with significant material savings and economic benefits. When the cross sections vary along the beam axis, thin-walled open section beams show relatively low bending and torsional stiffness coefficients. They tend to buckle out of the plane by deflecting laterally and twisting, as a compressive load is applied [1-3]. The solution to this coupled stability problem can be quite complex and difficult to obtain in analytic form. A number of theoretical and numerical approaches have been proposed in the literature to face this problem, including the Initial Parameter Method, Bifurcation an Dynamic Approaches, Energy Methods, Imperfection Sensitivity Analyses and Finite Element Approximations [4-7]. Worth of mention are the works by Elishakoff [8] on the behaviour of variable cross section beams, and by El-Mahdy [9] on the inelastic lateral-torsional buckling of symmetric beams with variable flange ratio.

We develop in the present paper a numerical investigation on the elastic flexural-torsional buckling of a simply supported thin-walled beam with uniformly varying C-cross section, through a numerical approach base on the Dirichlet variational method. The adopted method provides numerical estimates of the buckling load of the examined beam. It relies on the principle of stationary potential energy and employs the Rayleigh-Ritz discretisation procedure, with varying number of test functions. The given numerical result show that such an approach qualifies as a robust and versatile tool that allows to obtain reliable upper bounds of the true critical load, on considering the remarkable complexity of the mathematical problem ruling the lateral-torsional buckling problem of beams with variable cross-section.

2. Mechanical model

Let us consider a simple supported thin-walled beam made of S420 steel. We assume that that such a beam has C-section with height and width that vary along the length according to the following laws:

$$\begin{aligned} B'(z) &= -0.01z + 28.5 \\ H'(z) &= 0.10z + 87.0 \end{aligned} \qquad (1)$$

while the cross-section thickness δ remains constant ($\delta = 4$ mm).

The beam material is linear elastic with Young's modulus $E = 210\ GPa$; it is subjected to an axial compressive load, F, at the centroid G_A of the free end cross section (as shown in figure 1); it is defined with respect to a right-handed rectangular coordinate system $Gxyz$ (axes z coincides with the longitudinal centroidal axis of the member, while x and y coincide with the two principal axes of the cross section).

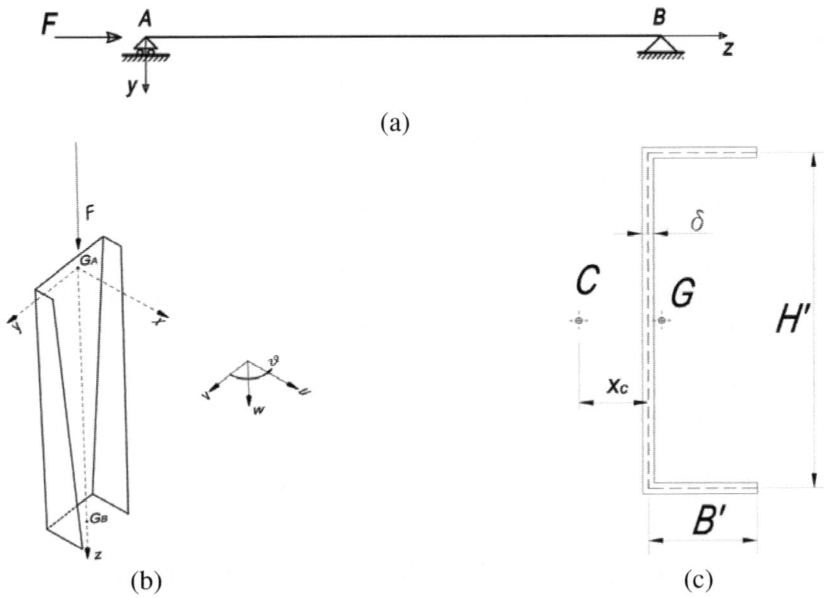

Figure 1. Theoretical model: load arrangement and constrain conditions (a); basic notation (b, c).

2.1. First order torsional and flexural equilibrium equation of thin-walled beam

According to the Vlasov torsion theory, the differential equations of equilibrium describing the flexural-torsional behaviour of the analysed beam [1] are

$$\begin{cases} \left(E\, I_y\, u''\right)' + F\, u' - F\, \hat{y}_G\, \vartheta' = 0 \\ \left(E\, I_x\, v''\right)' + F\, v' + F\, \hat{x}_G\, \vartheta' = 0 \\ C_\omega\, \vartheta'''' + C_\omega'\, \vartheta''' - \left(W_z - \dfrac{F\, I_c}{\mathcal{A}}\right) \vartheta'' - \left(W_z - \dfrac{F\, I_c}{\mathcal{A}}\right)' \vartheta' - F\, u''\, \hat{y}_G + F\, v''\, \hat{x}_G = 0 \end{cases} \tag{2}$$

in which the apex represents the z-derivative operation; $\mathcal{A}$, I_x and I_y are the cross-section area and the second moments of area about the x and y axes, respectively; I_c is the polar moment of area about shear centre; W_z, and C_ω are St. Venant warping rigidity and warping constant, respectively; $\hat{x}_G$ and $\hat{y}_G$ are the coordinates of the centroid with respect to the coordinate system $G\hat{x}\hat{y}\hat{z}$, which is obtained by a translation of the reference $Gxyz$ coordinate system in the shear centre of the free end. The in-plane displacement of an arbitrary centroid point with coordinates x and y are

$$\begin{cases} u = u_c - (y - y_c)\, \vartheta \\ v = v_c + (x - x_c)\, \vartheta \end{cases} \tag{3}$$

where u_c and v_c are the displacements of the shear centre, ϑ is the angle of twist of the cross section, and x_c and y_c are the shear centre coordinates. Due to the variability of the cross sections along the beam and their symmetry with respect to the z axis, the location of the shear centre will be determined by the following equations

$$\begin{cases} x_c = -\dfrac{B'^2 H'^2 \delta}{4\, I_x} + \dfrac{H' N\, \delta}{2\, T_y \mathcal{A}^2} \dfrac{dB'}{dz} \dfrac{B'\delta}{2} - \dfrac{M_x}{\delta} \dfrac{d}{dz}\left(\dfrac{6\, B'\, H'}{6\, B'\, H'^2 + H'^3}\right) \dfrac{B'\delta}{2} \\ y_c = 0 \end{cases} \quad (4)$$

where T_y denotes the shear force in y direction, M_x is the bending moment about x axis, and B' and H' are the lengths depicted in figure 1.

2.2. *Total Potential Energy method and Rayleight-Ritz discretization procedure*

Let $\mathcal{E}$ be the total potential energy which characterises the equilibrium condition in the analysed elastic system. The Dirichlet variational principle assumes that the stationarity of the second variation of $\mathcal{E}$

$$\partial_2\, \mathcal{E} = L_2 + W \quad (5)$$

is associated to a critical position of equilibrium. In equation (5), L_2 and W are the potential of applied forces and the strain energy, respectively.

The use of Total Potential Energy method leads to approximate values of the critical buckling load as an upper bound for the real value. The accuracy of the solution depends on using approximate deformation shapes that are as close as possible to the exact deformation shapes and also satisfy the boundary conditions of the problem.

According to Rayleigh-Ritz discretisation, a class of kinematically admissible configurations can be describe by the finite series

$$s_j = \sum_{i=1}^{m} q_{j,i}\, \psi_{j,i} \qquad j = 1, 2, 3 \quad (6)$$

where ψ-terms are arbitrarily chosen coordinate functions (i.e. a system of orthogonal functions) and $q_{i,j} = q_{i,j}(u, v, \vartheta)$ is a corresponding set of $3m$ Lagrangian multipliers.

Substituting equation (6) into the energy expression (5) the total energy $\mathcal{E}$ becomes a quadratic function of the Lagrangin multipliers and the minimum condition provides a system of $3m$ linear homogeneous equations (necessary and sufficient conditions of equilibrium) from which the critical buckling force F_c can be calculated without solving the nonlinear equilibrium equations (2).

$$\frac{\partial \mathcal{E}}{\partial q_{i,j}} = 0 \qquad j = 1, 2, 3 \qquad i = 1, \dots, m \quad (7)$$

3. Numerical results

Our numerical results assume the following class of kinematically admissible configurations

$$\begin{cases} u_c = \displaystyle\sum_{i=1}^{m} u_i \sin\left(\frac{m\,\pi\,\hat{z}}{l}\right) \\ v_c = \displaystyle\sum_{i=1}^{m} v_i \sin\left(\frac{m\,\pi\,\hat{z}}{l}\right) \\ \vartheta = \displaystyle\sum_{i=1}^{m} \vartheta_i \sin\left(\frac{m\,\pi\,\hat{z}}{l}\right) \end{cases} \quad (8)$$

which satisfy the boundary conditions of the analysed problem, namely

$$\begin{cases} u(0) = u''(0) = u(l) = u''(l) = 0 \\ v(0) = v''(0) = v(l) = v''(l) = 0 \\ \vartheta(0) = \vartheta''(0) = \vartheta(l) = \vartheta''(l) = 0 \end{cases} \quad (9)$$

Table 1 shows the obtained lower eigenvalues of the characteristic equation (i.e. the critical value) by taking one, two or three terms of the equation series (8). The results given in table 1 highlight an asymptotical convergence of the employed numerical scheme, with limiting value of the buckling load approximatively equal to 15.6 kN. On increasing the number of waves m up to 3, the numerical solutions exhibit asymptotic convergence, and progressively reducing in amplitude predictions of the buckling load.

Table 1. Computational results.

	F_c [kN]
One wave ($m = 1$)	16.4
Two waves ($m = 2$)	15.7
Three waves ($m = 3$)	15.6

4. Conclusions

We have conducted a numerical investigation of nonlinear stability of thin-walled C-section with uniformly varying inertia along the beam axis. First, we have derived the equilibrium equations of the analysed problem from the Vlasov theory [1-3]. The mathematical formulation of this problem leads to a homogeneous system of three linear ordinary differential equations with non-constant coefficients, whose exact solution is challenging to obtain in closed form. Therefore, we have numerically predicted an upper bound of the buckling load by means of the Dirichlet variational approach and the Rayleigh-Ritz algorithm. The numerical results presented in Sect. 3 demonstrate that the critical load corresponding to coupled flexural-torsional buckling asymptotically converges with the increasing order of test functions. This remarkable result leads us to conclude that the proposed numerical approach is a robust tool for obtaining technically reliable predictions of the true buckling load of beams with non-uniform cross-section. It is worth noting that such approach can be usefully generalised to the stability analysis of a variety of mechanical systems, with special focus on mechanical metamaterials, hierarchical metamaterials and structures equipped with curved members [10]–36]. An experimental validation of the proposed theory is also addressed to future research, through the fabrication and testing of physical models at a reduced scale, on using, e.g., sustainable additive manufacturing techniques [37-44]. Future extensions of the present research will also include an experimental study of the effects played by stiffening diaphragms on the lateral-torsional behaviour of thin-walled beams with non-uniform cross-sections.

Acknowledgments

AA, IM and FF acknowledge financial support from the Italian Ministry of Education, University and Research (MIUR) under the 'Departments of Excellence' grant L.232/2016. MM acknowledges financial support from the Department of Structures for Engineering and Architecture, University of Naples Federico II.

References

[1] Mascolo I and Pasquino M 2016 *Curved al Layer. Struct.* **3443(1)** 146-53.
[2] Mascolo I, Fulgione M and Pasquino M 2018 *Int. J. Mansory Research and Inn.* (In press).
[3] Vlasov V Z 1959 *Thin-walled elastic beams* (Moscow: Fizmatgiz).
[4] Ková M 2015 *Slovak J. Civil Eng.* **23(1)** 23-32.
[5] Fraternali F, Marino A, Elsayed T and Della Cioppa A 2011 *Mech. Adv. Mater. Struct.* **18** 225-43.
[6] Reddy J N 2002 *Energy Principles and Variational Methods in Applied Mechanics 2nd* (New York: John Wiley).
[7] Guarracino F and Walker A 1999 *Energy Methods in Structural Mechanics: A Comprehensive Introduction to Matrix and Finite Element Methods of Analyzis* (Thomas Telford).
[8] Elishakoff I 2000 *Arch. Comput. Method E.* **7** 451–526.

[9] El-Mahdy GM, El-Saadaway MM 2015 *Thin-Walled Struct.* **87** 149–57.
[10] Amendola A, Carpentieri G, De Oliveira M, Skelton R E and Fraternali F 2014 *Compos. Struct.* **117** 234-243.
[11] Modano M, Mascolo I, Fraternali F and Bieniek Z 2018 *Front. Mater.* **5** 5.
[12] Mascolo I, Amendola A, Zuccaro G, Feo and Fraternali F 2018 *Front. Mater.* **5** 16.
[13] Fraternali F, Marino A, Elsayed T and Della Cioppa A 2011 *Mech. Adv. Mater.Struc.* **18** 225.
[14] Fraternali F, Lorenz C D and Marcelli G 2012 *J Comput. Phys.* **231** 528.
[15] Schmidt B and Fraternali F 2012 *J. Mech. Phys. Sol.* **60** 172.
[16] Fraternali F, Carpentieri G, Modano M, Fabbrocino F and Skelton R E 2015 *Compos. Struct.* **134** 247.
[17] Carpentieri G, Skelton R E and Fraternali F 2017 *Struct. Multidiscipl. Optim.* **55(2)** 449.
[18] Carpentieri G, Modano M, Fabbrocino F, Feo L and Fraternali F 2017 *Meccanica* **52** 1561.
[19] Cimmino M C, Miranda R, Sicignano E, Ferreira A J M, Skelton R E and Fraternali F 2017 *Comp. Part B: Eng.* **115** 275-81.
[20] Amendola A, Benzoni G and Fraternali F 2017 *Comp. Part B: Eng.* **115** 117-23.
[21] Fraternali F and Amendola A 2017 *J. Mech. Phys. Sol.* **99** 259-71.
[22] Fabbrocino F, Modano M, Farina I, Carpentieri G and Fraternali F 2017 *Compos. Struct.* **169(1)** 167-72.
[23] Amendola A and Fraternali F 2018 *Compos. Struct.* **191** 145–53.
[24] Blesgen T, Fraternali F, Raney JR, Amendola A and Daraio C 2012 *Mech. Res. Commun.* **45** 58-63.
[25] Ngo D, Fraternali F and Daraio C 2012 *Phys. Rev. E* **85** 036602-1-10.
[26] Raney J R, Fraternali F, Amendola A and Daraio C *Compos. Struct.* **93** 3013.
[27] Fraternali F 2011 *Mech. Res. Commun.* **38** 176.
[28] Orefice A, Mancusi G, Feo L and Fraternali F 2017 *Compos. Struct.* **160** 1126–7.
[29] Ascione L and Fraternali F 1992 *Comput. Struct.* **45** 985-99.
[30] Singh R, Singh N, Fabbrocino F, Fraternali F and Ahuja I 2016 *Comp. Part B: Eng.* **105** 23-9.
[31] Singh R, Bedia P, Fraternali F and Ahuja I 2016 *Comp. Part B: Eng.* **106** 20-7.
[32] Fraternali F and Reddy J N 1993 *Int. J. Solids Struct.* **30** 3337-55.
[33] Fraternali F, Senatore L and Daraio C 2012 *J. Mech. Phys. Sol.* **60** 1137–44.
[34] Fabbrocino F, Farina I, Berardi V P, Ferreira A J M and Fraternali F 2015 *Comp. Part B: Eng.* **83** 297-305.
[35] Fraternali F, De Chiara E and Skelton R E 2015 *Smart Mater. Struct.* **24** 105032.
[36] Fraternali F and Bilotti G 1997 *Comput. Struct.* **62** 837-69.
[37] Farina I, Fabbrocino F, Colangelo F, Feo L and Fraternali F 2016 *Comp. Part B: Eng.* **99** 305-11.
[38] Singh R, Singh S and Fraternali F 2016 *Comp. Part B: Eng.* **98** 244.
[39] Boparai, KS, Singh, R, Fabbrocino, F, Fraternali, F 2016 *Comp. Part B: Eng.* **106** 42.
[40] Singh R, Kumar R, Ranjan N, Penna R and Fraternali F 2018 *Compos. Struct.* **184** 704–13.
[41] Singh R, Kumar R, Feo L and Fraternali F 2016 *Comp. Part B: Eng.* **101** 77-86.
[42] Farina I, Fabbrocino F, Carpentieri G, Modano M, Amendola A, Goodall R, Feo L and Fraternali F 2016 *Comp. Part B: Eng.* **90** 76-85.
[43] Amendola A, Smith C J, Goodall R, Auricchio F, Feo L, Benzoni G and Fraternali F 2016 *Compos. Struct.* **142** 254–62.
[44] Singh R, Kumar R, Mascolo I and Modano M 2018 *Comp. Part B: Eng.* **14** 132-40.

IOP Conf. Series: Materials Science and Engineering **473** (2019) 012012 doi:10.1088/1757-899X/473/1/012012

Staging and Pretensioning of Cable-Stayed Bridges

M Modano[1], I Mascolo[2] and F Fraternali[2]
[1]Department of Structures for Engineering and Architecture, University of Naples Federico II, Italy
[2]Department of Civil Engineering, University of Salerno, Italy

Abstract. This paper describes a study of cable-stayed bridges (CSBs), with a focus on the computation of the optimal pre-tensioning cable forces to be applied in the construction phase and the service behavior. We distinguish the case of a structure built using temporary supports from that of a structure built through the so-called cantilever method. We formulate an optimization procedure aimed at determining the cable pre-tensioning forces that ensure a target stress distribution along the longitudinal axis of the structure. The goal of this procedure is to make the bending diagram distribution as much uniform as possible over the bridge length, in order to avoid undesired stress localization effects. The proposed procedure is based on an influence matrix approach, and makes use of a single pre-tensioning cycle. It can be applied to optimize the mechanical response of the structure in any phase of its life cycle.

1. Introduction

One of the most common problem related to the design of cable-stayed structures is concerned with the determination of the initial cable forces, and the pre-tensioning sequences to be applied to such elements. An optimal pre-tensioning sequence is useful to control the stress and strain states exhibited by the structure, in correspondence with the construction phase, and also in the long-term behavior. The procedure presented in this work aims at directly solving the problem of cable pretension in cable-stayed structures, through a sensitivity analysis approach that can be applied to different construction techniques. Alternative approaches to the analyzed problem are available in the literature [1, 2], but there is a lack of analytical and explicit solutions. Available methods are typically based on iterative algorithms [3], which call for a progressive adjustments for the cable's forces during the construction phase. Such approaches require recursive cable tightening operations, and may be affected by a number of technological, structural, and economical issues. The present procedure instead makes use of an influence matrix method [1-6], and uses a single pre-tensioning sequence, in order to achieve a given bending moment distribution along the longitudinal axis of the structure.
As compared to previous available works on the same topic [3-5], this paper presents new results on the application of the proposed optimization procedure to cable-stayed bridges realized through two different assembly procedures, namely: bridges whose construction phase employs temporary supports, and bridges built through the so-called cantilever method.

2. Construction phases and basic principles

The structure's operational behavior is variously influenced by geometric, static and mechanical characteristics, constructive modalities, construction phases and techniques. The tools and procedures used to monitor and control the behaviour of these structures are then an important factor for the overall construction. The procedure can be applied to any type of cable-stayed bridge (CSB) once the design, mechanical system characteristics, and construction sequences and techniques are known. The method proposed in this paper is applied to a basic structure built with two different construction

IOP Conf. Series: Materials Science and Engineering **473** (2019) 012012 doi:10.1088/1757-899X/473/1/012012

procedures. In both cases, the method solves the optimisation problem and can then be considered a valid aid for the construction of CSBs.

2.1. Theoretical model and main hypothesis

The bridge idealised diagram includes a semi-fan, a central suspended span and a self-supported cable-stayed system. Towards the longitudinal direction, the bridge is symmetric with respect to the mid-span and it includes two towers and two sets of cables. The deck comprises three spans (i.e. the central span is150m long, the two approaching spans are both 75m long) and assembled through semi-precast elements which are12.50m long. The towers are made of pre-cast sections of reinforced high-strength concrete (C45/55). The deck is made of steel-welded beams (S355) and reinforced concrete sections (concrete slab (C25/30), 26cm thick). The concrete slab and the steel beams are connected using metallic bolts. The cables have a total diameter of 22 cm. The mono-dimensional model of the structure was developed to solve the elastic problem. The two-dimensional orthonormal system (OYZ), at this stage considered a mechanical model, represents the longitudinal profile of a continuous decked, semi-fanned, cable-stayed bridge with symmetrical spans and a cable inclination which varies between 65° and 25°. The geometric characteristics of the model described above are represented in the single-lined model shown below.

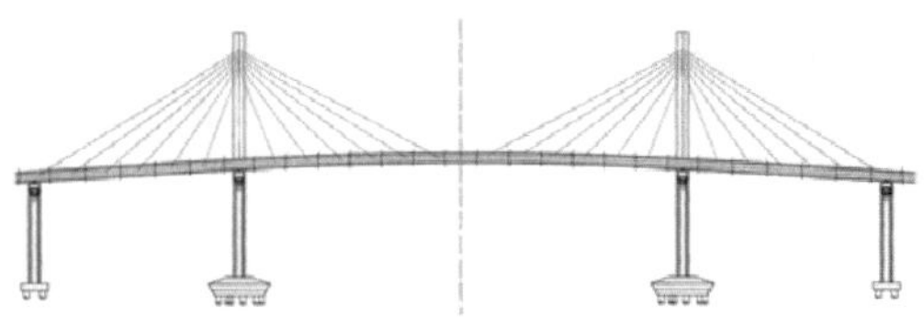

Figure 1. Full configuration of the CSB.

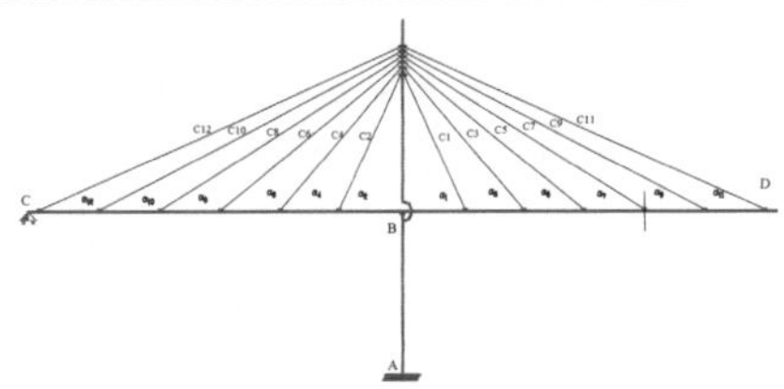

Figure 2. Idealisedmodel of the structure.

In addition to FEM modelling, the Principle of Virtual Works (PVW) was applied to develop the single-line models to assess the behavioral significance of the structure's main mechanical parameters. The main hypothesis are:

- the pile is completely fixed to the base (point A);
- the internal span has a roller-type support at the final point C;
- the deck and tower are connected via an internal hinge;
- the cables have a symmetrical configuration in respect to the tower. They follow an odd ascending numeration on the central span (points B to D) and an even descending numeration on the internal span (points B to C);
- each normal plane section has a rigid range of motion, excluding warping deformation. The tower and the deck are considered mono-dimensional elements, in accordance with Eulero-Bernoulli's beam theory.

2.2. Sensitivity analysis

In order to apply the optimization procedure, different models of the structure were analyzedand a sensitivity analysis was carried out. This analysis aimed to define a domain in which the behaviour of the structure is placed. Many virtual models of the structure were created by varying the stiffness of the main elements (deck, pile, cables, antenna).

- High-stiffness deck (HSD) - Low-stiffness cables, pile, and antenna
- Low-stiffness deck (LSD) - High-stiffness cables, pile and antenna
- High-stiffness tower [pyle and antenna together] (HST) - Low-stiffness deck and cables
- Low-stiffness antenna (LSA) - High-stiffness cables, deck, and pile
- High-stiffness cables (HSC) - Low-stiffness deck, pile, and antenna
- Low-stiffness cables (LSC) - High-stiffness deck, pile, and antenna

2.3. Case studies
Two models were analyzed to identify the behavior of the real structure. The first model was calculated for the construction phase related to the temporary supports method. The second model relates to the cantilever method. In both cases, comparative analyses were carried out together with the results obtained from the sensitivity analysis. It is useful to remember that the sensitivity analysis was developed to identify the best possible configuration of the stresses induced on the structure.

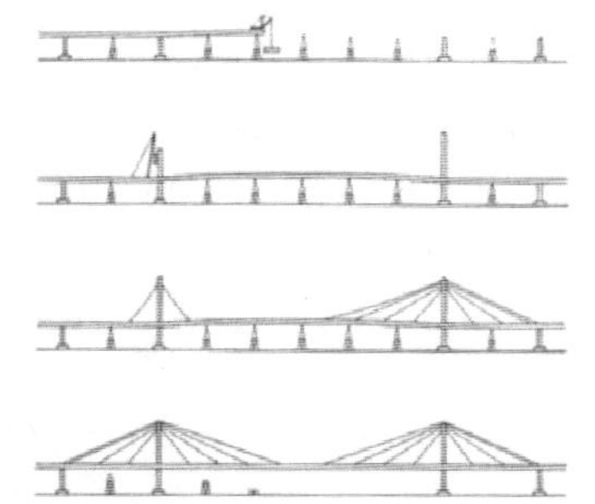

Figure 3. Construction stage: Temporary supports method.

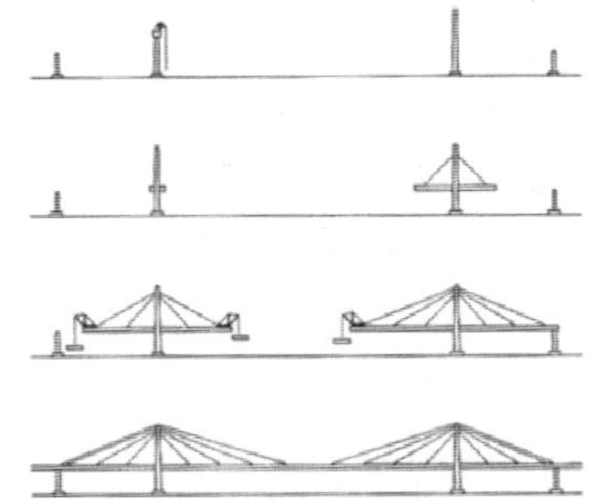

Figure 4. Construction stage: Cantilever method.

3. Proposed optimization procedure
The optimization of the initial cable forces was carried out by elaborating an analytical method based on the structure's global equilibrium conditions. Taking an adequate parameter as a reference, the optimal entity of the initial cable stresses was determined to obtain a specific distribution of the stress state in the final assembly configuration. The proposed method leads to a linear elastic analysis under the hypotheses of small displacements, with a consequent linear sum of effects, with the aim of reaching a predetermined distribution of bending moments on the deck. This method acts directly on the static entities and indirectly on the elastic deformations, nullifying the vertical displacement in the cable-to-deck connection. The problem has been formulated in terms of structural optimization. After a restrained structural system has been assigned and the precise load conditions defined, an optimal entity of the cable stressing sequences is sought to obtain a predetermined distribution of bending moments on the deck. The starting function consists in the distribution of the cable stresses corresponding to a specific initial configuration (C_0), and the target function represents the distribution of cable stresses in the ideal configuration (C_d). For unit influence matrix D, whose coefficients represent the generic a-dimensional element a (i,j), which is the stress level reached in the i-th cable when the j-th cable is tightened by a unitary amount, the closed-form solution of the problem may be found in the following equilibrium relation:

$$[D^T]\{X\}=\{\Delta T\} \tag{1}$$

The problem is reconstituted through the resolution of a linear algebraic system whose unknowns $X=\{x_1, x_2 \cdots x_n\}$ are the optimal pretension stresses to be applied to each cable in the C_0 configuration until the optimal stress configuration C_d is obtained under the same control load.
By inverting the D matrix, it is possible to identify the group of pretension stresses to be applied to the cables in order to guarantee the desired conformation.

4. Numerical results
The results of the two case studies - CSB built with temporary supports and CSB built using cantilever method - are perfectly contained in the domain defined by the sensitivity analysis. Notably, the cantilever construction determines a more severe stress and strain state. In both cases, the method allows us to obtain a better configuration of the bending moment on the deck and a more uniform distribution of axial forces in the cables.

The optimal distribution was chosen from the multiple stress configurations and identified as an objective function. The solutions with compressive stresses in the cables were discarded. In the following figures are shown the diagrams of the moments obtained on the structures built with the two different construction methods. Figure 5 shows the bending moments on the deck in significant (virtual) configurations used in Sensivity Analysis: HSD – orange; LSD – light blue; Real config. of struct. – blue; HSC – green; LSC – red. Figures 6, 7 show cables axial forces in most significant (virtual) configurations used in Sensivity Analysis and in the real configuration of structure. The most important are: HSD – light blue; LSD – orange; Real config. of struct. – grey. The diagram of the bending moment on the deck obtained after applying the optimization procedure is shown in figure 9. that is bending moment diagram after applying the pretension stresses in the cables.

Figures 10 and 11 show the improved distribution of axial force in the cables following the optimisation procedure. After pre-tensioning, the axial stresses in the cables become more uniform, improving the functioning of the structure.

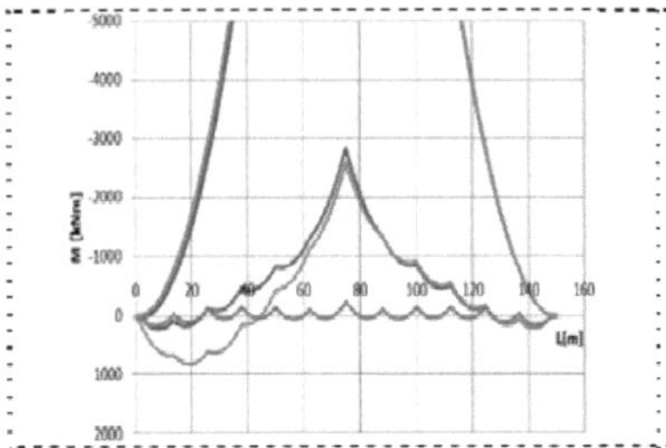

Figure 5. Sensitivity analysis: Envelope of bending moment.

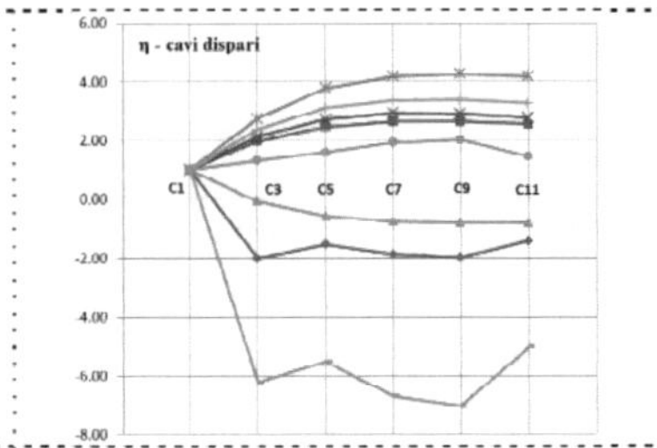

Figure 6. Sensitivity analysis: A-dim. even cables axial force.

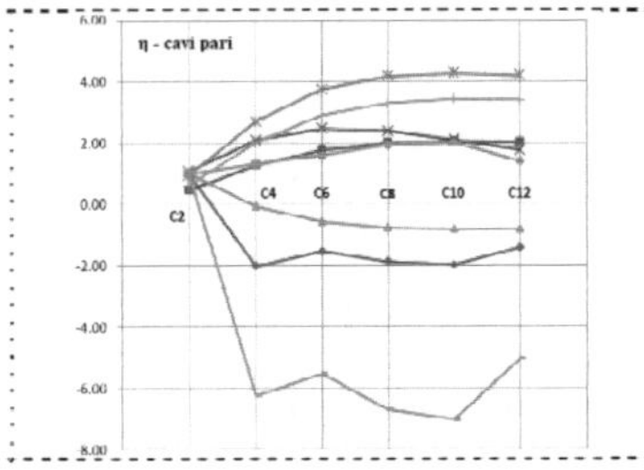

Figure 7. Sensitivity analysis: A-dim. odd cables axial force.

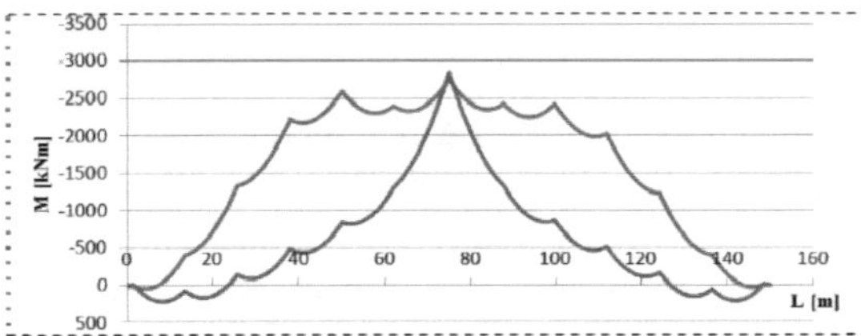

Figure 8. Bending moments deck – Temp.Supp. (red) Cant.Meth. (blue).

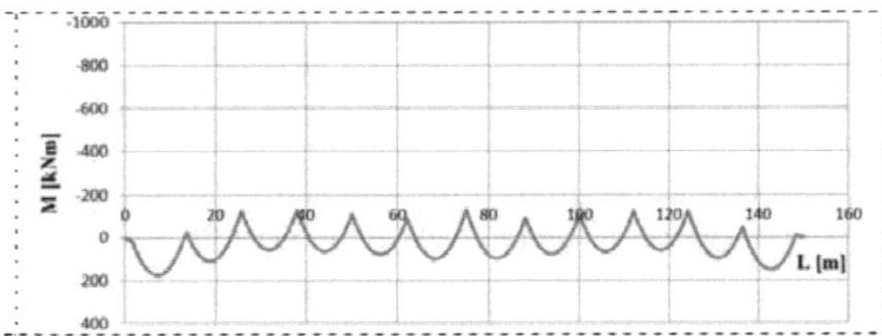

Figure 9. Bending moments deck – after optimization procedure.

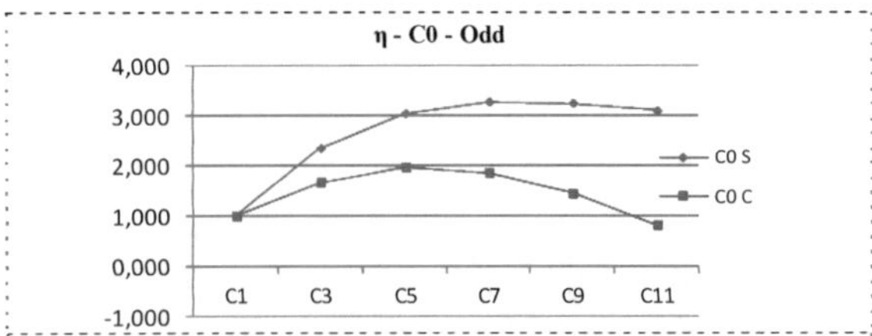

Figure 10. A-dimensionalised axial force in odd cables (right span) before and after the optimization procedure.

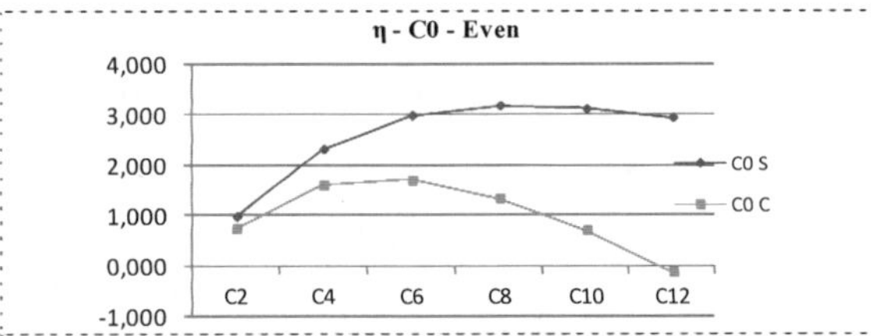

Figure 11. A-dimensionalised axial force in even cables (left span) before and after the optimization procedure.

The optimal configuration can be achieved for any given initial configuration: "temporary support" as well as "cantilever methods". In both cases (T.S. and C.M.), the axial forces in the cables tend to be uniform, limiting the stress variations - compared to the 1st cable in span - from 3,8 to 1,8. In both cases there is also a drastic reduction of the bending moment on the deck. Bending moments are driven to the values of continuous beam on simple supports. In fact, the pretension in the cables generates positive bending moments that reduce negative moments on the deck.

5. Concluding remarks

We have presented an approach to the optimal pre-tensioning design of cable-stayed bridges which aims at achieving a target bending moment distribution over the deck. The proposed methodology is based on the matrix of influence method, and determines the cable forces that permit to avoid bending moment concentrations. It can be applied to any construction phase of the structure, and can be easily generalized to account time dependent phenomena due, e.g., to viscosity effects [7, 8]. The given results highlight that the optimization procedure illustrated in Ref. [3] can be suitably generalized to cover different construction techniques of cable stayed bridges. Such a generalization allows us to strengthen the applicative potential of this procedure in real-life examples. It is worth observing that the design technique proposed in the present work can also be applied to masonry structures [5], arch bridges with suspended decks, and all the prestressable structures [3, 7-14]. We address an experimental validation of this technique to future work, through the fabrication and testing of reduced-scale models [15-34], making use of innovative materials [15-34] and sustainable additive manufacturing techniques [15-34].

Acknowledgments

MM acknowledges financial support from the Department of Structures for Engineering and Architecture, University of Naples Federico II. FF and IM acknowledge financial support from the Italian Ministry of Education, University and Research (MIUR) under the 'Departments of Excellence' grant L.232/2016.

References

[1] Jadhav P B, Ganesh M and Vinayagamoorthy M 2017 *I. J. CI. E. T.* **8(4)** 252-64.
[2] Kim HK, Lee MJ, Chang SP 2002 *Engineering Structures* **24(12)** 1547-59.
[3] Fabbrocino F, Modano M, Farina I, Carpentieri G and Fraternali F 2017 *Compos. Struct.* **169(1)** 167-72.
[4] Carpentieri G, Modano M, Fabbrocino F and Fraternali F 2015 *Proc. COMPDYN (Crete)* (Athens: National Technical University of Athens) 1731.
[5] Carpentieri G, Modano M, Fabbrocino F, Feo L and Fraternali F 2016 *Proc. ECCOMAS* (Crete) vol 2 (Athens: National Technical University of Athens) 3386.
[6] Fraternali F, Marino A, Elsayed T and Della Cioppa A 2011 *Mech. Adv. Mater.Struc.* **18** 225.
[7] Amendola A, Carpentieri G, De Oliveira M, Skelton R E and Fraternali F 2014 *Compos. Struct.* **117** 234-43.
[8] Modano M, Mascolo I, Fraternali F and Bieniek Z 2018 *Front. Mater.* **5** 5.
[9] Mascolo I, Amendola A, Zuccaro G, Feo and Fraternali F 2018 *Front. Mater.* **5** 16.
[10] Fraternali F and Bilotti G 1997 *Comput. Struct.* **62** 837-69.
[11] Amendola A and Fraternali F 2018 *Compos. Struct.* **191** 145–53.
[12] Fraternali F, Lorenz C D and Marcelli G 2012 *J. Comput. Phys.* **231** 528.
[13] Schmidt B and Fraternali F 2012 *J. Mech. Phys. Sol.* **60** 172.
[14] Carpentieri G, Skelton R E and Fraternali F 2017 *Struct. Multidiscipl. Optim.* **55(2)** 449.
[15] Carpentieri G, Modano M, Fabbrocino F, Feo L and Fraternali F 2017 *Meccanica* **52(7)** 1561.
[16] Singh R, Kumar R, Feo L and Fraternali F 2016 *Comp. Part B: Eng.* **101** 77-86.
[17] Singh R, Kumar R, Mascolo I and Modano M 2018 *Comp. Part B: Eng.* **14** 132-40.
[18] Fabbrocino F, Farina I, Berardi V P, Ferreira A J M and Fraternali F 2015 *Comp. Part B: Eng.* **83** 297-305.
[19] Amendola A, Benzoni G and Fraternali F 2017 *Comp. Part B: Eng.* **115** 117-23.
[20] Fraternali F and Amendola A 2017 *J. Mech. Phys. Sol.* **99** 259-71.
[21] Orefice A, Mancusi G, Feo L and Fraternali F 2017 *Compos. Struct.* **160** 1126–7.
[22] Ascione L and Fraternali F 1992 *Comput. Struct.* **45** 985-99.
[23] Singh R, Singh N, Fabbrocino F, Fraternali F and Ahuja I 2016 *Comp. Part B: Eng.* **105** 23.
[24] Singh R, Bedia P, Fraternali F and Ahuja I 2016 *Comp. Part B: Eng.* **106** 20-7.
[25] Blesgen T, Fraternali F, Raney J R, Amendola A and Daraio C 2012 *Mech. Res. Commun.* **45** 58.

[26] Ngo D, Fraternali F and Daraio C 2012 *Phys. Rev. E* **85** 036602-1-10.
[27] Raney J R, Fraternali F, Amendola A and Daraio C 2011 *Compos. Struct.* **93** w013.
[28] Fraternali F 2011 *Mech. Res. Commun.* **38** 176.
[29] Fraternali F, Carpentieri G, Modano M, Fabbrocino F and Skelton R E 2015 *Compos. Struct.* **134** 247.
[30] El Sayed T, Mota A, Fraternali F and Ortiz M 2008 *J. Biomech.* **41(7)** 1458.
[31] El Sayed T, Mock W, Mota A, Fraternali F and Ortiz M 2009 *Comput. Mech.* **43(4)** 525.
[32] Cimmino M C, Miranda R, Sicignano E, Ferreira A J M, Skelton R E and Fraternali F 2017 *Comp. Part B: Eng.* **115** 275-81.
[33] Mascolo I and Pasquino M 2016 *Curved al Layer. Struct.* **3443(1)** 146-53.
[34] Mascolo I, Fulgione M and Pasquino M 2018 *Int. J. Mansory Res. Inn.* (In press).
[35] Fraternali F, Senatore L and Daraio C 2012 *J. Mech. Phys. Sol.* **60** 1137–44.
[36] Fraternali F and Reddy J N 1993 *Int. J. Solids Struct.* **30** 3337-55.
[37] Boparai, KS, Singh, R, Fabbrocino, F, Fraternali, F 2016 *Comp. Part B: Eng*. **106** 42.
[38] Singh R, Kumar R, Ranjan N, Penna R and Fraternali F 2018 *Compos. Struct.* **184** 704–13.
[39] Fraternali F, De Chiara E and Skelton R E 2015 *Smart Mater. Struct.* **24** 105032.
[40] Farina I, Fabbrocino F, Colangelo F, Feo L and Fraternali F 2016 *Comp. Part B: Eng.* **99** 305.
[41] Farina I, Fabbrocino F, Carpentieri G, Modano M and al. F 2016 *Comp. Part B: Eng*. **90** 76-85.
[42] Singh R, Singh S and Fraternali F 2016 *Comp. Part B: Eng*. **98** 244-9.
[43] Amendola A, Smith C J, Goodall R, Auricchio F, Feo L, Benzoni G and Fraternali F 2016 *Compos. Struct.* **142** 254–62.

Research on the Residual Stress and Influence Factors of Butt Welding Channel Section of BS700 High Strength Steel

L Gao, Q Wang, L Y Bai, and X H He*
[1] College of Filed Engineering, Army Engineering University of PLA, 210007, Nan Jing, Jiangsu, China

E-mail: gcbhxh@qq.com

Abstract. Taking the butting weld structure of channel which was fabricated by BS700 as study object, the residual stress and influence factors of welding box section which was butted by two channel section were studied. The study results show that the main welding residual stress of butt weld channel steel is longitudinal residual tension stress, and it affect the stability and fatigue performance. Longitudinal residual distribution of two pieces of weld are located at the district of tension stress, the residual stress of welding foot in the second weld have bigger peak value. The varieties of section length have influence on transverse residual stress, but have little influence on longitudinal residual distribution. Changes of thickness of wall in channel section have marked influence on residual stress value of transverse and longitudinal weld. So it is necessary to choose appropriate thickness of channel for reducing the influence of residual stress on channel steel structure.

1. Introduction

The use of high-strength steel instead of the traditional steel structures can effectively achieve the purpose of improving the overall property of the structure, saving steel and reducing the cost of the project [1-2]. In the past decades, China has developed rapidly in designing the steel structures, and the high-strength steel commonly used in China is the material of Q460 steel [3], but the application of higher grade steel is still lack. So the steel structures of high-strength steel, especially for high strength steel products, still have many problems to be solved in China [4-5]. In recent years, BS700 high-strength steel have been independently developed and produced by China and have been successfully applied [6-8]. In this paper, BS700 butt welding channel steel structure was taken as the research object, and the welding residual stress and its influence factors of the structure was studied in order to promote the application of BS700 high strength steel in the steel structure engineering of China.

The finite element simulation software ANSYS was used to simulate the welding process of butt welded channel steel structure with different dimensions. In the model, the von Mises yield criterion and the thermo-mechanical coupled finite element simulation technique were used. The corresponding conclusions were obtained by comparing the calculation results of each plan. The parametric modelling method is adopted for the model size, which can easily modify the computational model.

In the process of designing the dimensional parameters of channel steel specimens, the spans of the influencing factors are equal: the length of the section of side channel steel with equal section length is taken as 70mm, 1000mm, and 130mm, respectively, and the length-width ratio of the section of side channel steel with non-equal-section length is respectively taken as 1:1.2, 1:1.4, and 1:1.6. The wall thickness of channel steel is 4mm, 6mm, and 8mm, respectively. To reduce the number of calculations and time, different pipe sizes are combined into eight model solutions. The design plans of the butt welding channel steel dimensions are shown in table 1.

Table 1. The design plans of the butt welding channel steel size parameters.

Plan number	1	2	3	4	5	6	7	8
The length of section/h	70	100	100	100	100	130	100	100
The thickness of section/b	70	100	100	100	120	130	140	160
The wall thickness Of channel steel/d	6	4	6	8	6	6	6	6

2. The numerical analysis model

The finite element model uses eight-node three-dimensional solid elements SOLID70, SOLID90, and surface effect element SURF152, with a weld width of 10 mm and a transition area width of 10 mm. In order to increase calculation accuracy, the weld seam is divided into denser meshes and the unit width is 1mm. The width of the transition area is 2mm, and the width of the remaining area is 4mm. The welding finite element model (such as plan 3) is shown in figure 1.

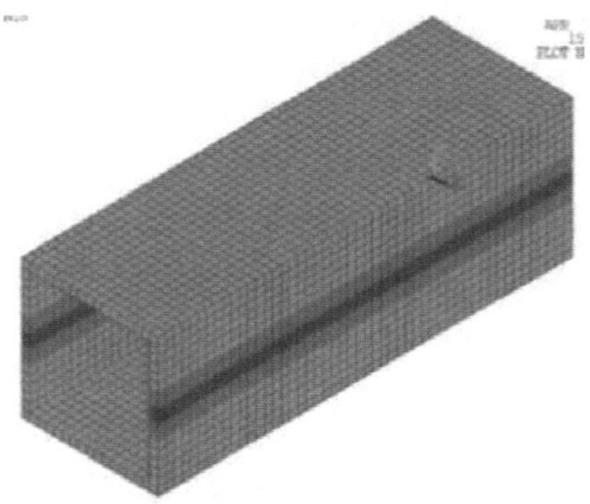

Figure 1. The finite element model.

The finite element model which is established in the course of research is a thin-walled structure with a wall thickness changing from 4mm to 8mm [10]. A Gaussian heat source model is used for the Gaussian distribution model in the welding finite element simulation. When simulating, the heat source is applied to the weld surface in the form of heat flux which is calculated by equation (1).

$$q(r) = \frac{3Q}{\pi R^2} e^{-\frac{3r^2}{R^2}} \tag{1}$$

In the equation: $e = 2.71828$, Q is input heat, $Q = \eta UI$, η is efficiency, U is welding voltage, I is welding current, R is arc effective heating radius. The initial environment temperature is 20 °C. While loading the temperature field, the convection density is loaded on the surfaces of SOLID70 and SOLID90, and the heat flux density is loaded through the SURF152.

3. Butt welding channel residual stress distribution

The transverse and longitudinal residual stress distributions are shown in figure 2 and figure 3.

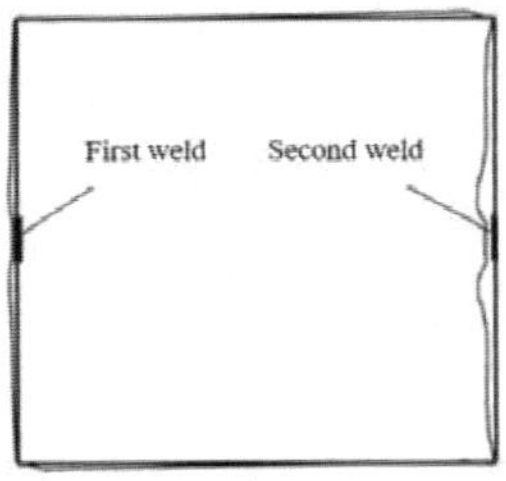

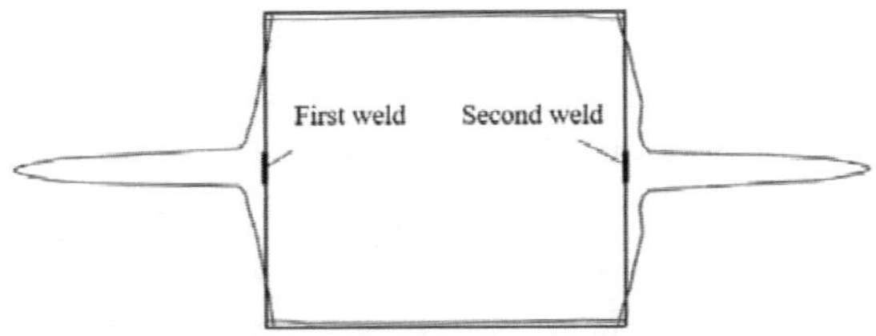

Figure 2. Transverse residual stress distribution. **Figure 3.** Longitudinal residual stress distribution.

It can be seen from figure 2 and figure 3 that the level of residual tensile stress of the welding channel steel is much higher than the residual compressive stress. The peak value of the welding residual stress on the long side of the steel channel section is located at the weld toe position. The peaks of the residual stress on the upper and lower edges of the channel steel section are mainly concentrated in the corners of the channel steel section. The transverse and longitudinal residual stress comparison curves of the first and second weld seams are shown in figure 4 and figure 5.

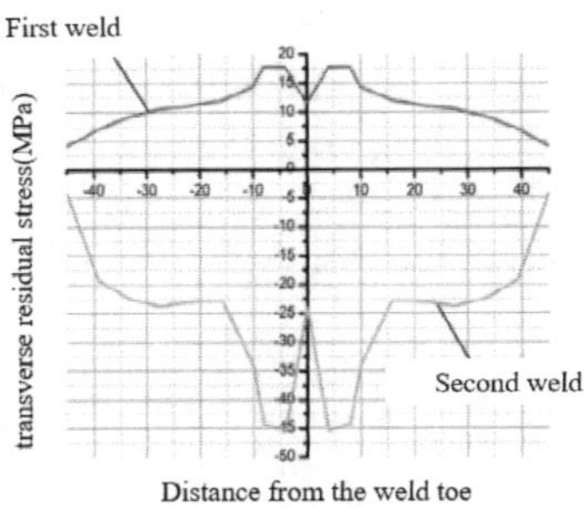

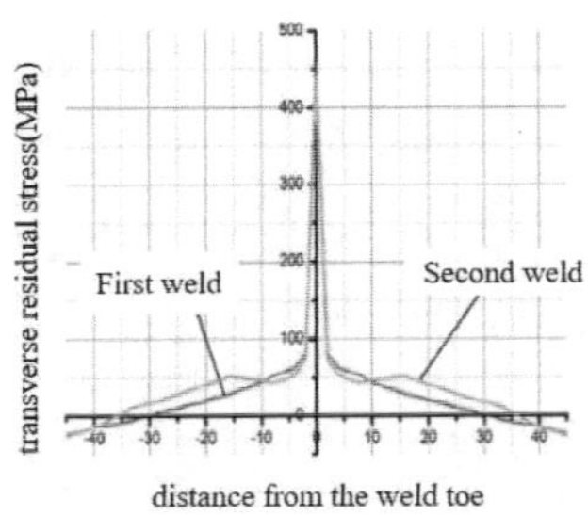

Figure 4. transverse residual stress comparison. **Figure 5.** longitudinal residual stress comparison.

From figure 4, it can be seen that the peak position of the residual stress at the weld position of the butt welded channel steel structure is not the weld toe, but about 5mm away from the weld toe, and as the distance from the weld toe increases, the transverse residual stress shows the trend of increasing first and then decreasing. It can be seen from figure 5 that the peak position of the longitudinal welding residual stress is at the weld toe position, and the peak value of the residual stress at the weld toe of the second weld is higher than the first weld. With the increase of the distance from the weld toe, the longitudinal residual stress near the weld seam gradually decreases. When the distance from the weld toe exceeds 10mm, the longitudinal residual tensile stress value near the first weld slightly increases and higher than the second.

4. The influence factors on welding residual stress of welded channel steel

4.1. Influence law of section length of butt welded channel steel on structural residual stress distribution

In this paper, the influence of the section length of welded channel steel on the structural residual stress is studied through plan 1, 3 and 6. The variation of the side length of each plan is equal: 70mm, 100mm and 130mm respectively. Except for the side length, other parameters of the finite element model are the same, aiming to eliminate the influence of other factors on the analysis result. The transverse and longitudinal residual stress curves calculated by different models are shown in figure 6.

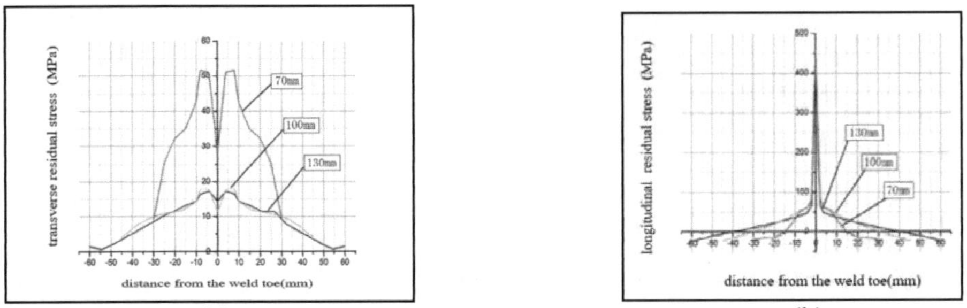

(a) (b)

Figure 6. comparison curves of transverse and longitudinal residual stress of different side lengths.

As can be seen from figure 6(a), the transverse residual stress near the weld seam is always presents as tensile stress, and the peak location of transverse residual stress is 8mm away from the weld toe. The peak of transverse residual stress decreases with the increase of section length of channel steel. When the section length of the channel steel is reduced from 100mm to 70mm, the peak value of transverse residual stress increases by 33.877MPa, which is much higher than the increment of peak.

As can be seen from figure 7(b), the longitudinal residual stress near the weld seam is tensile stress, and its peak is mainly concentrated near the weld seam. With the increase of the section length of the channel steel, the peak of the longitudinal residual tensile stress at the near weld decreases gradually: 458.36MPa, 456.50MPa and 429.57MPa respectively. The longitudinal residual stress in plan 1 shows a significant drop down process.

4.2. The influence law of thickness of channel steel on structural residual stress distribution

The influence law of the thickness of butt welded channel steel on the structural residual stress was studied. The thickness of channel steel was changed by 4mm, 6mm, and 8mm respectively. The transverse and longitudinal residual stress curves which were calculated by different models are shown in figure 7.

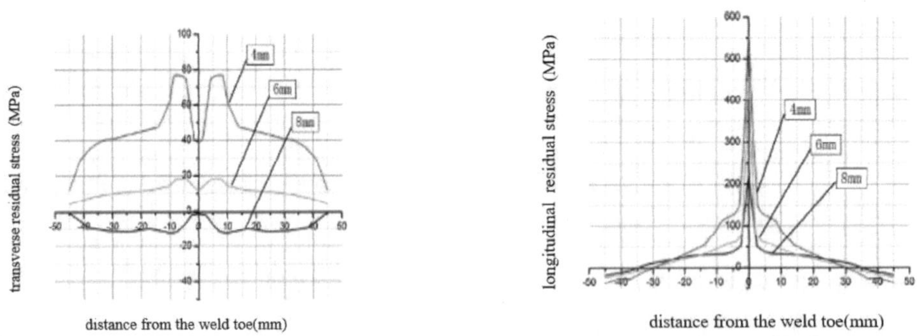

(a) (b)

Figure 7. comparison curves of transverse and longitudinal residual stress of different thickness

As can be seen from figure 7(a), with the increase of the thickness of the channel steel, the variation law of transverse residual stress in the calculation model of each plan is gentle, and when the thickness of the channel steel is 4mm, the transverse residual stress has the most changes with the increased distance from the weld toe. With the increase of the thickness of the channel steel, the transverse residual stress near the weld area changed from the tensile stress region to the compressive stress region, and when the thickness of the channel steel reaches 8mm, the transverse residual stress distribution is completely transformed into compressive stress. However, with the change of the thickness of channel steel, the peak position of the transverse residual stress remains unchanged at about 8mm to the weld toe.

As can be seen from figure 7(b), the peak value of the longitudinal residual stress decreases gradually with the increase of the thickness of channel steel: 537.36MPa, 439.72MPa and 232.81MPa in turn. Meanwhile, the smaller the thickness of the channel steel, the greater the gradient of the longitudinal residual stress with the increased distance from the weld toe. From the vibrational peak of residual stress at weld toe, we can see that the deviation of peak value of longitudinal residual stress in plan 2 and plan 3 is 97.64MPa, while the deviation of peak value of longitudinal residual stress in plan 3 and plan 4 is 206.91MPa, which indicates that the greater the thickness of the channel steel, the greater the reduction amplitude of the peak value of longitudinal residual stress with the increase of thickness.

5. Summarize

Through the above analysis, we can draw the following conclusions:

(1) The welding residual stress distribution in the butt welded channel steel structure is mainly longitudinal residual tensile stress, the peak position of tensile stress is weld toe, the peak of tensile stress can reach 537.36MPa, which seriously affects the stability and fatigue performance of the butt welded channel steel structure.

(2) The transverse and longitudinal residual stress distribution of butt welded channel steel structure is different between two weld seam because of the welding sequence, the transverse residual stress distribution in the vicinity of the first welding line is presented as compressive stress, and the transverse residual stress distribution near the second welding line is tensile stress. Although the longitudinal residual stress distribution of two welding lines is in the tensile stress region, the longitudinal residual stress at weld toe of the second welding line is higher.

(3) The variation of length in the constant-section side-length mainly influences the distribution of transverse residual stress in the butt welding channel steel structure, and has little influence on the longitudinal residual stress distribution in the structure. While in the structure of butt welding channel, the stress of the structure performance is mainly longitudinal residual stress.

(4) The change of the thickness of the channel steel will significantly affect the transverse and longitudinal welding residual stress value, the smaller the thickness of the channel steel, the higher the transverse and longitudinal welding residual stress in the structure. When the thickness of the channel steel reaches a certain value the transverse residual stress value near the weld area is converted to compressive stress and the peak value of longitudinal residual stress at the weld toe can be reduced remarkably.

Acknowledgments

This research work was financial supported by the Major State Basic Research Development of China (973 Program, Grant No.2014CB046801) and Postdoctoral Science Foundation of China (Grant No.2017M623403)

References

[1] Pocock G. High strength steel use in Australia, Japan and the US 2006 *Structural Engineer* **84**(21) P 27-30

[2] Cui Wei. Recent research advances and application of high strength steel structures 2011 *Building Structure* **41** P 27-30

[3] Zhou Sihong，Zhu Zhongyi，Qi Wuhui, et al. Structural design on the project of Phoenix International Media Center 2011 *Building Structure* **41** P 56-62

[4] Shen Zuyan. Development of the Chinese code for design of steel structures 2010 *Journal of Building Structures* **6** P 1-6

[5] Shi Gang, Ban Huiyong, Shi Yongjiu, Wang Yuanqing. Overview of Research Progress for High Strength Steel Structures 2013 *Engineering Mechanics* **30** P 1-13

IOP Conf. Series: Materials Science and Engineering **473** (2019) 012013 doi:10.1088/1757-899X/473/1/012013

[6] Xiao yi, Ma minli, Zhou daiyi. Microstructure and mechanical properties of welded joint of BS700MC low carbon micro-alloy steel for automobile beam 2012 *Materials for mechanical engineering* **38** P 15-20

[7] Ma Hongyong. Research on Hot-stamping of BS700 high strength steel track shoe 2011 (Harbin: Harbin institute of technology) p 89-116

[8] Gou Mingkang, Tao Li. The Application of High Strength Steel ($\sigma s \geq 700$MPa) To Movable Bridge 2002 *Steel Structure* **17** P 6-9

[9] Fang wanhong. Research on welding structur 2008 (Beijing: China machine press) p 67-86

[10] Bai linyue, Jiang kebin, Gao lei. Finite element analysis of residual stress distribution of butt-welded pipeline 2015 *Steel Construction* **30** P 77-81

[11] Curfs C,Kirstein O,Studeraj,et al. Residual stress measurements in Australia:Present and future 2005 *Mater Sci Forum* P 490-491

[12] Sinha V K,Godaba V S. Residual stress measurement in worked and heat treated steel by X-ray diffractometry 2008 *Materials Science and Engineering* A488 P 491-495

IOP Conf. Series: Materials Science and Engineering **473** (2019) 012014 doi:10.1088/1757-899X/473/1/012014

Synchronic Excitation in Footbridges due Human-Induced Forces in Lima Peru

G Huaco[1] and L Vasquez[2]

[1] Professor, Peruvian University of Applied Science UPC, Lima, Peru
[2] Project Manager, Ensoft, Austin Texas, USA

Email: pccighua@upc.edu.pe

Abstract. Structural analysis of human-induced induced vibrations in pedestrian bridges due to the passage of pedestrians is presented. An antecedent of this phenomenon is the Millennium Bridge located in the city of London, which had to close to the public, shortly after its inauguration due to unexpected lateral vibrations produced by the passage of people across the bridge. The present research analyses this effect on a pedestrian bridge in Lima, for transverse, lateral, longitudinal and torsion movements, due to the passage of pedestrians walking with a period similar to the modal periods of the structure corresponding to each direction of movement, finding displacement and acceleration responses. A software Puente.exe was developed to find the response of the peruvian footbridge due the synchronic excitation.

1. Introduction

Some bridges have had serious structural problems because extra vibrations were induced by pedestrians. The displacement and acceleration induced by these vibrations must be taken into account for the proper design of those structures. In the current report, those effects are analyzed and revised due to vibration lateral, longitudinal, transverse and / or torsion presented on a footbridge located in Lima, Peru. This footbridge is loaded by external forced induced due the pedestrians.

Pedestrians, while walking, generate forces which are transmitted to the footbridge. People passing through a bridge transmit vertical forces synchronized with a period of walking, lateral and longitudinal forces due to friction between the shoe with the slab of the bridge, and torsion due a possible eccentricity between the axis of the bridge and the line through the pedestrian. Those events could create troubles, because the community would not rather use those footbridges by insecurity. The investment to build the footbridge would be down.

One case of this phenomenon is the Millennium Bridge. The bridge is located in London, England and opened to the public on June 10, 2000, on which an estimated 80,000 to 100,000 people crossed it. Like any structure, the Millennium Bridge has been designed to an expected level of movement, however, when large groups of people crossed the bridge, lateral movements occurred more than expected. When the pedestrian bridge was opened to use and pedestrians walked on a slab, there were unexpected lateral acceleration, which were every sensitive to pedestrians. It was necessary to close the bridge for further evaluation and reinforcement. The maximum displacement of the platform was approximately 70mm. [2]

There are other bridges having different structures than the Millennium Bridge and those have experienced significant lateral vibrations when they have been congested, such as The Road Bridge at the port of Auckland in 1975. These cases have not been published extensively and therefore the phenomenon is not well known.

2. A Synchronic excitation – tolerance of vibration

When people walk, it creates a pattern of repeated forces making the mass of the structure vibrate up and down. This phenomenon is additional to the gravity load by the weight of pedestrians. The phenomenon creates a vertical force fluctuation of about 250 N. which is repeated with each step. There is also a small oblique force caused by the shaking of our body because our legs are slightly apart. This force of about 25 N. goes to the left when it rests on the left foot and right when it rests on the right foot and repetitions occur with each successive step [3].

The balance of the footbridge under the pattern of load explained was not significantly affected by vertical movements; however, movements are not tolerated by perpendicular forces. If the walking surface begins to oscillate, pedestrian must put his foot a little more separated from the other to seek stability and strength (lateral) oblique increases. This tendency towards synchronization has the effect that every step that begins to work increasing the movement of the slab. While the movement progresses, it moves with the same period of motion of the surface. Also tend to widen the position, thereby increasing the lateral force. The effect explained increase with a greater number of people. The individual patterns of response vary but most of the crowd interacts with the surface and develops a lateral move in sync. This phenomenon is referred to as synchronous lateral excitation [3]. Some design codes recommend use vibration tolerance, in order to design more stiffness structures members.

Approximate levels of acceptability for vibration acceleration are defined as a function of the human perception to vibrations. A common human problem is that motion causes the pedestrian to become concern about the safety of the structure, where danger of structural collapse is not an issue. There is no enough number of design codes which treat this. Some examples of design codes are and the Ontario Bridge code. [1]

The British bridge design code takes pedestrian response to vibrations into account by the acceleration serviceability expressed by the following equation:

$$A_{\text{limit}} = 0.5\, f_1^{0.5}\ [\text{m/s}^2] \tag{1}$$

This formula works for fundamental natural frequencies f1 (in Hz) less than 5 Hz.

The Ontario bridge code [1] the serviceability acceleration limit is given by:

$$A\ \text{limit} = 0.25\, f_1^{0.78}\ [\text{m/s2}] \tag{2}$$

This level of vibration is not provided for in the design of the bridge. This excessive movement due to vibration may occur in other bridges, existing or under construction, for a lateral frequency below 1.3 Hz and with a sufficient number of pedestrians. [2] This frequency was measured by non-destructive vibration tests.

3. Measurement of the force from the walking pedestrian

The vibration induced by the people is a mobile force caused by walking pedestrians. The force induced by them on the bridge has been measure by dynamics tests, measuring the displacement and the acceleration on the bridge slab. According to several studies [1, 5, 6] the frequency of this vibration is 2 Hertz with a standard deviation of 0.175. For this investigation the variation is considered as that in figure 1. To simulate the horizontal effects, the friction force between the person and the audience is considered by a force function based on that shown in figure 1, and using a friction coefficient = 0.2μ. Horizontal Force is shown in the figure 2.

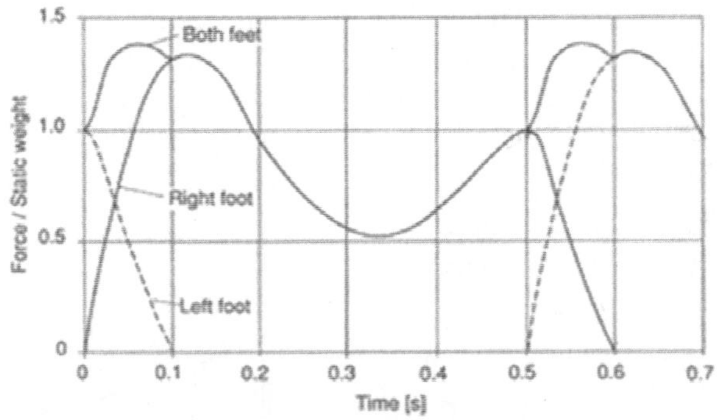

Figure 1. Forcing function resulting from footfall overlap during walking with a step rate of 2 Hz [1].

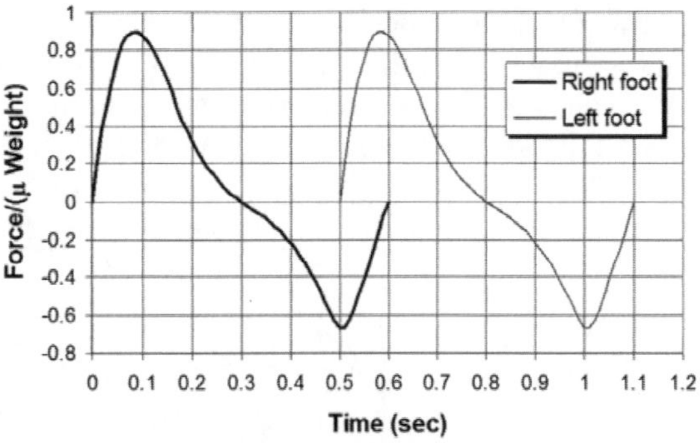

Figure 2. Frictional forcing function considered.

4. A peruvian footbridge case and puente software developed

One Peruvian pedestrian bridge was analyzed in order to study this phenomenon of vibration. This bridge is located in the city of Lima, Peru. It consists of two 21mts spans; each one is articulated by neoprene supports at both ends to 2 meters for each column, as figure 3 shows. The two extreme columns are circular in cross section with a height of 7.00 m., and the central pillar has oval cross section.

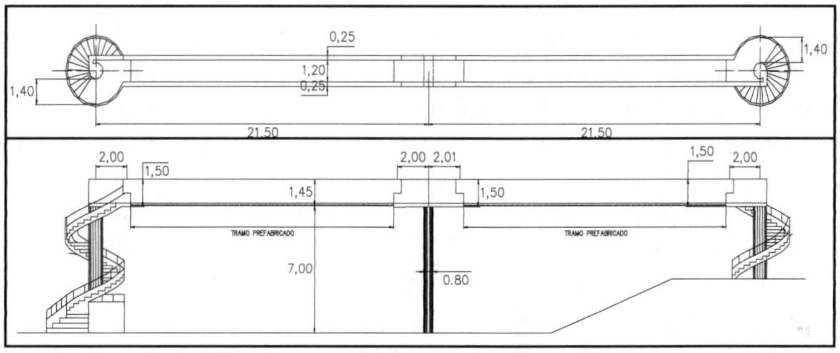

Figure 3. Peruvian Footbridge's scheme.

A linear elastic analysis was performed in order to calculate the dynamic properties of the current footbridge such as vibration modes. The software used was SAP2000 [7] making a structural model on this. The three-dimensional analytical model of the bridge was developed for obtain the following dynamic characteristics:

Table 1.- Dynamics properties of the structure.

Mode	Frequency (Hz)	Period(sec)
M1 Transversal	2.239	0.4466
M2 Lateral	2.253	0.4439
M3	3.293	0.3037
M4	3.336	0.2997
M5 Vertical	4.012	0.2492
M6	5.313	0.1882
M7 Twist	7.473	0.1338

The dynamic properties were calculated. The first mode of the system is associated with the fundamental mode of the bridge and is a movement across the axis of the bridge. The second mode is associated with lateral movement, the sixth mode is predominantly vertical movement, and the seventh shows the torsion mode. Those properties are used to calculate the acceleration tolerance.

Table 2.- Tolerate acceleration of footbridge.

Frequency (Hz)	**Acceleration tolerance (cm/sec^2)**	
	Equation 1	Equation 2
2.239	74.815	47.067
2.253	75.045	47.295
4.012	100.155	74.407
5.31	115.217	92.713

The current results were calculated using classical and conventional analysis methods. However, the synchronic excitation is not considered on those types of structural analyses. It is necessary to make a more complete analysis in order to consider the synchronic excitation,

PUENTE.EXE software was developed in order to analyze the effects of vibration caused by pedestrian loads. The software has a graphic interaction with the user (figure 4). This software is based on the modal displacements of the structure obtained by SAP2000, obtaining values of displacements, velocities and accelerations 6 degree of freedom (DOF) each node of the bridge deck. PUENTE.EXE works with different number of people walking on the bridge, as well as the distances between pedestrians and separation between their legs. This software considers the maximums vibration effects occurred under synchronic excitation, vibration effects such as resonance.

When the period of the applied force is similar to the natural period of structure, resonance is reached. This event produces the maximum amplification of the structural response. Looking to find the maximum responses of the structure (resonance), analyzed the bridge was subjected to periods associated with transverse movements, lateral, vertical and torsion. It shows the effect of amplification of responses when the period of the loads due to pedestrian approaches the period of the structure. Increasing the mass or if the structural configuration of the bridge is different, the period of pedestrians can reach some of the periods of the modes of the structure, presenting the phenomenon of resonance. The value of damping for this work was 0%, no viscous properties on concrete material for this specific case was considerate.

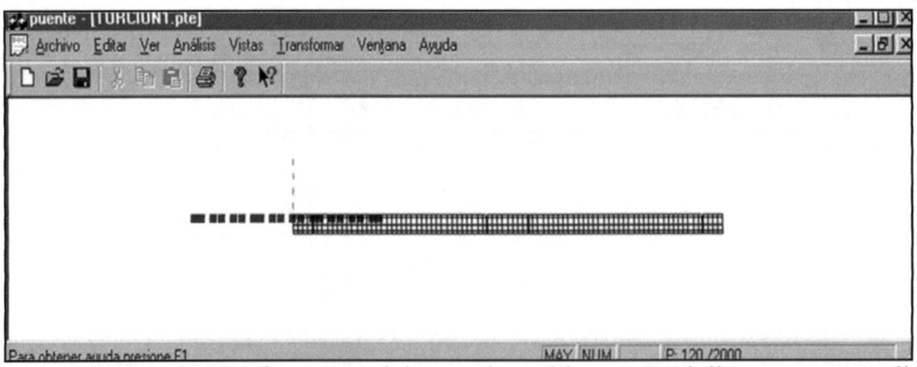

Figure 4. PUENTE.EXE software graph interaction with user. Dash lines represent a line of pedestrian walking on the footbridge. Torsion mode shape is expected for this eccentric line of load due pedestrian location.

PUENTE.EXE calculates responses to cycle and moving different quantities of punctual loads. The structure was analysed for the walking of one person and 50 people, and unit load (force / self-weight pedestrian = 1), to calculate responses of different weights of different pedestrians, the answer will be the value obtained by the program PUENTE.EXE amplified by the weight of the pedestrian. The results expected have cycle patterns of displacement and accelerations.

5. Results of vibration analysis

It is presented the results of the analysis performed by PUENTE.EXE per each case of fundamental mode shape of the footbridge. Each mode shape represents each type of vibration case would appear on case load pattern (due pedestrian loads) carry out on the footbridge. Displacement and acceleration were calculated considering the synchronic excitation explained. The results are listed below.

5.1 To the first mode of vibration, there is the increase of responses (displacements and accelerations) over time pedestrian travel through the bridge deck. The bridge reaches resonance in the transversal direction to start 13seg the pedestrian walk. The bridge is moving freely. The structure has a zero damping to seek maximum responses (Vibration Free System). For the walking of one person with about 60 kilos, the peak displacement is 0.254cm and an acceleration of 50 gals, but 50 people have 12cm and 2390gals. Maximum responses appear on left extreme of the footbridge.

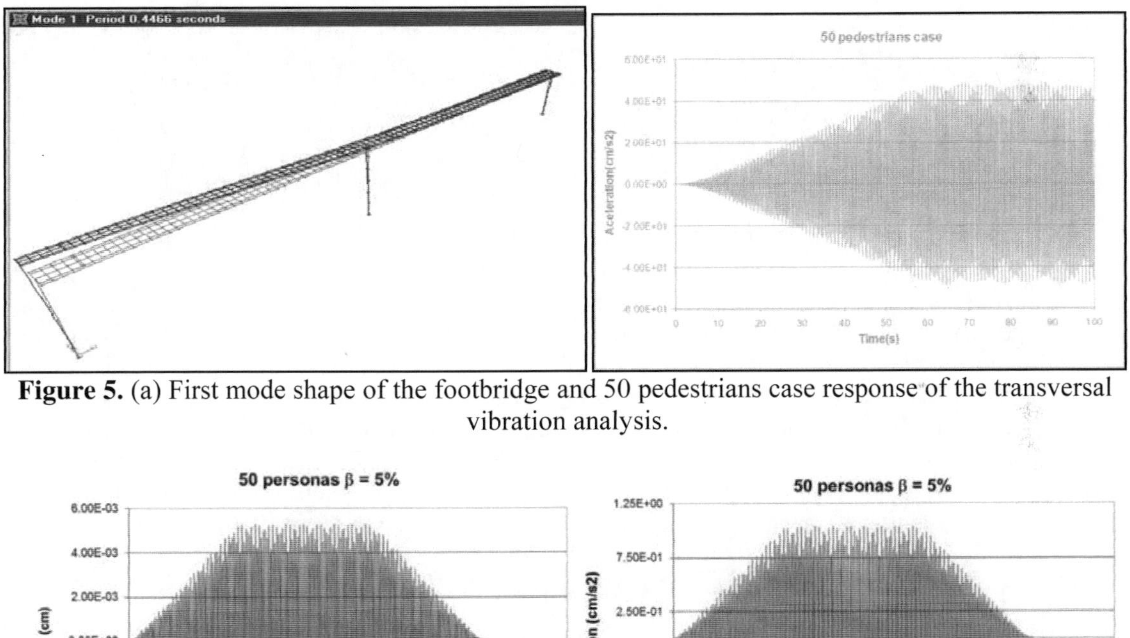

Figure 5. (a) First mode shape of the footbridge and 50 pedestrians case response of the transversal vibration analysis.

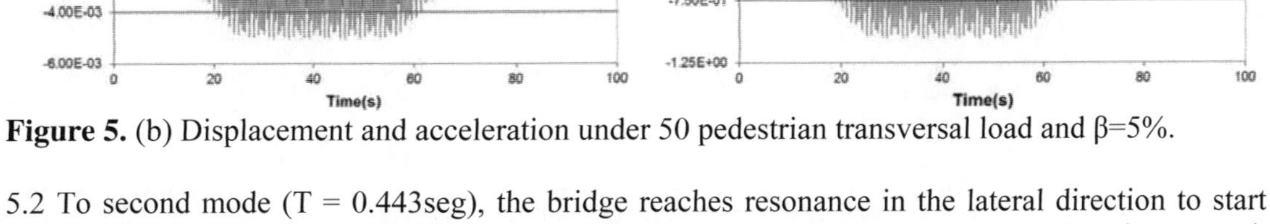

Figure 5. (b) Displacement and acceleration under 50 pedestrian transversal load and β=5%.

5.2 To second mode (T = 0.443seg), the bridge reaches resonance in the lateral direction to start 25seg the pedestrian walk. The bridge is swinging freely. The structure has a zero damping to seek maximum responses (Vibration Free System) like transversal direction. For the passage of a person of about 60 kilos, the peak displacement is 0.242cm and an acceleration of 48 gals, but 50 people have

12cm and 2420gals. Maximums response effects appear at 5.50m and 27.00m from the left extreme of the footbridge.

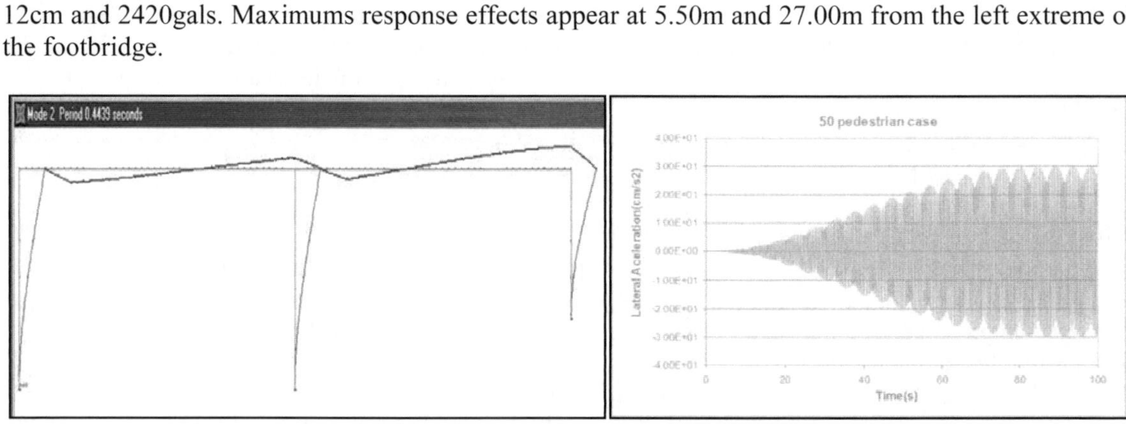

Figure 6. (a) Second mode shape of the footbridge and 50 pedestrians case response of the lateral vibration analysis.

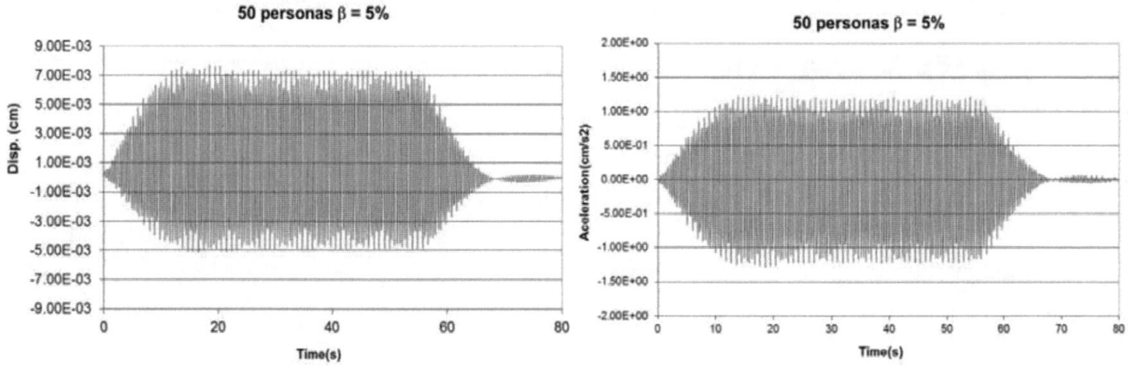

Figure 6. (b) Displacement and acceleration under 50 pedestrian lateral load and β=5%.

5.3 To the fifth mode (T = 0.188seg), the bridge reaches resonance 13seg vertical movement to start the walk from the pedestrian. The bridge is swinging freely. The structure has a zero damping to seek maximum responses (Vibration Free System). For the walking of a person of about 60 kilos, the peak displacement is 0.856cm and acceleration of 59 gals, but for 50 people is 4.28cm and 2920gals. Maximum response effects appear at 11.00m and 32.50m from left extreme of the footbridge.

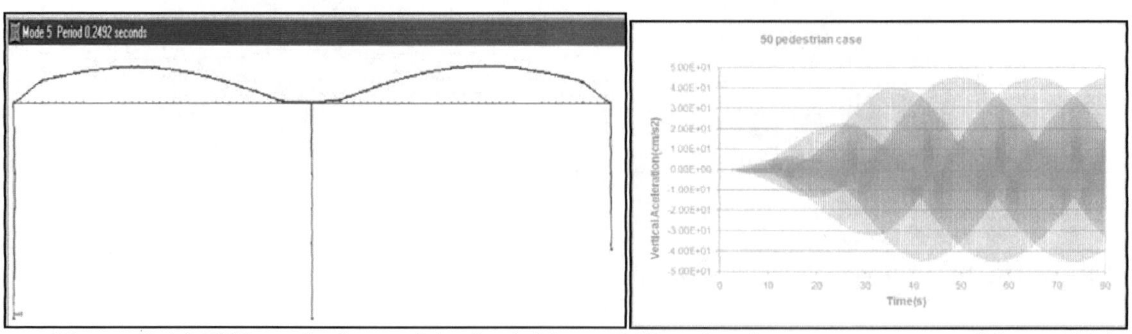

Figure 7. (a) Fifth mode shape of the footbridge and 50 pedestrians case response of the vertical vibration analysis.

IOP Conf. Series: Materials Science and Engineering **473** (2019) 012014 doi:10.1088/1757-899X/473/1/012014

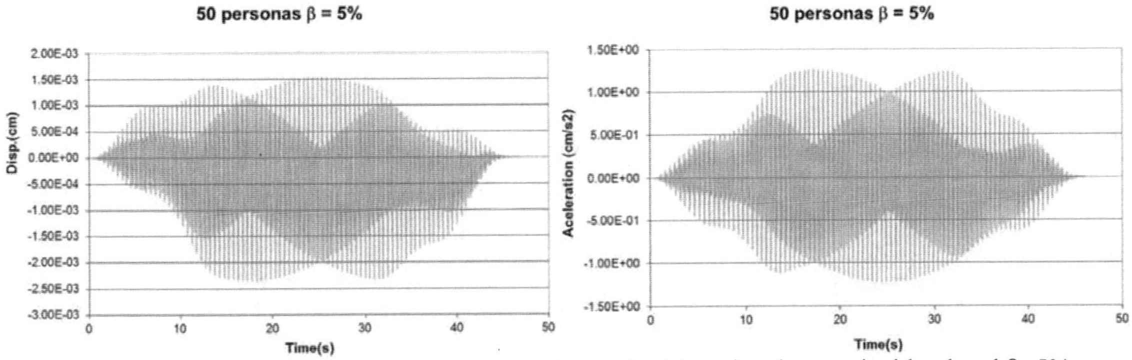

Figure 7. (b) Displacement and acceleration under 50 pedestrian vertical load and β=5%.

5.4 To the seventh mode (T = 0.134seg) the bridge reaches resonance 11seg vertical movement to start the walk from the pedestrian. The bridge is swinging freely to twist movement. The structure has a zero damping to seek maximum responses (Vibration Free System). For the passage of a person of about 60 kilos, the maximum displacement is reached 4.16E-4cm and acceleration of 0944 gals, but for 50 people has 0.0152cm and 33gals. Maximum response effects appear at 11.00m and 32.50m from left extreme of the footbridge as vertical response case.

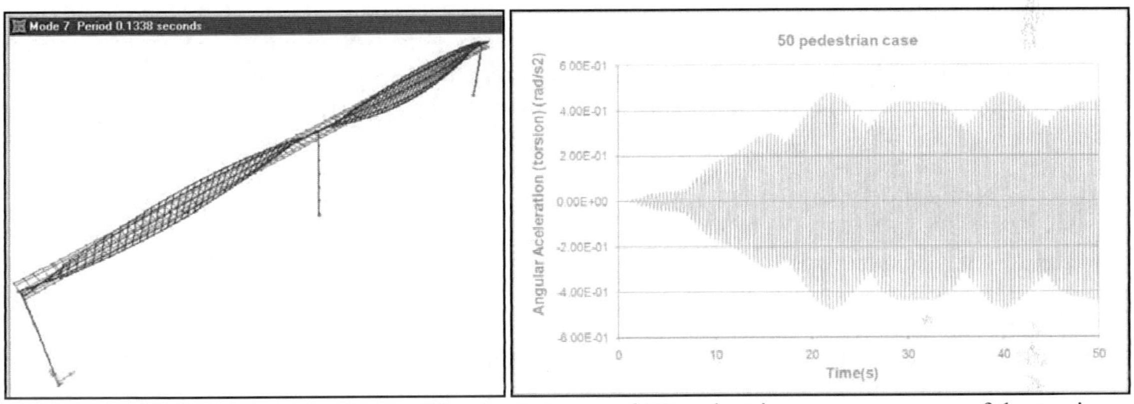

Figure 8. (a) Seventh mode shape of the footbridge and 50 pedestrians case response of the torsion vibration analysis.

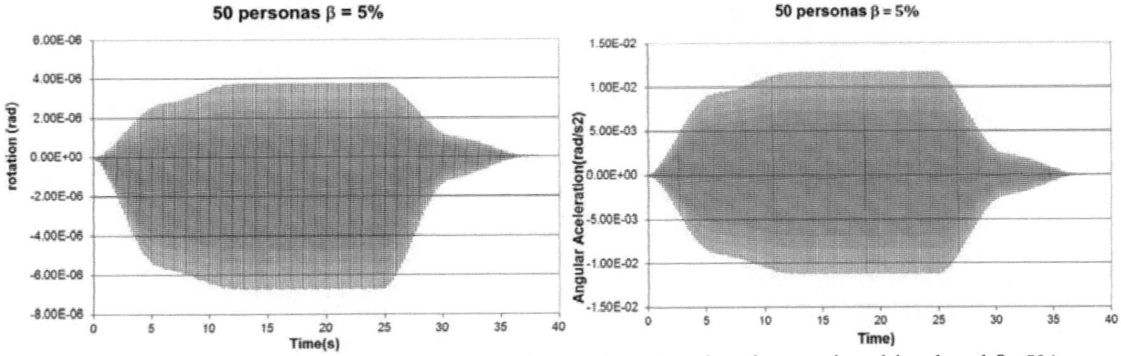

Figure 8. (b) Displacement and acceleration under 50 pedestrian torsional load and β=5%.

Table 3. Summary of Results.

Weight Pedestrian		60 kg			
			Damping β=0%		
		TRANSVERSAL	LATERAL	VERTICAL	TWIST
	Mode Period	0.4466	0.4439	0.2492	0.1338
1 pedestrian	Displacement(cm)	0.254	0.242	0.856	0.00042
	Aceleration(gals)	50	48	59	0.944
50 pedestrians	Displacement(cm)	12	12	4.28	0.0152
	Aceleration(gals)	2390	2420	2920	33
			Damping β=5%		
		TRANSVERSAL	LATERAL	VERTICAL	TWIST
	Mode Period	0.4466	0.4439	0.2492	0.1338
1 pedestrian	Displacement(cm)	0.055	0.0146	0.0127	0.0000429
	Aceleration(gals)	9.06	2.93	5.39	0.0713
50 pedestrians	Displacement(cm)	0.461	0.317	0.0913	0.0004
	Aceleration(gals)	77.6	63.2	75.4	0.703

6. Analysis of the results

The resonance can be reached if the frequency period of walking people on the footbridge is similar or has the same value of the bridge's natural period. The effects produced by the pedestrian bridge must be considered. The bridge may not necessarily be pedestrian, but any other modal having a period that is similar to the passage of people (low stiffness, greater mass, characteristics of pedestrian bridges, or pedestrian platforms in the vehicular bridge wings). It is necessary consider and make vibration analysis.

Due vibration analysis performed on the Peruvian footbridge, it was analyzed the responses of the bridge are not considering the damping of the concrete, this in order to observe the phenomenon of amplification of dynamic response. For design calculations must take into account the damping. The program PUENTE.EXE also considers this parameter. The results have been compared with the tolerance limit presented.

It is found that due transversal, lateral and vertical excitation the footbridge must have troubles on vibration, especially on the acceleration. Values of acceleration results calculated for 0% damping and show that they are exceeding the tolerance limits, on 20 more times than tolerate values. Accelerations calculated reach 2400 gals and the tolerance limits is less than 115 gals. For 5% damping case, concrete material damping, the acceleration response of the footbridge is close to tolerance values. The footbridge may have serious troubles on its use. Bridge designer must consider synchronic excitation on the design.

7. Conclusion

Due the many real examples on the world, the bridge designer must take into account this phenomenon, whose care is largely the perception of pedestrians. There are precedents that show these vibrations as shown. As Peruvian footbridge case, acceleration performed on the structure can reach and exceed human perception tolerance. These tolerance limits of acceleration must be compared with the vibration analysis response explained on the present research.

References

[1] Bach H et.al *Vibration Problems in Structures* Birkhauser, Berlin, 1995

[2] Dallard P et.al *London Millennium Bridge* Pedestrian-Induced Lateral Vibration. Journal of Bridge Engineering. November/December (2001) Pag. 412-417

[3] Fitzpatric, T., et.al. "Linking London, The Millennium Bridge". The Royal Academy of

Engineering, June 2001.
[4] Shun-Ichi N., “Model for Lateral Excitation of Footbridges by Synchonus walking” Journal of Structural Engineering ASCE, January 2004.
[5] Tilly, G, Cullington, D., Eyre, R., Dynamic Behaviour of Footbridges, IABSE Surveys S-26/84, pp 13-24.
[6] Wheeler, J., “Prediction and Control of Pedestrian Induced Vibration in Footbridges”. Journal Structural Division ASCE Vol.108, NoST9. September 1982. Pp.2045-2065.
[7] Computers and Structures, Integrated Software for Structural Analysis and Design, SAP2000, Inc 2009. California. U.S.A.

Expression Forms and Application of Ceramic Materials in Mural Paintings in Architectural Environment

Lei Baojie[1], Kim Chul Soo[2]

[1]Henan Institute of Science and Technology, Dept. of Industrial Design, Pukyong National University
[2]Dept. of Industrial Design, Pukyong National University

Abstract. With increasing humanistic spirit, health concepts and environmental awareness in human beings, ceramic materials, with its strength, are increasingly applied in murals in architectures. As innovative breakthroughs are constantly taking place in terms of ceramic materials and workmanship, greater satisfaction of people's aesthetic needs for architectural environmental decoration are being realized. Focusing on the expression forms of ceramic materials in architectural murals, the paper analyses the practical application of modern ceramic materials in architectural murals from the perspective of color and texture. The paper also explores the innovative development of ceramic murals in hope of promoting the application of ceramic materials in relevant fields.

1. Introduction
Traditional ceramics are made by smashing, mixing, moulding, and high-temperature calcination of clay and natural minerals. Environmentally friendly and non-toxic, such materials have the features of high hardness, colorfastness, moisture resistance, and corrosion resistance with no deformation. In fact, knowledge of the expression forms, technological processes of ceramic murals and their compatibility with the architectural environmental is necessary for the study of ceramic materials in architectural murals.

2. Expression forms of ceramic materials in architectural murals
Ceramic materials have complex process techniques and a great number of expression forms when applied in architectural murals. The selection of ceramic materials may vary according to different requirements of concrete architectural environments. The major expression categories of ceramic murals include overglazed color, underglazed color, tri-color pottery of the Tang Dynasty, fancy glaze, embossed form, and composite materials.[1]

Table 1. Properties Comparison of Different Expression Forms of Ceramic Murals.

Variety	Calcination Temperature	Features of Glaze Color	Suitable Space
Overglazed Color	720~800(9)°C	Stable, Bright-colored	Interior, Exterior
Underglazed Color	1230~1260°C	Simple, Classic	Exterior
Tri-colored Pottery in the Tang Dynasty	900°C	Bright-colored, Florid	Interior
Fancy Glaze	1200~1300°C	Profound, Antique	Interior, Exterior
Embossed	1300°C	Antique, Monolithic	Exterior

2.1. Overglazed color murals

Overglazed color murals are painted on ceramic plates or tiles with specially made pigments, low temperature ceramic pigments and diluting agents. Using simple calcination method, overglazed color decoration has stable performance, rich, steady and vigorous color expression. Varied expression techniques such as sketching contours, dotting stroke, chapped stroke, chopping, dyeing, and brushing can render a vivid and fascinating image. Overglazed color murals are the major expression forms of architectural murals.

2.2. Underglazed color murals

Painting mineral pigments on a green body or unglazed ware, using traditional workmanship, underglazed drawings are performed on the semi-finished goods. With application of transparent glaze, underglazed color decoration is made after high-temperature calcination. More endurable than overglazed color, underglazed color figure is bright, classic and monolithic. With sparkling and wet color surface, it features as corrosion resistant, wear resistant, environmental friendly, non-toxic and easy to clean. Underglazed color decoration has higher requirements on technological processes, and is thus more suitable for small-size architectural murals.

2.3. Tri-colored glazed murals with the tang dynasty style

Tri-color pottery of the Tang Dynasty is made by direct painting with pigments of multiple color low temperature glaze on unglazed wares before low temperature calcination. With time-honored workmanship, it is increasingly becoming a creative form of ceramic murals. Traditional tri-color glazed pottery of the Tang Dynasty mainly exhibits itself in a three-colored glaze, i.e. yellow, green and brown. Modern tri-color glazed pottery of the Tang Dynasty style is formulated by using colored painting bounded by lead powder. With plain color and natural furnace transmutation, it has simple and unadorned beauty.

2.4. Fancy glaze murals

Fancy glaze is manufactured by high temperature calcination using high temperature color glaze as pigments. Wonderful artistic effects can be produced through the inter-infiltration of colored glazes during the high temperature calcination. With fascinating and varied transmutation, monolithic glaze color, antique style, mysterious beauty and harmonious artistic features, fancy glaze are increasingly becoming a popular ceramic mural material.

2.5. Embossed ceramic murals

Embossed ceramic murals have the features such as good overall effect with clear distinction and stereoscopic impression. Embossed murals on unglazed wares have been quite popular since recent years. No glazes are applied on the ware in this way to protect the unadorned feature. Shoveling, chopping and hollow engraving by using sweep blade, slanting knives and blunt shovel are performed on the unpainted clay body to produce plain, uncomplicated, monolithic and powerful charm.

2.6. Comprehensive decorative ceramic murals

Comprehensive decorative ceramic murals are a particular form that employs multiple ceramic expression techniques for artistic representation. For example, the ceramic mural that combines overglazed color figure with high temperature fancy glaze, or underglazed color figure with embossed form.

In terms of the workmanship, comprehensive ceramic murals are an innovative integrated expression form that draws upon the strength of each type of ceramic murals to avoid the usage of single technique and material.[2]

3. Analysis of ceramic materials' application in architectural murals

3.1. Application of colors

The colors of ceramic murals are rich and varied, with different glaze colors to reflect different color

IOP Conf. Series: Materials Science and Engineering **473** (2019) 012015 doi:10.1088/1757-899X/473/1/012015

styles. The rich glaze color is the unique quality of ceramic murals. In the process of calcination, the mysterious glaze color reflects a unique artistic beauty. The glaze color produces unique artistic charm in the furnace, which endows the cold and bland building environment with emotional touch. The fancy change of glaze color of ceramic murals, blended with the architectural environment, evokes the resonance of people's hearts. This emotional color touches the human mind, and exudes a natural and mysterious artistic atmosphere, which naturally arouses people's aesthetic interest. While it touches a chord in human heart, it also paves the way for a broader development of architectural environment.

Between 1957 and 1959, Joan Miró created two earthenware glazed murals called "Wall of the Moon" and "Wall of the Sun" for the UNESCO building in Paris (figure 1) (figure 2). The two pieces are characterized by the unique charm of ceramic glaze, with abstract shapes and bright colors, which renders strong yet soft visual effect of the picture. The works are filled with the sense of freedom and innocence, which gives people fancy feelings and highlights the yearning for freedom and peace. The beauty of nature and harmony is revealed, and the color style of the works is in harmony with the UNESCO architectural environment.

Figure 1. Joan Miró, "Wall of the Moon".

Figure 2. Joan Miró, "Wall of the Sun".

3.2. Application of texture

The texture of ceramic murals is characterized by unevenness, criss-crossing, rough and smooth texture structure, which reproduces rich and varied mural features as well as changing visual effect. The texture of ceramic art is diverse due to factors such as glaze color, clay, shape and craftsmanship. Artists take advantage of the characteristics of ceramic materials in the production process, using techniques including tapping, carving, embossing, scribing, and pinching, so as to realize a variety of texture effects. [3]

The ceramic mural work "Dongping River Catalogue" in the Shiwan Ceramics Museum in Guangdong uses the stacking method (figure 3) to carry out hand-moulding and embossed decoration on clay mud, which realized lively and vivid textures of the characters and landscapes.

Repetitive pasting of single piece of clay is adopted in the ceramic mural named "A Gentle Breeze" in Shanghai Pudong Airport (figure 4), forming a rhythmic texture effect full of modern flavor.

Figure 3. Zhang Wenzhi, "Dongping River Catalogue".

Figure 4. Zhu Legeng, "A Gentle Breeze".

Figure 5. Zhang Guofan, "Folk Dance".

3.3. Material application

Ceramic murals involve artistic processing of clay materials through special workmanship, so as to serve the architectural environment with the artistic beauty of ceramic materials. The properties of ceramics are unmatched by other mural materials.

"River of Life", created by Chinese potter Zhu Legeng, uses clay as material to recreate the lines of life and rhythm for modern urban space. Ceramic materials that are usually static in the traditional art are changed into a myriad of rhythmic ceramic lines in the mural. Many flowers form a flowing ceramic river, reflecting the beauty of ceramic materials (figure 6). In another ceramic mural by Zhu Legeng called "A Book from Heaven", clay was fully employed through kneading and tapping to form a natural crack, making the mural natural and monolithic. The wrinkles of the mud pieces and the cracks formed through furnace transmutation make the mural painting seemingly relaxed and free, more vividly expressing the theme.

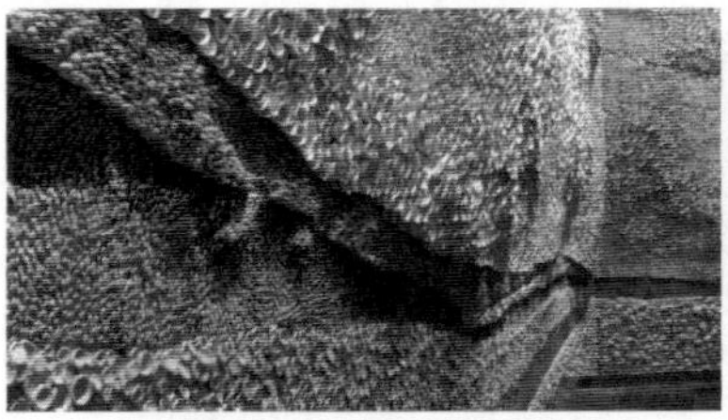

Figure 6. Zhu Legeng, "River of Life".

Figure 7. Zhu Legeng, "A Book from Heaven".

Figure 8. Zhu Legeng, "Nirvana".

3.4. Application of comprehensive materials

With the rapid development of ceramic art workmanship, ceramic mural materials, coupled with their special features in quality, color and texture, are increasingly being used with other materials to form a new mural expression form that produces original and harmonious artistic effect. The integrated use of different materials has created a new artistic beauty of ceramics, brought new vitality to ceramic murals and haven been increasingly satisfying people's aesthetic requirements for ceramic murals.

Glass material workmanship was applied on the ceramic plate of a glass embossed ceramic mural work called "Water Splashing Folk Customs", which makes the mural style simple and beautiful, reflecting the distinct ethnic style (figure 9). The marble ceramic embossed mural work named "Now and the Future" uses marble and ceramic, two different materials, to produce a layered, dynamic and powerful visual effect (figure 10). The glass ball mosaic ceramic mural entitled "Unlimited" was made by pasting glass balls on the surface of ceramic mosaic wall, rendering vigorous and splendid design (figure 11).

Figure 9. Lang Sen, "Water Splashing Folk Customs".

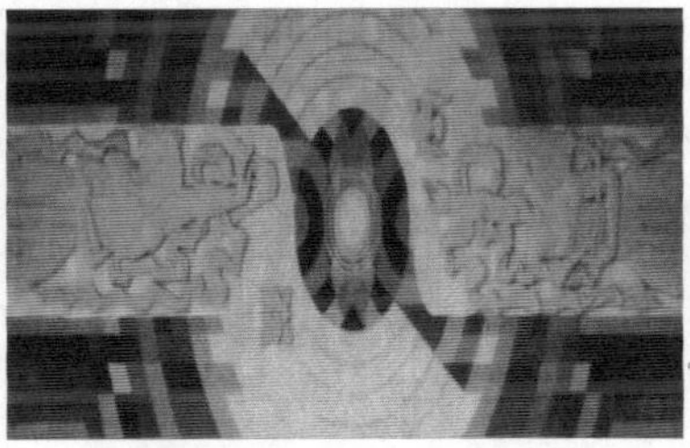

Figure 10. Zhang Songnan, "Now and the Future".

Figure 11. Cui Bingliang, "Unlimited".

In summary, with particular artistic characteristics, ceramic murals can serve as a unique form in the architectural environment, in which the characteristics of color, texture and material play an important role in the formation of ceramic mural style. Making full use of the influencing factors of ceramic materials in the design of architectural murals can create impressive mural paintings with artistic beauty.

4. Prospect of ceramic materials' application in architectural murals

With rapid social and economic development, and the improvement of people's ideological and cultural standards, diversified development trend is taking place in the field of ceramic murals.

Modern ceramic materials have developed rapidly along with the advancement of architectural mural art. On the basis of ceramic media, artists have created art works with greater modern aesthetic awareness. Ceramic mural creation participates in architectural design with active openness and becomes an important artistic carrier in environmental art. In the future application of ceramic materials in architectural environment, the following aspects should be emphasized.

First, people are constantly pursuing the optimization of urban environment and creating a quiet, harmonious and relaxed living space. This requires ceramic mural artists to keep up with the development of the times, constantly adapt to people's new aesthetic viewpoints, and make ceramic material murals fashionable and relatively endurable. Second, with the changes and development of ceramic mural materials, artists need to constantly explore the potential of such materials, introduce new workmanship and new technologies, and better grasp the expression language and characteristics of ceramics. Third, while maintaining the characteristics of local culture, the study and exchange of international mural art should be strengthened to integrate advanced design concepts and techniques, and promote the rapid development of ceramic murals.[4]

5. Conclusion

As ceramic mural materials boast high adaptability to surrounding environment and have close affinity to nature, many architects have begun to adopt such materials as a way to pursue and express their understanding about beauty. Ceramic materials used in architectural murals are environmentally and ecologically friendly, and are easy to clean, waterproof, fireproof, anti-corrosive, and anti-oxidant. These advantages represent the unique artistic charm of architectural murals. Ceramic murals can effectively decorate the physical space through the expressions of color, shape, and texture, and can enhance the cultural and artistic atmosphere of the entire architectural environment. The permanence, affinity, contingency and aesthetics of ceramic murals make it unique in the architectural environment. Used in public space mural paintings, modern ceramic materials play an important role in architectural environment, as they can turn architectural space into a distinctive and ideal medium for human interaction in artistic life. With the boom in modern urban architecture and the evolvement of the idea "living in art", ceramic mural materials will be applied extensively in modern architectural space, and become the driving force behind the fusion of artistic life and architectural techniques.

Reference

[1] Song Z J 2013 *Ceramic Mural Glaze Form Language Study*. Chongqing: Southwest University.
[2] Huang X R 2014 *Application of Ceramic Murals in Indoor Environment Design*. Jiangxi: Jingdezhen Ceramic Institute.
[3] Ma L 2016 *Application of Ceramic Murals in Architectural Environment*. Jiangsu: Suzhou University.
[4] Hou Y M 2004 *Chinese Mural Painting for 100 Years*. Beijing: China Building Industry Press.

IOP Conf. Series: Materials Science and Engineering **473** (2019) 012016 doi:10.1088/1757-899X/473/1/012016

Stabilization of Pavement Granular Layer using Foamed and Emulsified Asphalt under Critical Low Temperature Conditions

J Sánchez[1], N Shoji[1] and G Lazo[2]

[1] Student, Universidad Peruana de Ciencias Aplicadas
[2] Professor, Universidad Peruana de Ciencias Aplicadas

Abstract. One of the efficient techniques that is currently used in pavement engineering to improve the stress–strain response conditions of the pavement structure includes the stabilization of its structural layer components. There are several laying techniques that include the usage of stabilizing agents such as emulsified and foamed asphalt. In these technologies, the temperature of the environment is observed to play a critical role. This condition is often considered to be the main factor that encourages the usage of foamed asphalt for stabilization. Further, the aforementioned technique specifies that, at temperatures of approximately 10 °C or lower, the asphalt particles are not effectively inserted in the foamed granular material mastic; they are instead merged with other asphalt particles. This causes an agglomeration of coalesced asphalt, prevents adequate foaming, and hinders the pavement layer from compaction. In such a situation, the pavement will exhibit a structural strength deficiency. Therefore, this study experimentally investigates the environmental temperatures at which an efficient layer that is stabilized using foamed asphalt may be obtained both in the laboratory as well as in the field. Further, this study proposes a new limit. Apart from exceeding this temperature limit, the study also offers an alternative with respect to the usage of asphalt emulsions to stabilize the granular layer. This technique comprises the dispersion of asphalt particles in an aqueous medium; however, when this technique is applied at low temperatures, the low temperatures do not allow the system to reach a critical condition of inapplicability, which is observed when the asphalt foam is used for stabilization. Further, the mechanical behaviors of the foamed pavement at temperatures of lower than 10 °C are discussed. Additionally, this study exhibits the results of stabilized layers that use a slow-setting asphalt emulsion (CSS-1h) as a solution to the temperature problem that is associated with the usage of the foaming technique. The investigation is performed based on a project that is conducted in the extremely low temperature areas of the Peruvian Andean highlands, which are located at an altitude of 4,000 to 5,000 meters above the sea level.

1. Introduction

The usage of foamed asphalt and asphalt emulsions as stabilization techniques for the granular layers of pavements has been increasing. Both foamed asphalt and asphalt emulsions efficiently increase the structural strength of the material that is measured in the compact condition. These methods, which were initially introduced in the United States market during the 1950s, exhibited various advantages such as low energy consumption, reduction in emissions, and improvement in the support capacity of the mixture after compaction [1]. Currently, based on modern techniques, we can observe that both the methods provide excellent stabilization alternatives [2].

Recent studies have denoted that the usage of foamed and emulsified pavements exhibit better resistance to the damage caused due to moisture and fatigue as well as permanent deformation as compared to that exhibited by the traditional granular base layers [3]. Both stabilizing techniques are observed to add asphalt to the granular material in different manners.

Foamed asphalt is a physical process during which a thermal shock is observed when hot asphalt comes in contact with water, which results in the formation of asphalt bubbles filled with water vapor [4]. These bubbles coming into contact with the aggregates results in expansion. This will be effective if they fall into the mastic, which is a fine fraction that comprises sand, Portland cement, and water, and form connectors that increase the cohesion of the mixture [5].

However, low temperatures cause the optimum asphalt and water temperatures that are required for the occurrence of thermal shock to drop rapidly. This causes the foaming quality indicators, such as expansion ratio and half-life, to exhibit values that are lower than the expected values, thus reducing the foaming quality. Further, these low temperatures cause the aggregates to cool down in such a manner that the asphalt particles are not able to connect in the aforementioned fine fraction, which makes it impossible to form ductile connectors as a consequence of poor foaming [6].

Additionally, emulsions do not require asphalt to be heated to high temperatures of 160 °C–180 °C, which is a temperature range that is similar to that observed in the case of foaming. Further, the asphalt emulsion forms an electrochemical process comprising asphalt, water, and an emulsifying agent. The asphalt particles are dispersed in water because of the temporary effect of the emulsifier. To combine the asphalt particles with the granular material, the breaking and curing processes must be conducted. The first process that can be referred to as emulsion breaking occurs when the water is separated from the asphalt and begins to evaporate. In the second process, or system curing, the water completely evaporates, leaving the asphalt bound in the aggregates [7].

The objective of this study is to define the temperature conditions under which foamed asphalt and asphalt emulsions can be effectively used, which can be further used to validate the application of these techniques with respect to quality parameters such as indirect tensile strength (ITS), loss of stability, and the compaction and deflectometry controls. The experiments that are associated with this study are performed in a construction project that is located between the cities of Tacna and Puno, Peruvian highlands, at very low temperatures. Additionally, the highest sections of this road are observed to be located 4,800 meters above sea level.

2. Background: design of a mixture stabilized with foamed asphalt

This study began using the initial laboratory test results that were obtained for the foamed asphalt design based on the specifications from the Wirtgen Manual. Because the granular material that was to be stabilized did not exhibit a plasticity index, 1% of the Portland cement was used as the active filler. Based on the foaming laboratory design, it can be determined that the factors that achieved the most efficient foaming, in terms of both expansion ratio and half-life, corresponded to the usage of asphalt cement 85/100 penetration grade at a temperature of 160 °C and in the presence of 2% water. Using this combination, samples were manufactured with 1.5%, 2%, and 2.5% asphalt content to evaluate the combination that provided the optimal stabilization conditions for the granular material. The specimens in a compact condition were subjected to the Marshall press to determine their ITS in dry and saturated conditions, establishing 2% of residual asphalt as the optimum value for the foamed mixture.

However, despite the fact that the design of the mixture that was stabilized with foamed asphalt was successful in the laboratory during the early stages, progressive difficulties were reported in preparing and laying the foamed mixture during the construction process in the field. The conditions of the construction work increased in altitude and decreased in temperature; thus, unfavorable temperature conditions were achieved, and these results were further verified in the laboratory.

While achieving a temperature of 7 °C or lower, which is characteristic of the Peruvian highlands in this sector at which large extensions are observed at an altitude of 4,000 meters above the sea level, the foaming quality was observed to be considerably diminished by the adverse climate. Thus, during the foaming process, water and asphalt entered the expansion chamber, which further resulted in the production of thermal shock at lower temperatures than those recommended in the design. Further, asphalt bubbles were dispersed over very cold aggregates, which caused poor foaming and stabilization.

(a) (b)

Figure 1. (a) Asphalt particles coalesced in the field. (b) Specimens manufactured using the foamed material with coalesced asphalt connectors.

Such circumstances caused the inefficient insertion of asphalt particles as connectors in the mastic. This resulted in the formation of coalescing asphalt masses, which are colloquially referred to as gummy asphalt, as depicted in figure 1a.

In figure 1b, the base foaming material is taken to the laboratory, where it was compacted and subjected to the Marshall press to determine its ITS using the ITS test. The obtained ITS values were observed to be lower than those recommended for the technique: 200 kPa, 100 kPa, and 50% ITS in dry, saturated, and retained strength, respectively.

Table 1. Structural strength of the foamed mixture in a compact condition.

Sector km	Density (g/cm^3)	Humidity (%)	Strength (kPa)		Retained Strength (%)
			Dry	Wet	
126+000 to 127+000	2,116	7.9	180.9		45.4
	2,016	7.0	171.3		
	2,073	7.9	187.3		
	2,106	7.4		80.3	
	2,093	7.0		88.0	
	2,011	8.2		76.8	
Average	2,069	7.5	179.8	81.7	

3. Method: design of the stabilized mixture with asphalt emulsions

In the first stage, the quality control of the granular material that was to be stabilized was performed. Aggregates were procured from the Tasabaya Quarry, which is located 86 km away from Tacna, Peru. The results of the laboratory tests that were performed on aggregates, as presented in table 2, were obtained based on the specifications of the ASTM (American Society for Testing and Materials) and the Ministry of Transport and Communications [8].

Table 2. The results of the granular material laboratory tests.

Test	Standard	Result	MTC Specification
Fine Material (%)	ASTM D 6913 / MTC E 202	8.70	Maximum 10%
Liquid Limit (%)	ASTM D 4318 / MTC E 110	18.0	
Plasticity Index (%)	ASTM D 4318 / MTC E 111	NP	Maximum 9%
Sand Equivalent (%)	ASTM D 2419 / MTC E 114	34.0	
Los Angeles Abrasion (%)	ASTM D 131 / MTC E 207	31.5	Maximum 50%
Loss with Magnesium Sulphate for Coarse Aggregate (%)	ASTM D 88 / MTC E 209	14.96	Maximum 18%
Loss with Magnesium Sulphate for Fine Aggregate (%)	ASTM D 88 / MTC E 209	2.90	Maximum 15%

The Illinois methodology was used to design the granular material that was stabilized using emulsions with an objective of determining the optimum content of residual asphalt using a slow-setting cationic emulsion (CSS-1H), which exhibited a penetration of 75 mm at 25 °C. The product was observed to be compliant with the ASTM D2397-13 specifications. The initial step was to evaluate the theoretical residual asphalt content, which was 2%. Using this value, Marshall briquettes were prepared at 75 strokes per face at different moisture contents to determine the optimum water content for compaction, which was 8.6%. Briquettes were prepared again with an optimum moisture content and at different residual asphalt contents of 1%, 2%, and 3%. Half of these samples were subjected to the Marshall press under dry conditions, and the other half under saturated conditions. The Marshall dry and wet stability values were found, which are demonstrated in figure 2 (a). Therefore, an optimum residual asphalt content of 2% was selected, because the highest Wet Marshall Stability of 1052.5 kgf and a loss of stability of 49.4% was achieved with this value. Also, the highest Dry Bulk Density of 2,162 g / cm3 was obtained, which is demonstrated in figure 2 (b) with a percentage of air voids of 13.55%. These results allowed the stabilization of the granular layer of the pavement with asphalt emulsion to have the highest resistance under critical conditions of the environment to which it would be subjected.

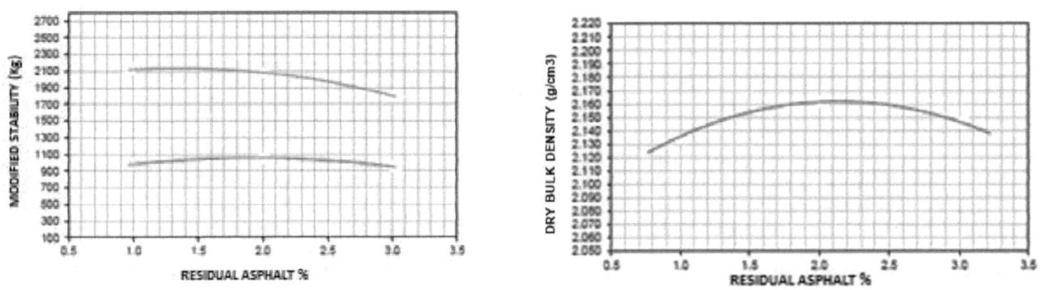

(a)

(b)

Figure 2. (a) Dry (top line) and wet (bottom line) Marshall stabilities.
(b) The dry bulk density at different contents of residual asphalt.

4. Results

The laboratory design of the granular layer that was stabilized with asphalt emulsions at 2% residual asphalt was observed to be efficient at low temperature climate conditions, since this technique was not incidentally dependent on the climatic conditions that were observed at temperatures of 10 °C and more. This was similar to that observed in the case of the foamed asphalt, which required adequate foaming and adequate insertion of the connectors in the foamed mastic with aggregates at equally controlled temperatures. Further, the asphalt was temporarily dispersed in water due to an emulsifying agent to ensure stabilization with emulsions. This aqueous solution was maintained at a temperature of 30 °C.

This can be verified using field results that involved compaction degree controls that were performed every 50 meters, accepting a minimum compaction of 98% Maximum Dry Density. This was completed with the results of the deflectometry, using the Benkelman Beam every 25 meters alternating lane, to determine if the deflection was recoverable at the passage of the load, compared to the admissible deflection value of 0.95mm. These controls can be observed in figure 3 (a) and (b).

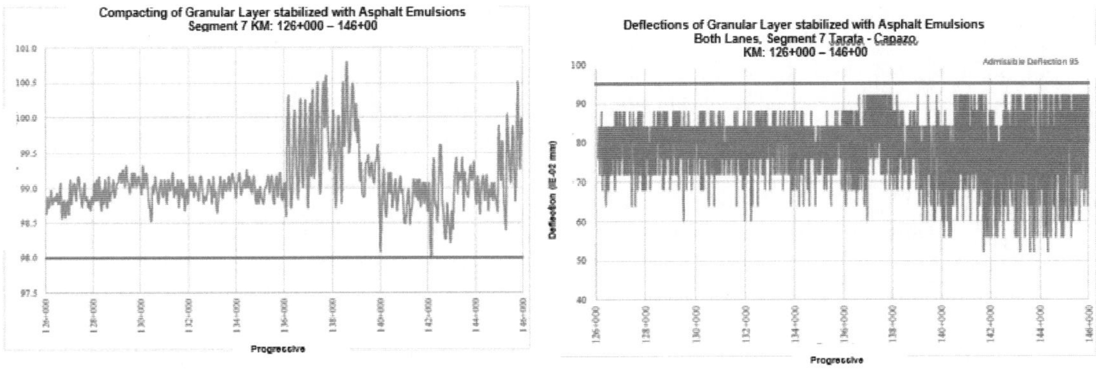

(a) (b)

Figure 3. (a) Degree of compaction of the layer stabilized with asphalt emulsions.
(b) Pavement deflections measured with the Benkelman beam.

Similarly, the stock material that was emulsified in the field can be taken to the laboratory to verify whether the production complied with stability losses of less than 50%, as depicted in table 3.

Table 3. The results of the granular material laboratory tests.

Nmbr	Dry density (g/cm^3)	Dry load (kgf)	Wet load (kgf)	Loss of Stability (%)
1	1,832	2181.47	1444.35	33.8
2	1,867	1731.59	988.14	42.9
3	1,921	1375.50	926.90	32.6
4	1,821	1443.47	975.55	32.4
5	1,941	1302.53	781.40	40.0

Finally, the versatility of the stabilization of granular materials with asphalt emulsions as compared to the stabilization of granular materials with foamed asphalt was demonstrated in low temperature conditions (5 °C). This can be verified when the quality parameters of the finished product are satisfied from 126 + 000 km to 146 + 000 km, when compaction grades are greater than 98%, deflections are less than 0.95 mm, and loss of stability is less than 50% are observed.

5. Conclusions

The climate temperatures are fundamental in stabilizing the pavement granular materials with foamed asphalt, since proper foaming is not achieved at low temperatures. In the case of the project analyzed in this study in the Peruvian Puno highlands (one of the few places in the world where large extensions, not only steps, are located at an altitude of more than 4,000 meters above the sea level), efficiently foamed asphalt mixtures could be achieved at a temperature of 7 °C. The quality of the foaming was poor below this temperature. Further, stabilization with foamed asphalt could not be achieved along with cold aggregates, which generated massive coalesced asphalt agglomerations that were immersed in the foamed granular mass, as evidenced in both the laboratory and the field.

Before this thermodynamic environmental issue was observed, the stabilization of granular layers with asphalt emulsions achieved a structural contribution that was equal to the results obtained using granular layers that were stabilized with foamed asphalt. Further, no changes were observed in the thickness of the stabilized layer, and the dosage of the asphalt cement that was required to stabilize the granular material was not affected, which ensured that the residual asphalt would remain 2% in both the techniques. The important factor was the variation that was observed during the construction procedure; the emulsions worked and managed the asphalt in a manner that effectively controlled the adversities encountered at low temperatures, thereby facilitating the stabilization of the granular layer with asphalt emulsions.

Both techniques exhibited discontinuous ductile ligament stabilizations, which further provided higher structural capacity, less deterioration from fatigue and less failure due to permanent deformation, thus creating a new group of efficient materials in the stabilization of pavements.

6. References

[1] Yin F, Arámbula E, Newcomb D 2015 *Int. J. Pavement Eng.* **18** 358–66.

[2] Bernucci L, Goretti da Mota L, Ceratti J, Barbosa S 2008 *Asphalt paving* (Rio de Janeiro: Petrobrás) 501.

[3] Arimilli S, Nagabhushana M, Jain P 2017 *Road Materials and Pavement Design.*

[4] Wirtgen Group 2012 *Wirtgen Cold Recycling Technology* 1st edition (Germany) 367.

[5] Lazo G, Morales E and Vásquez P 2017 *Contribution of foamed bitumen in the structural contribution of pavements in height as a solution for the Puno-Tacna road conservation project, Capazo-Mazocruz stretch* (Lima: Tesis UPC).

[6] Halles F 2013 *Stiffnes evolution mechanism of recycled mixes with foamed bitumen and cement* (Chile: Pontificia Universidad Católica de Chile).

[7] Asphalt Insitute and AEMA 1999 *Basic Handbook of Asphalt Emulsions* (USA) 12.

[8] Ministerio de Transportes y Comunicaciones 2013 *General Technical Specifications for Construction* (Lima: MTC)

Analysis of Temperature Field of Tunnel Surrounding Rocks in Freezing-Thawing Environment

Jie Zhang[1], Nan Liu[2]

[1] School of transportation civil engineering, Shandong Jiaotong University, China
[2] School of transportation civil engineering, Shandong Jiaotong University, China

Abstract. On the background of large scale rocks engineering in cold regions, Temperature Field and Stress Field of Tunnel Surrounding Rock under the freeze-thaw environment are numerically analysed. According to the single temperature field equation of tunnel surrounding rock and Da Ban mountain tunnel project in Qinghai province as an example, large scale finite element software is used to simulate temperature distribution under freeze-thaw cycles, which is from the rising temperature to the decreasing temperature and from the rising temperature to the decreasing temperature again. The frost heaving force of the tunnel in the cold region is introduced and the mathematical model about rocks temperature field is also researched. Finally, finite element method is used to calculate the frost heaving force of rock tunnel in the cold regions. The results show that in the freezing and thawing environment, the area near the inner wall of the tunnel is more likely to be affected by the ground temperature and the effect of open ventilation and convection, and the temperature and stress changes more violently. In the design and construction of the cold regions, the influence of the frost heave force of the surrounding rock must be considered. Because freeze-thaw circles could cause the destruction of the tunnel structure, convective heat transfer will occur between tunnel mouth and walls and the environment, resulting in large temperature difference. Therefore, in order to reduce freeze-thaw disasters in the cold regions, we should consider the Protection measures about the hole and the wall. In this way, it reduces the range of freeze-thaw circles and prevent the convective heat transfer effectively.

1. Research background

Seasonal frozen soil is widely distributed in Northeast China, North China, northwest and other regions. About 2 million 150 thousand square kilometers permafrost is distributed in the northeastward and Xiao Hinggan Mountains, the western mountains and Qinghai-Tibetan Plateau. Among this, the Qinghai-Tibetan Plateau permafrost area is 1 million 490 thousand square kilometers, accounting for 70% of the total permafrost area in China, which is rich in minerals, land, biology and tourism resources [1]. Thus, as can be seen, the construction and utilization of resources in cold regions play a very important role in our national economy.

This paper is based on the tunnel engineering of rock mass in cold region, taking the section K106+025 of the exit section of the Qilian Mountains in Qinghai Province as the object of analysis. Numerical simulation of single temperature field and thermal mechanical coupling of tunnel surrounding rock using large general finite element software ANSYS put into use, it analyses the interaction and changing rule of rock temperature and stress field. Measures for restraining freezing and thawing disasters are put forward. Therefore, it provides a scientific basis for quantitatively evaluating the engineering stability of rock mass in cold regions.

2. Fundamental theory

After the tunnel through the cold zone, the original stable thermodynamic conditions were destroyed instead of a new thermal system with no direct solar radiation and open ventilation convection. The new heat exchange system results in the new freeze-thaw characteristics of the surrounding rock mass, and creates the conditions for forming permafrost [2]. According to the annual variation rule of tunnel wall temperature based on the data of meteorology, geology and hydrogeology of the Qilian Mountains in Qinghai Province, and on this basis the freeze-thaw condition of surrounding rock and the condition of permafrost formation after tunnel penetration are preliminarily analysed with different annual average wall temperature.

$T(t,x,y)$ is the temperature value which is at the t time point at (x,y). $T(t,x,y)$ satisfies the heat conduction equation with the following phase transition [3]:

$$C_f \frac{\partial T_f}{\partial t} = \nabla(\lambda_f \nabla T_f), \quad 0 < t < D, x \in S_f(t) \tag{1}$$

$$C_u \frac{\partial T_u}{\partial t} = \nabla(\lambda_u \nabla T_u), \quad 0 < t < D, x \in S_u(t) \tag{2}$$

$$T_f[t,\xi(t)] = T_u[t,\xi(t)] = T_0, \quad 0 \le t \le D \tag{3}$$

$$\left(\lambda_f \frac{\partial T_f}{\partial t} - \lambda_u \frac{\partial T_u}{\partial t}\right)\Bigg|_{x=\xi(t)} = L\frac{d\xi}{dt}, \quad 0 \le T \le D \tag{4}$$

Where, $S_f(t)$, $S_u(t)$ are freeze-thaw districts; λ_f, λ_u are coefficient of heat conduction in freeze-thaw state; C_f, C_u are volumetric heat capacity in freeze-thaw state; $x=(x,y)$, $\xi(t)$ are phase transition interface of freeze-thaw; T_0 is Critical temperature of rock freezing;

$L = \gamma_d \times (W - W_u) \times 80$ is latent heat of water; γ_d is dry density of rock; W is moisture content; W_u is Unfrozen water content [4].

According to the altitude (3800*m*) and the annual mean temperature (-3°C) of the Daban Mountain tunnel, the air density is calculated out: ρ=0.774*kg/m*^{3}; Because of water vapor in the atmosphere, the specific heat at constant pressure is taken as C_p=1.847*KJ/(kg·*°C);

The dynamic viscosity coefficient of air is taken as μ=9.218×10^{-6}*kg·s*;

Thermal conductivity: λ=2.0×10^{-2}*W/(m·*°C)

The thermal diffusivity of air is obtained: α =1.3788×10^{-5}*m*^{2}*/s*;

The kinematic viscosity: υ=1.19×10^{-5}*m*^{2}*/s*

The thermal conductivity of rock mass is treated according to the condition of broken rock. Dry bulk density of rock: γ_d=1835kg/m^3. The moisture content and unfrozen water are respectively:

W =9%, W_u=1.5%, λ_u=1.4338*W/(m·*°C*)*, λ_f =1.8247*W/(m·*°C*)*; The specific heat capacity of rock is taken as 0.8*kJ/(kg*°C*)*.

$$C_f = \frac{(0.8 + 4.128W_u)}{1+W} \times \gamma_f, C_u = \frac{(0.8 + 4.128W)}{1+W} \times \gamma_d \tag{5}$$

The initial boundary value of the temperature is taken as:

$$T(0,x,R_0) = [f(x) - R_0] \times 0.03 \tag{6}$$

$$T(0,x,r) = \begin{cases} [f(x) - r] \times 0.03 \ldots\ldots R < r \le R_0 \\ A - B \ldots\ldots\ldots\ldots\ldots\ldots\ldots r \le R \end{cases} \tag{7}$$

Here *f(x)* is the distance of permafrost lower limit to tunnel vault. The geothermal gradient is 3%.

The natural annual average temperature of the outside of the tunnel is A=-3°C, and the annual temperature variation amplitude is B = 12°C [5].

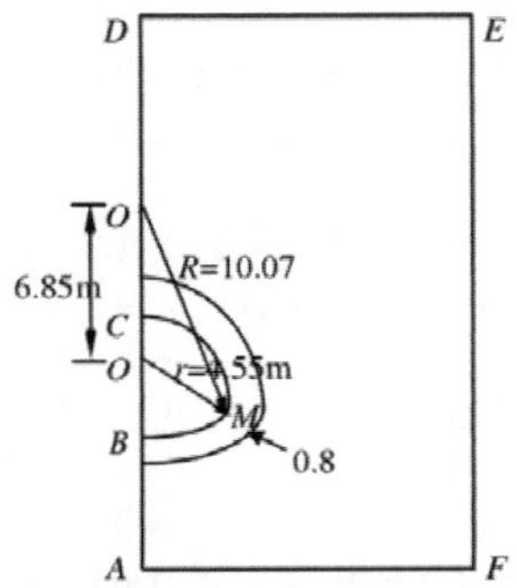

Figure 1. Physical model of single temperature field.

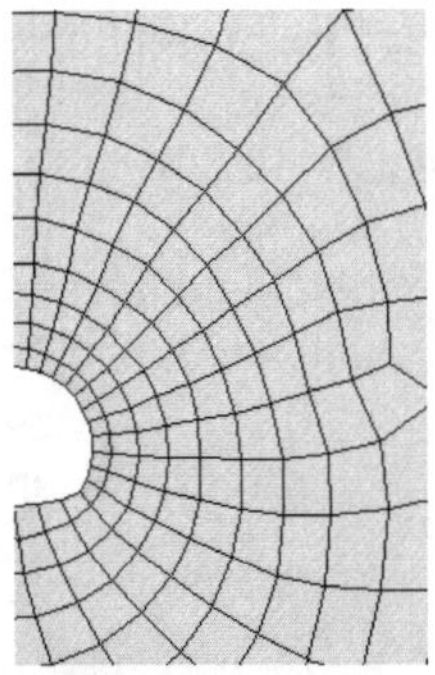

Figure 2. Model grid partition diagram.

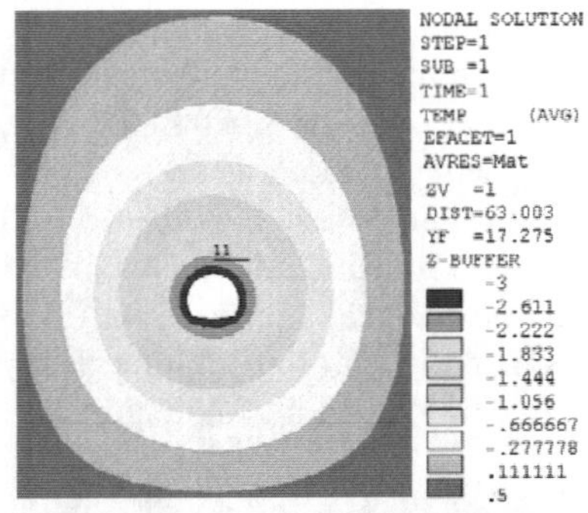

Figure 3. Temperature field of tunnel surrounding rock temperature at -3°C.

The annual average temperature of the Daban Mountain tunnel is -3 ℃. The section of K106+025 at the tunnel exit is taken as the computational model (figure 1). In the vicinity of the section, the groundwater is abundant, and the seepage is very large. The daily discharge is more than 200*m* in the tunnel. At this section, the tunnel tip is located at the lower limit of permafrost 100*m*. OD=74.55*m*, DE is located in the lower limit of permafrost 20*m*, OA=40*m*, AF=DE=40*m*. According to the symmetry, CD and AB are adiabatic boundary, and the BC side is communicated with the atmosphere. From the climatological data of Daban Mountain [6], the temperature variation curve can be obtained as follows:

$$T_a = -3 + 12\sin\left(\frac{2}{8760}\pi t_k + \pi/6\right) \tag{8}$$

The physical model is shown in figure 1.

3. Numerical simulation of temperature field of tunnel surrounding rock in cold region

The finite element software ANSYS is used to simulate the single temperature field of the surrounding rock of the tunnel in cold region. The thermal unit PLANE55 is selected and the characteristic of materials parameters are selected as the thermal conductivity. According to the default form adopted by the ANSYS meshing tool, the model is divided into 1784 nodes and 1679 four node nonparametric elements. The result of the model grid partition is shown in figure 2.

In fact, the process of thermal analysis of ANSYS first divides the object to be treated as a finite element (including a number of nodes), and then based on the equation of heat balance of the heat transmission science classical theory, the heat balance equation at each node are under certain boundary conditions and initial conditions is solved. Thus the temperature of each node is calculated. And then other relevant quantities are solved. In the thermal analysis of ANSYS, the thermal analysis is divided into two categories: steady thermal analysis and transient thermal analysis, not by the familiar heat transfer mode as the classification standard, but it is based on the different temperature field properties. Steady state thermal analysis is used to determine the effect of a stable thermal load on the system or components. Before the transient thermal analysis is carried out, steady state thermal analysis is used to determine the initial temperature distribution. Through steady-state thermal analysis, the parameters such as temperature, thermal gradient, heat flux and heat flux density caused by stable thermal load can be determined. The values of these parameters do not vary with time. That is to say, the temperature field of the system is constant in steady-state heat transfer analysis.

Below is the temperature distribution of the surrounding rock of tunnel wall under the action of temperature: -3°C →10°C →-20°C →10°C →-3°C. That is to say, the temperature distribution of tunnel surrounding rock is under the effect of heating up, cooling, warming up and freezing and thawing cycle, and there is no insulation layer. From the chart, we can see the temperature distribution trend of the surrounding rock and its roughly frozen depth, from which we can get the seasonal freeze-thaw change law.

From the figure 3, it can be found that when the wall temperature of the surrounding rock reaches -3°C, the temperature gradient of the tunnel wall is large, and the average temperature changes at 0.5 *m*. At around 0°C, the surrounding rock will start to freeze, and the freezing depth will be around 10*m* when the temperature changes.

In the same way, as is shown in figure 4, when the temperature of the inner wall of the surrounding rock rises to 10°C, the depth of melting is about 10*m*, and the depth belongs to the frozen area for many years, and the effect is not significant.

When the wall temperature drops to -20°C, the freezing depth has reached 12*m*, which has affected the freezing depth for many years.

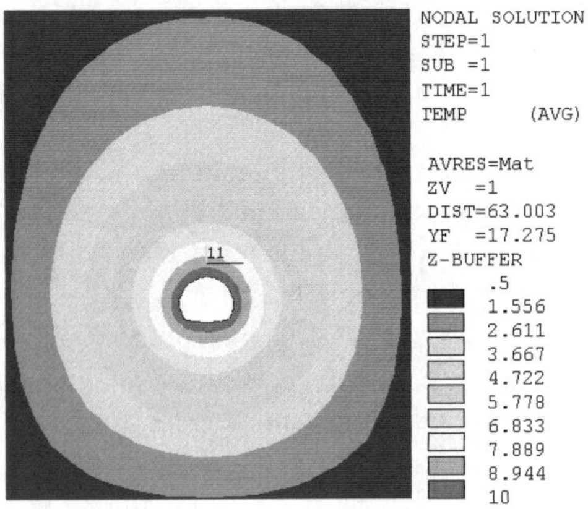

Figure 4. Temperature field of tunnel surrounding rock temperature at 10°C.

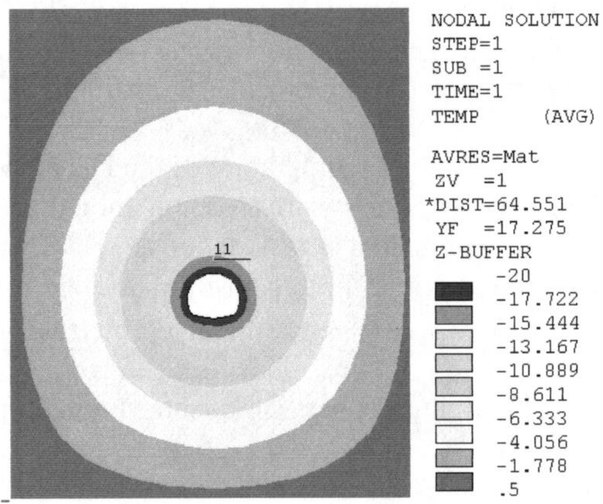

Figure 5. Temperature field of tunnel surrounding rock temperature at -20°C.

4. Conclusion

According to the single temperature field equation of tunnel surrounding rock and Da Ban mountain tunnel project in Qinghai province as an example, large scale finite element software is used to simulate temperature distribution under freeze-thaw cycles. The temperature field distribution of tunnel surrounding rock under the action of temperature is: -3°C →10°C →-20°C →10°C →-3°C, which is from the rising temperature to the decreasing temperature and from the rising temperature to the decreasing temperature again:

(1) The temperature field of surrounding rock gradually changes with the surrounding rock depth, and the temperature changes significantly near the wall.

(2) When the wall temperature of the surrounding rock reaches -3°C, the temperature gradient of the wall will be larger, and the temperature will change at the average 0.5*m*. When it is near 0°C, the surrounding rock will begin to freeze, and the freezing depth is about 10m after the temperature change is stable. When the inner wall temperature of the surrounding rock rises to 10°C, the surrounding rock has melted. The depth of melting is about 10*m*, and the depth can be in the frozen area for many years, and the impact is not significant. When the temperature of the inner wall of the surrounding rock falls to -20°C, once again, the freezing depth of the surrounding rock will reach 12*m*, which will affect the freezing depth for many years. Therefore, engineering measures should be taken to effectively control the temperature range.

(3) For the calculation model considered in this paper, because freeze-thaw circles can cause the destruction of the tunnel structure, convective heat transfer will occur between tunnel mouth and walls and the environment, resulting in large temperature difference. Therefore, in order to reduce freeze-thaw disasters in the cold regions, we should consider the protection measures about the hole and the wall. It is suggested that a protective door in the hole should be added, and insulating layer on

the wall should be filled to minimize the heat-exchange between the temperatures of the inside and outside of holes and surrounding rocks. In this way, it reduces the range of freeze-thaw circles and prevent the convective heat transfer effectively.

Acknowledgments
This work was supported by the College fund of Shandong Jiaotong University.

References

[1] Zhang H M 2010 *Study on the Damage Mechanical Charateristics and Moisture-heat-stress Coupling of Rock under Freeze-thaw Condition* Doctoral Dissertation of Xi'an University of Science and Technology chapter **6** 70–85.
[2] Xu G M 2006 *Study on Mechanical Charateristics of Rock at Low Temperature Damage Due to Freezing and Thawing and Mutiphysical Coupling Problems of Rock in Cold Regions* Doctoral Dissertation of Chinese Academy of Sciences chapter **5** 61–84.
[3] Chen X S 1998 *J. China Coal Soc.* **23** 53-7.
[4] Yang G S 2007 *Rock Soil Mech.* **3** 11-9.
[5] Li N, Cheng G D and Xu X Z 2001 *Adv. Mech.* **31** 95-102.
[6] Zhang X F, Lai Y M 2002 *J. Chin. Railway Soc.* **24** 92-96.

IOP Conf. Series: Materials Science and Engineering **473** (2019) 012018 doi:10.1088/1757-899X/473/1/012018

Experimental Study on Cutting Forces Developed during Dry Turning of a CoCrWNi Alloy

M R Dijmărescu[1]

[1] University POLITEHNICA of Bucharest, Manufacturing Engineering Department, 313 Splaiul Independentei, 060042 Bucharest, Romania

Abstract. This scientific study focuses on the investigation of the cutting forces developed during dry longitudinal turning with TiAlN PVD coated inserts of a CoCrWNi alloy used in medical applications, especially in orthopaedics. The present paper is organised in two main parts: the first part presents the experimental procedure in terms of material, input data, necessary technological means, physical experiments and data registration while the second part presents the results and their subsequent interpretation. The study reveals the variation of the turning cutting forces in relation to the cutting regime parameters.

1. Introduction

The machinability of metallic materials is one of their most important properties. One of the most important machinability parameters is represented by the cutting forces developed during machining, cutting forces which are instrumental for the optimization of the turning process quality, being a significant factor in determining the necessary machining power, proper dimensioning of the cutting tools, determining the tool wear evolution, etc. [1-3]. Subsequently, an adequate understanding and in-depth knowledge of this parameter is important for choosing a machine tool that complies with the power requirements and can perform the necessary processes while maintaining a low level of energy consumption.

Most studies on metallic biomaterials machinability focus on stainless steels and Titanium alloys, while Co-Cr based alloys still present unexplored areas in this field of research [1, 4-6]. The majority of scientific studies on Co-Cr based alloys used in medical applications focus on the machinability of CoCrMo alloys; no specific studies related to the cutting forces developed during turning CoCrWNi alloys using TiAlN coated inserts were found.

The aim of this research paper is to determine by experimental investigation methods the cutting forces developed during dry turning of a CoCrWNi alloy and to present the influence exerted by the variation of the cutting regime parameters on these forces.

The main steps of the conducted research program were: determination of certain effective characteristics of the analysed alloy and of the cutting regime parameters, preparation of the material samples, physical experiment (dry longitudinal turning and data registration), experimental data processing and analysis of the influence exerted by the variation of the cutting regime parameters on the cutting forces.

IOP Conf. Series: Materials Science and Engineering **473** (2019) 012018 doi:10.1088/1757-899X/473/1/012018

2. Experimental procedure

This section provides details on the material and the equipment used to perform the physical experiment. The experimental process is presented in terms of input data (material samples, cutting regime parameters, etc.), technological means, and output data (cutting forces graphs and values).

2.1. Material and methods

CoCrWNi alloy is a Co-Cr based metallic biomaterial with a high resistance to wear, high corrosion resistance and a high degree of hardness used in medical applications, especially for hip and knee implants [7, 8]. The CoCrWNi samples used for the experiment had a 6.5 mm diameter.

The technological equipment used during the experimental process consisted of: a precision lathe (type SN 320), PCLNR 2525M-12X-JHP [9] lever lock tool, rhombic CNMG 120404-VL TiAlN PVD coated inserts [10], a Kistler stationary dynamometer with piezoelectric transducers, a signal amplifier and a computer with DynoWare software (DynoWare Type 2825A-02).

The values of the cutting parameters were established depending on the material characteristics, the cutting tool manufacturer recommendations and the capabilities of the machine tool. The specific range of the values taken into account for the processing experiment are shown in table 1.

The Kistler dynamometer was assembled on the lathe carriage and the cutting tool body was fixed on it with clamps. The machine carriage dimensions prevented the dynamometer from being positioned with its axis system in correspondence with the machine-tool axis system, see figure 1.

Therefore, the relation between these two systems is as follows: $X_{dynamometer} = -Y_{machine\text{-}tool}$, $Y_{dynamometer} = X_{machine\text{-}tool}$ and $Z_{dynamometer} = Z_{machine\text{-}tool}$. This position will determine negative values on the resulted graphs for F_p.

Table 1. Ranges of the cutting regime parameters.

Depth of cut, a_p, (mm)	Feed rate, f (mm/rev)	Cutting speed, v_c, (m/min)
0.25 – 0.5	0.03 – 0.07	20.4 – 32.7

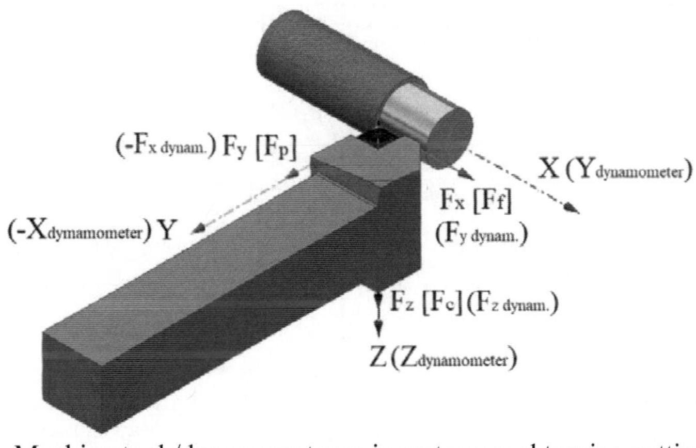

Figure 1. Machine tool /dynamometer axis systems and turning cutting forces.

2.2. Experimental data

Thirteen dry longitudinal turning experiments were conducted using the aforementioned parameters and conditions. During each experiment, for 20 seconds, graphs and numerical values for F_c ($F_{z\ dynamometer}$), F_f ($F_{y\ dynamometer}$) and F_p ($F_{x\ dynamometer}$) cutting forces were recorded. The cutting forces signal variation graphs for two representative turning experiments are shown in figures 2 and 3. The cutting parameters used and the maximum registered values of the cutting forces for these two experiments are presented in table 2.

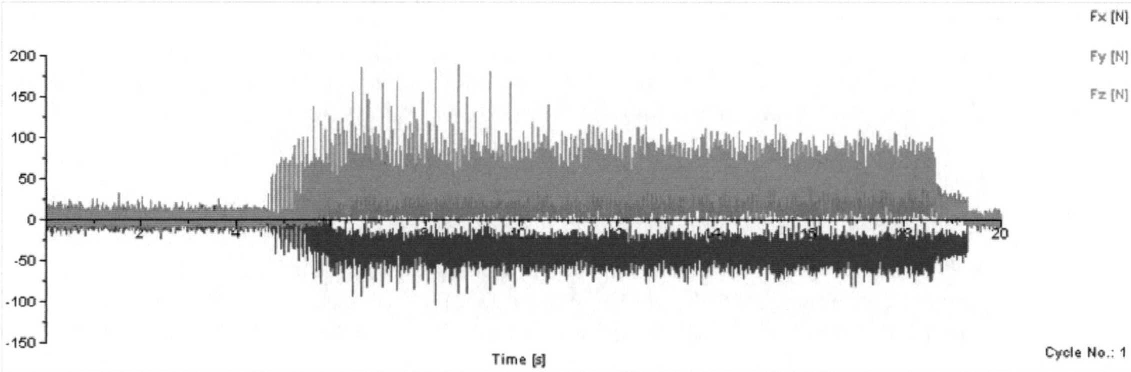

Figure 2. Turning cutting forces variation for experimental case no. 3.

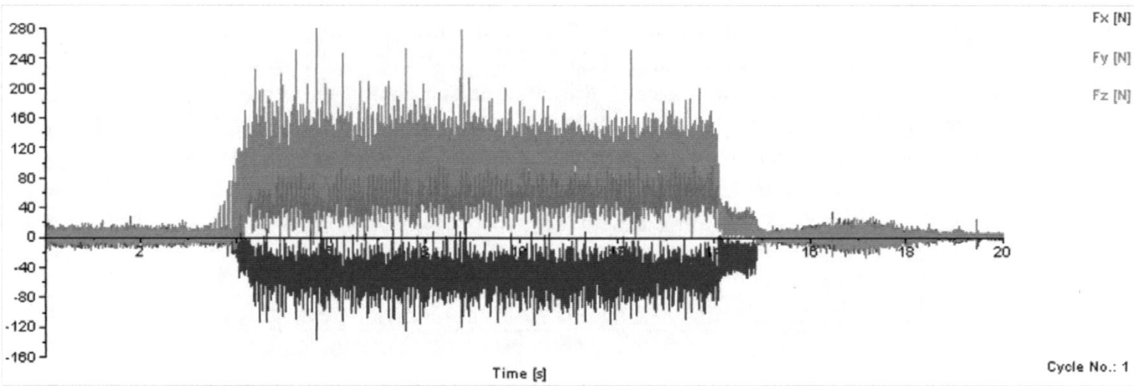

Figure 3. Turning cutting forces variation for experimental case no. 10.

Table 2. Cutting parameters ranges and measured cutting forces for representative experiments.

Exp. no.	Cutting regime parameters			Experimental cutting forces maximum values			Observation
	a_p, (mm)	f (mm/rev)	v_c (m/min)	F_c (N)	F_p (N)	F_f (N)	
3	0.05	0.375	25.5	188.690	102.875	86.884	median values of the cutting parameters
10	0.05	0.5	25.5	279.388	135.620	128.510	largest values of the turning cutting forces

3. Results and discussion

Based on the measured values of the turning cutting forces, the resultant cutting force (F) was calculated using equation (1).

$$F = \sqrt{F_x^2 + F_y^2 + F_z^2} \ [\text{N}] \tag{1}$$

The variation of the turning cutting forces and their resultant force in relation to the cutting regime parameters are shown in figures 4-6.

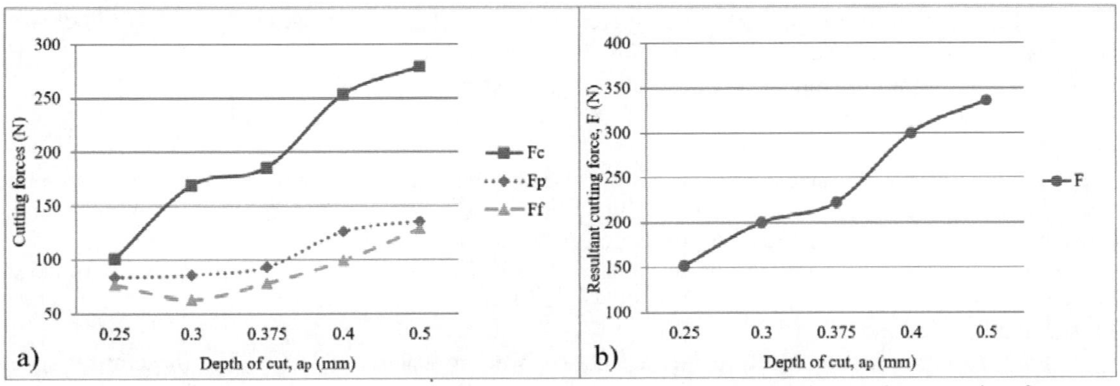

Figure 4. Variation of the (a) turning cutting force components and (b) resultant cutting force in relation to the depth of cut.

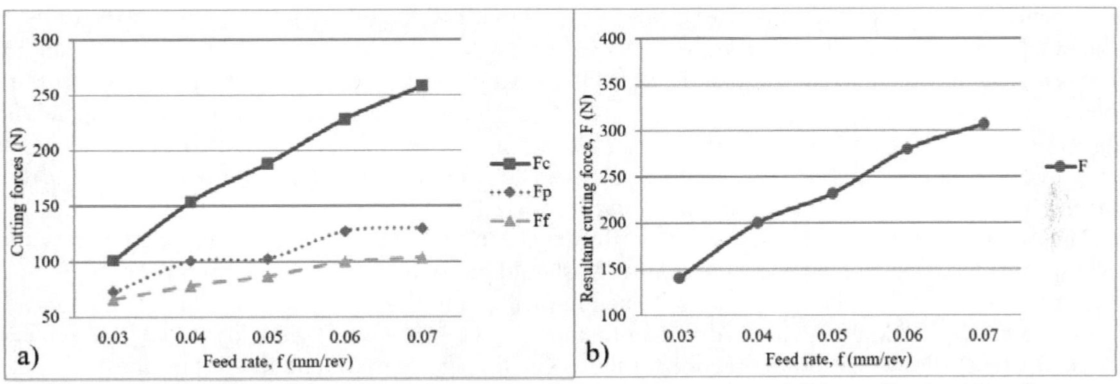

Figure 5. Variation of the (a) turning cutting force components and (b) resultant cutting force in relation to the feed rate.

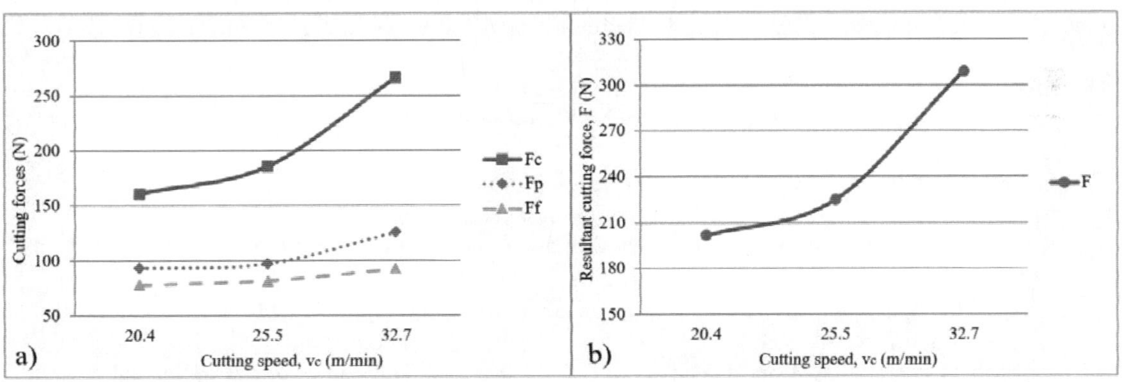

Figure 6. Variation of the (a) turning cutting force components and (b) resultant cutting force in relation to the cutting speed.

Analysing the variation graphs between the turning cutting forces and the cutting parameters, we can ascertain that:

- If f = 0.05 mm/rev and v_c = 25.5 m/min, when a_p increases by 100 %: F_f increases by 68.51%, F_p increases by 62.90%, F_c increases by 178.18% and resulted F increases by 122.43%.
- If a_p = 0.375 mm and v_c = 25.5 m/min, when f increases by 100 % – F_f increases by 51.22%, F_p increases by 74.86%, F_c increases by 126.38% and resulted F increases by 98.82% and

when f increases by 133. (3) % – F_f increases by 56.87%, F_p increases by 78.72%, F_c increases by 155.45% and resulted F increases by 117.90%.

- If f = 0.05 mm/rev and a_p = 0.375 mm, when v_c increases by 60.29%: F_f increases by 18.35%, F_p increases by 35.05%, F_c increases by 65.62% and resulted F increases by 53.14%.

The cutting parameters have a nonlinear influence on the turning cutting forces, the major influence being exerted by the depth of cut – F_c increases with 178.18% and F increases with 122.43% when a_p increases with 100%; while the lowest influence is exerted by the cutting speed.

The cutting force component that is influenced the most by the variation of the cutting parameters is the main cutting force (F_c). $F_{c\ max.}$ = 279.388 N and $F_{max.}$ = 336.103 N, when a_p = 0.5 mm, f = 0.05 mm/rev and v_c = 25.5 m/min.

Taking into account the aspects presented above, we can ascertain that raising the cutting speed will have a low influence on the cutting forces, but, at the same time, it will result in a shorter processing time.

4. Conclusions

The study focuses on the identification of the most significant parameters that influence the cutting forces during dry longitudinal turning. The results reveal the variation of the turning cutting forces in relation to the cutting regime parameters – depth of cut, feed rate and cutting speed - and expand the knowledge base developed by the existing studies made on Co-Cr alloys, with a particular focus on the CoCrWNi alloy. Subsequently, these results can be used as a basis for the optimization of the turning process quality for this specific alloy, when using TiAlN coated inserts.

The outlined results of the study are valid for the CoCrWNi alloy, selected cutting parameters and their specified variation ranges. Any extrapolation should be confirmed by further experiments.

Further research will focus on the optimization of the cutting parameters and the development of cutting forces prediction models for dry longitudinal turning of the analysed alloy, models which can be used in research and industrial practice for the evaluation of the machinability of this alloy.

Acknowledgment
This work has been funded by University POLITEHNICA of Bucharest, through the "Excellence Research Grants" Program, UPB – GEX 2017. Identifier: UPB-GEX 2017, ctr. no. 51/25.09.2017.

References

[1] Korka Z I, Micloşină C-O and Cojocaru V 2013 *Analele Universității "Eftimie Murgu" Reșița* **2** 25-32.
[2] Babu J R and Babu A 2014 *Proc. AIMTDR 2014* 459.
[3] Malagi R R and Rajesh B C 2012 *Int. J. Eng. Inn. Tech.* **2(1)** 37-43.
[4] Khorasani A M, Gibson I, Goldberg M, Nomani J and G. Littlefair 2016 *Sci. Adv. Mat.* **8** 1491-511.
[5] Ohkubo C, Watanabe I, Ford J P, Nakajima H, Hosoi T and Okabe T 2000 *Biomater.* **21(4)** 421-8.
[6] Jackson M J, Novakov T and Da Silva M B 2015 *Mater.* 231-71 (Switzerland, Springer).
[7] Park J B and Lakes R S 2007 *Biomaterials - An introduction* (New York, Springer-Verlag).
[8] ISO 5832-5:2005 *Implants for surgery - Metallic materials - Part 5: Wrought Cobalt-chromium-tungsten-nickel alloy.*
[9] *http://www.iscar.com/eCatalog/Item.aspx?cat=3603117,* accessed on 13.08.2018.
[10] *http://www.iscar.com/eCatalog/Item.aspx?cat=5567594,* accessed on 13.08.2018.

The Influence of Particle Flux Density and Particle Size Distribution in Surface Morphology of Cold Spray Coatings

K H Moonga[1, 2] and T C Jen[1]

[1]University of Johannesburg, Kingsway Campus, Johannesburg, South Africa
[2]University of Zambia, Great East Road Campus, Lusaka, Zambia

Abstract. In this article the coupling between the Cold Gas Dynamic Spray (CS) particle flight instance just before impact and resulting surface morphology after impact of a cluster of particles are investigated with a view to understanding the complex CS process for possible optimisation for the CS fabrication process in nano thin films. This study identifies two main variables for the analysis namely particle flux density and particle size distribution and examine their influence in the structural properties of the coatings, in particular, the deposition efficiency of the cluster of particles and its applications in functional properties of the coatings. This process determines the deposition parameters that characterize the surface roughness. Given r as a radial distance from the axis of a nozzle, and R as the maximum radial distance reached by the particles, the results indicate that increasing the flux density of a cluster beyond a certain limit (yet to be identified) creates a deformation behaviour closely related to that of a single particle, however this process lowers the deposition efficiency as r approaches R; where R is the radius of the nozzle particle flux distribution radius at impact surface and r is measured from the centre axis , (where r = 0), of the impinging particle jet to towards the peripheral (where r = R). A similar characteristic on the morphology of the surfaces is observed when particles have a varying particle size: the smaller particles (and indeed any other particle sizes) that hit the target substrate at the peripheral at an angular velocity and possibly with reduced velocity from the average particle flux jet velocity, eject outward due to the oblique impact imposed by the traverse velocity component. These findings have potential applications in fabrication of functional surfaces for various applications and provide us with control variables necessary for modifying the surfaces.

1. Introduction

The cold gas dynamic spray technology involves acceleration of particles to supersonic velocities (such as 400 – 1200m/s) for deposition in the manufacturing of thin films. The thin films so fabricated can be used for various applications; in our research we are more interested in applying this technology in nano thin films where the physical structure such as surface roughness and the material properties define device functionality [1].

Numerical methods assist in understanding the behavior of complex dynamical processes such as the CGDS process to variation in process variables. There are two main aspects involved in the systems analysis of a CGDS process: the fluid dynamics aspect [2, 3] which involve particle flight through the nozzles and outside the nozzle; and the solid mechanics aspect [4, 5] which involve the interaction of the particles and the substrate at impact (i. e the impact process). From a systems engineering aspect these two ends are coupled in the flight dynamics of the particles outside the nozzle, this involve the trajectories taken by the particles, the preferred trajectories for varying sizes; the preferred temperatures for varying particle size; the densities of the particles, and the preferred velocities for varying particle sizes. High pressure in the impact zone creates a bow shock which can deflect smaller size particles [6]. Further particles falling an appreciable distance away from the nozzle axis will be subjected to unbalanced traverse forces due to reduced pressure of axis which will give a traversal velocity component during their flight to impaction with underlying substrate material. Upon impact, depending on the adhesive or cohesive forces bonding or ejection of the particle will occur.

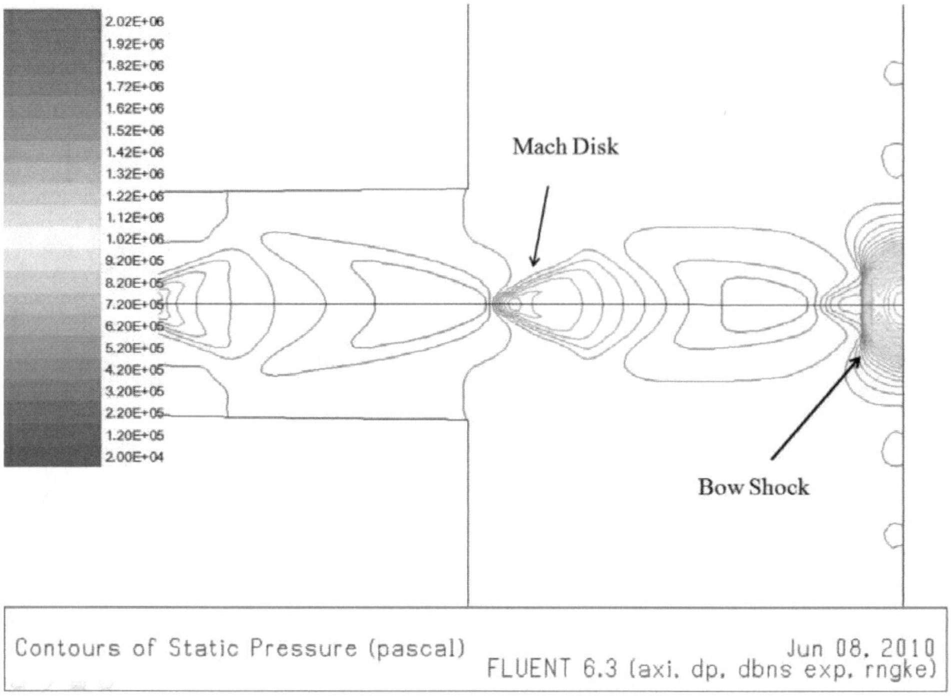

Figure 1. Pressure contour of N_2 outside the nozzle [6], bow shocks in high pressure CS systems.

2. Comparisons between modeling the cold spray process and practical applications of the CGDS

This study involves a numerical analysis of the impact process, and therefore the computational model begins with a capture of the particle instance just before impact of the first particle (see figure 2). As is the case in most numerical studies, this instance is normally simplified. In some instances, this is necessary in order to reduce the complexity of the problem setup. This simplification must be properly recorded as limitations of the computational model have effect of the extent to which the results will compare with practical applications. Usually it is acceptable to make these assumptions, and a comparison to any experimental data must discuss the variables involved in the experiments in comparison to those captured by the numerical methods. Sometimes the computational model may turn out to be more theoretical than is practical while in other instances the assumptions and simple models will be beneficial in determining the contribution of specific variables to the complex process under study. This process is illustrated in the block diagram of figure 2 where experimental results in [7] and numerical results in [4] are examined.

Due to limitations in numerical models, it is important to specify the limitations of the model doing so has the advantage that results obtained in numerical studies can be used to understand the experimental process variables captured in the numerical simulation instance.

In this paper we study a cluster of copper particles impacting on an aluminum substrate (figure 3 [4]) and analyze further possible scenarios under which it can be used in understanding the CGDS process. The assumptions in this model were that particles are of a uniform diameter, d_p; implying a zero standard deviation on the diameter. The substrate was considered to be initially smooth compared to the size of particles used. The velocity of impact is 500m/s in the axial direction and zero in the

IOP Conf. Series: Materials Science and Engineering **473** (2019) 012019 doi:10.1088/1757-899X/473/1/012019

traverse direction to the jet axis, the initial temperatures of the substrate and particles is defined to be 25 degrees Celsius, particles are assumed to be 10 um in diameter.

EXPERIMENT/CFD
Off nozzle process parameters

The properties of the particles Including materials for substrate and particle sizes (statistical distribution of particle size and shape); propellant gas pressures; etc.

Flight mechanics

Particle flow fields and velocity profile across off nozzle; preferred trajectories for various particle sizes

Interaction with the substrates The process parameters (Experimental) plan: the $x - y$ motion, nozzle size, feeder system, transport system, flux density, and irradiation techniques

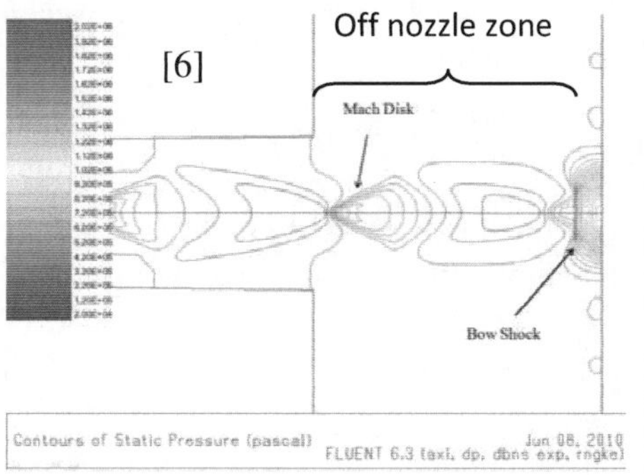

NUMERICAL PROCESS must capture the off nozzle process parameters

Input parameters in a cold spray process Including materials for substrate and particles

Defined by either: Continuum Material properties: definition of material properties through constitutive equations (Johnson cook Plasticity models; Viscoelastic, viscoplastic; Heat and Mass transfer);

Or Molecular dynamics: for example numerical optical glass material processing, ensembles; force fields

Interaction with the substrate defines process parameters (Numerical) plan: the $x - y$ motion, nozzle size, feeder system, transport system, flux density, irradiation techniques (defines initial velocities and initial temperatures)

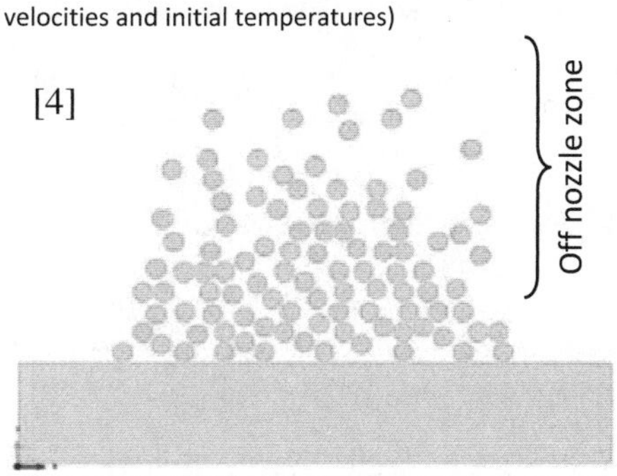

RESULTS: EXPERIMENTAL COATINGS

Output Parameters; film thickness, surface roughness, interface roughness, grain sizes, residual stresses, etc

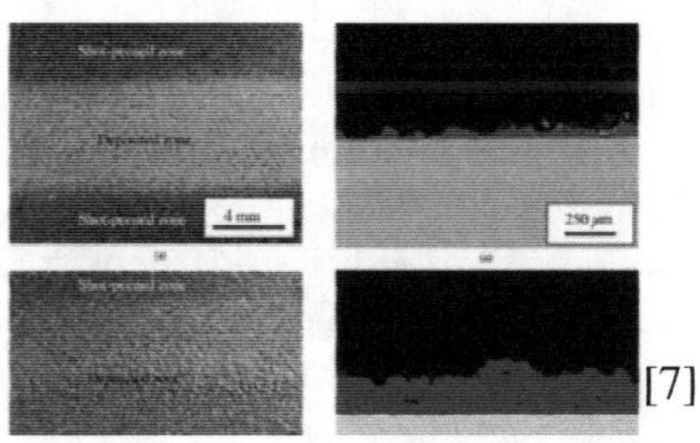

RESULTS: PREDICTED COATINGS

Output Parameters; film thickness, surface roughness, interface roughness, residual stresses, porosity

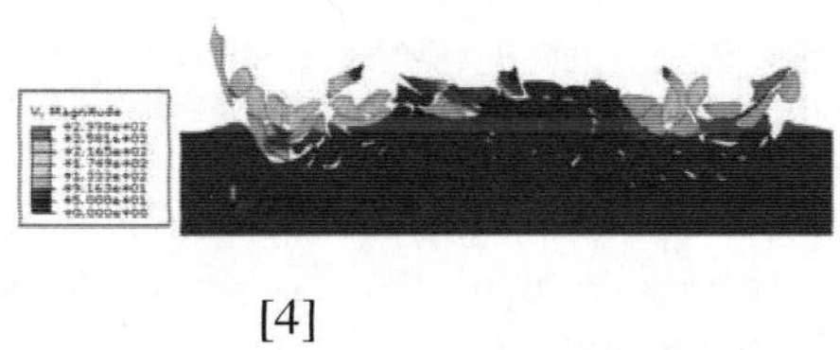

Compare outputs to numerical result taking into consideration limitations in numerical and computational models

Figure 2. Parameter Search and System Identification Process Flow Chart.

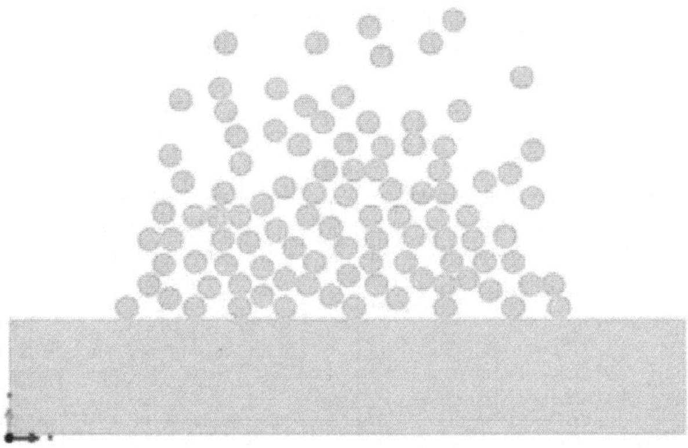

Figure 3. Computational domain, (meshing resolution of 1/20, where $d_p = 10\mu m$ particle diameter) showing a cross sectional length, part of a cross sectional length of the nozzle [4].

The simulations are based on the continuum material models using the Johnson-Cook constitutive equations [8] which define the yield flow stress, $\overline{\sigma}$ of the material as follows

$$\sigma^{-} = \left[A + B\left(\bar{\varepsilon}^{pl}\right)^{n}\right]\left[1 + C\ln\left(\frac{\dot{\bar{\varepsilon}}^{pl}}{\dot{\varepsilon}_0}\right)\right]\left(1 - \hat{\theta}^{m}\right) \quad (1)$$

where A, B, n, C, and m are material specific constants, $\bar{\varepsilon}^{pl}$ is the effective plastic strain, $\dot{\bar{\varepsilon}}^{pl}$ is the strain rate, $\dot{\varepsilon}_0$ is the reference strain rate, $\hat{\theta}$ is the dimensionless temperature parameter defined as:

$$\hat{\theta} \equiv \begin{cases} 0 & \text{for } T < T_t \\ \frac{(T-T_t)}{(T_m-T_t)} & \text{for } T_t \leq T \leq T_m \\ 1 & \text{for } T > T_m \end{cases} \quad (2)$$

where T is the current temperature, T_m is the melting temperature of the material, and T_t is the transition temperature which defines the temperature below which the resultant yield stress is considered to be temperature independent.

3. Analysis of the findings

In this model it was observed that the mean time to impact between two consecutive particles at the same axial location was less than the time for a single particle deformation. A rotational couple and oblique trajectory are prominent characteristics in the jetting region (figure 4). These material and particle jets have higher rebound energy than those with only longitudinal velocity components. In order to have a higher deposition efficiency, η_{beam} (close to 100% deposition efficiency), selection of propellant gas pressures and nozzle characteristics should be tailored to reduce the transverse velocity component while increasing the longitudinal velocity component. For the high flux density case, this could be done by reducing the flux density or increasing the longitudinal velocity component to such an extent that the adhesion energy counteracts the jetting components.

Figure 4. Rebound velocities and local deposition efficiency at a radial distance r from axis of nozzle.

Given r as a radial distance from the axis of a nozzle, and R as the maximum radial distance reached by the particles on the underlying substrate surface, the results indicate that increasing the flux density of a cluster beyond a certain limit (yet to be identified) creates a deformation behavior closely related to that of a single particle, however this process lowers the deposition efficiency $\eta_{film}(r)$ as r approaches R where R is the radius of the nozzle particle flux distribution radius at impact and r is measured from the center axis, (where $r = 0$) to towards the peripheral (where at $r = R$).

A similar characteristic on the morphology of the surfaces is observed when particles have a varying particle size: the smaller particles hit the target substrate at the peripheral at an angular velocity and possibly with reduced longitudinal velocity component and eject outward due to the oblique impact. The two scenarios (higher flux density on one hand and reduced flux density but with varying particle sizes on the other hand) are similar in terms of particles in the peripheral regions having traverse velocity components. These findings have potential applications in fabrication of functional surfaces for various applications and provide us with control variables necessary for modifying the surfaces. This analysis is also capable of prediction of agglomerate impacts of nano particles and likely morphology after impact of the agglomerate showing disintegrations of particles at the peripheral.

We therefore show that indeed the flux density and particle trajectories have a significant influence in the surface morphology. Importantly, and for future study, is to determine the settings for a 100% deposition efficiency (if at all it exists) which would yield a compact coating and also a smooth surface. It is the first time that this is being reported in the cold spray community about the influence of flux density and particle flight trajectories to surface properties of the cold spray coatings. In future, based on these findings, an analytical model for deposition efficiency can be formulated.

4. Conclusion

This paper described a simulation of the particle deposition process in a cold gas dynamic spray system and analyzing the findings show the influence of flux density to deposition efficiency and also the particle distribution and their preferred trajectories provide their contribution to reduced deposition efficiencies in cold spray systems. Deposition efficiency has been found to control the surface morphology by increased surface roughness with reduced deposition efficiency. In this analysis we took a systems approach to the problem. We also applied the concepts of data mining by collecting and analyzing observed structure in available experimental data. The numerical results were used to explain the underlying mechanics.

Acknowledgements

The authors would like to acknowledge the financial support from National Research Foundation (NRF) of South Africa. Kando H. Moonga would also like to acknowledge the financial support from The University of Zambia under the SSTEP Project 2016 - 2018.

References

[1] Voit W 2011 From Heat Shrink to Shape Memory to Flexible Bioelectronics: Why Radiation Doesn't Deserve a Bad Wrap. http://cirms.org/pdf/2011_conference_pdf/2011_10_19_CIRMS_Voit.pdf.

[2] Jen T C, Pan L, Li L, Chen Q and Cui W 2006 *Appl. Therm. Eng.* **26(5-6)** 613-21.

[3] Takana H, Ogawa K, Shoji T and Nishiyama H 2008 *Powder Tech.* **185(2)** 116-23.

[4] Moonga K H and Jen T 2017 *Int. J. Mech. Eng. Robot. Res.* **7(1)**

[5] Moonga K H and Jen T C 2017 *ASME Int. Mech. Eng. Congress Expo. (Adv. Manuf.)* **2** V002T02A017.

[6] Yen Y H 2010 The Effects of The Distance Between Nozzle and Substrate on Cold Gas Dynamic Spray Process *MS Thesis, (University of Wisconsin, Milwaukee).*

[7] Wu J, Fang H, Yoon S, Kim H. and Lee C 2005 *Appl. Surf. Sci.* **252(5)** 1368-77.

[8] Johnson G R and Cook W H 1983 A Constitutive Model and Data for Metals Subjected to Large Strains, Strain Rates, and High Pressures *In Proceedings of the 7th International Symposium On Ballistics.*

Multi-Response Optimization of Mechanical Properties of Hybrid (Fiberglass / Abaca Woven) in Polyester Matrix using Desirability Function based on DOE

J Paredes[1, a*], H Vaca[1, b], H Erazo[1, c] and C Pérez[1, d]

[1] Technical University of Ambato, Faculty of Civil and Mechanical Engineering, and "GI³M Research and innovation group in mechanical engineering", Av. Los Chasquis and Río Payamino, Ambato, Ecuador

Abstract. The study of the application and mechanical behavior of new composite and hybrid materials within the bodywork industry in its different parts (external and internal) has led to perform different characterizations and analysis of results which determined the material with the best mechanical properties in many cases depending on the subjectivity of the researcher. In the present study the optimization of the mechanical properties tensile and flexural hybrid composite made of natural fiber and synthetic fiber was made. To produce the hybrid fabric, abaca (AbF) and cotton (CF) over glass (FG), combined with a polymer matrix polyester resin was used. The configuration of the fibers and the type of drying (ambient and oven), were taken as input factors. The influence of these factors on the overall responses were analyzed using factorial design. A function of desirability was used for the optimization of experimental responses. The results indicated that the optimal factors to take advantage of the experimental responses or mechanical properties, were the configuration of 20% FV + 7.75% FAb (45°) + 2.25% FAI (135°) and the type of oven drying, with a value of global desirability of 0.8208.

1. Introduction

The study, application and the use of new composite materials of polymeric matrix have had a great development within the world industry, from the obtaining of prototypes, parts and pieces to the manufacture of large constructions, where the insertion of metallic reinforcements, particles and fibers (synthetic and natural) have tried to improve physical, chemical and mechanical properties. This has caused the increase in the range of materials and even increasing the market competitiveness of this type of elements [1, 2]. At the same time, the continuous interest in the preservation of environment and increase in use of green composites materials has led to an inclination to research composite materials and hybrids with natural fibers [3]. Research efforts in this field are aimed at finding possible plant fibers that match or exceed the characteristics of a synthetic fiber. In addition to the above, these fibers have been combined to improve the final properties at the lowest possible cost. [4, 5].

The application of these engineering materials in the secondary industry has grown in recent times, when metal parts began to be replaced by compounds with similar properties but obtaining a greater advantage in the reduction of structural weight and manufacturing costs. Synthetic fibers (carbon, glass, kevlar, hony comb, etc.) have been used mainly in composite materials within the automotive, body, naval and aeronautical industries [6-9]. The advantages of natural fibers in the manufacture of parts and pieces for several years have been investigated and applied, reaching an excellent behavior, physical, chemical and mechanical. These fibers can act alone in a polymeric matrix or, sometimes, combining with synthetic fibers forming the so-called hybrid materials, which have great results in mechanical resistance and economic. This is the case of the Mercedes Benz E-Class cars of the company Daimler AG, where it was possible to improve the original material of the panels of doors to the replacement by epoxy resin embedded in sisal fiber, obtaining a notable weight reduction of 20% [10].

In Ecuador there are several projects related to composite materials and the use of vegetable fiber, such as the developed in the project PIC-08-493 "Development of new materials for structural and industrial applications" [1] in which aspects such as the type of matrix, the morphology of fibers, surface treatments and mechanical properties. This has favored the use of greenmaterials such as abaca fiber, jute, coconut, cabuya, among others, in production activities of the industrial sector, thus benefiting the creation of new sources of employment with the possibility of promoting sustainable economic development in areas rural [11]. Ecuador is the second largest producer of abaca after the Philippines. The areas of Quinindé, Santo Domingo de los Tsáchilas, Quevedo and La Maná have a combined production of 10,000 tons per year. The abaca plant is an herbaceous structure belonging to the musaceae family (musa textiles), it is native to the Philippines and it grows in humid tropics. The fibers come from the sheath of the leaves that make up the pseudostem of the plant, their cells are characterized by being long and thin [12]. These fibers have a length ranging from 1 to 2 m or more with a diameter of 01.1 - 0.28 and are chemically formed by cellulose, hemicellulose, pectin, lignin. There are 5 grades of abaca fibers that are increasing according to their grade of quality, being grade 1 those of better quality and properties [5]. Nowadays, the experimental designs assisted with statistical tools allow an appropriate analysis of the mechanical behavior of the materials, as well as to predict said behavior for a given application [13, 14]. The use of DOE in the industry is for the purpose of solving a problem or checking an idea through statistical tests. This corresponds to effectively identifying which tests should be performed and how to do them, obtaining data that, when analyzed statistically, provide objective evidence to answer the questions posed [15]. Studies such as the analysis of the flexural strength of mild steel A36 by design of experiments, determine in the field of metalworking the best welding process under specific conditions [16]. Applied to thermodynamics DOE allows to optimize the accuracy of a certain temperature reading, as mentioned by F. Barari et al. [17]. In the development of composite materials E.H.Agung et al. [18] chemically modifies a high impact polystyrene composite reinforced with abaca fiber and uses a Box-Behnken design with desirable function to choose the acceptable combination of weight percentages of its components. Suardi et al. [19], evaluates the statistical properties of the diameter of abaca fibers and identifies a statistical method (weakest link, Weibull, t-test) to suppress the dispersion of tensile strength.

The present work used an approach based on the design of DOE experiments which seeks to analyze statistically and with the help of the desirability function [18][20][21] the experimental results of the proposed combinations (configuration and type of drying)) in order to establish the optimal combination of the input factors for the tensile and flexural mechanical properties of the hybrid composite material made of abaca (AbF) and cotton (CF) plus glass (FG), combined with a matrix polyester resin polymer.

2. Methodology

2.1. Materials and Equipment

2.1.1. Materials. The woven fibers of abaca, cotton and fiberglass are adhered to the polymeric matrix to form a hybrid composite material. Unsaturated polyester resin was used as the thermostable polymer matrix. For this study, quality abaca fiber 1 with a diameter of approximately 0.252 mm was used and obtained from the localities of Quevedo and Santo Domingo (Ecuador). The following table summarizes the materials used in the preparation of the composite:

Table 1. Materials.

Material	Brand	Characteristic	Country of origin
Fiberglass	Jushi	Synthetic fiber of 375 gr/m^2	China
Abaca fiber	-	Natural fiber	Quevedo - Ecuador
Polyester resin	Andercol 836	Unsaturated resin	Medellin
Cobalt octoate	Andercol	12% w/w	Colombia
Peroxide metal - ethyl ketone	-	Polymerization initiator	Ecuador
Desmoldable wax	Simoniz	Paraffin paste wax	Colombia

2.1.2. Equipment. For the laboratory tests of tensile and flexural, was used the Universal testing machine Metro Test 50 kN for polymeric, ceramic and composite materials. ASTM D3039 regulations, ASTM D7264 were used to prepare specimens for each test. An infrared curing furnace available at the Faculty of Mechanical Engineering, Technical University of Ambato was used to dry the specimens.

2.2. Experimental process

By means of the compression stratification process, the test pieces were obtained with the following volumetric fractions: 70% matrix (polyester resin) and 30% reinforcement (natural fiber plus glass fiber). An A 36 steel mold was fabricated to make the specimens based on the minimum number of specimens in accordance with ASTM D standards. Different configurations were made for the hybrid compound. Cotton fiber plus abaca (CF + AbF) with orientations of 45 °, 90 ° and 135 ° plus fiberglass (FG) with random orientation. Seven cases of studies that are shown in Table 2 were analyzed.

Table 2. Cases of hybrid material.

Cases	Configuration
Case 1	20% FG + 7.75% AbF (0 °) + 2.25% CF (90 °)
Case 2	10% FG + 17.75% AbF (0 °) + 2.25% CF (90 °)
Case 3	20% FG + 7.75% AbF (45 °) + 2.25% CF (135 °)
Case 4	10% FG + 17.75% AbF (45 °) + 2.25% CF (135 °)
Case 5	23.25% AbF (0 °, 45 °, 135 °) + 6.75% CF (90 °, 135 °, 225 °)
Case 6	10% FG + 20% Long Abaca Fiber
Case 7	10% FG + 20% Short Abaca Fiber

Two types of curing were carried out: ambient and oven. The specimens were subjected to the curing process at an ambient temperature of 18 °C to 21 °C and cured in an infrared drying oven, at a temperature of 80 °C. Specimens of the dimensions of 250x25x3 mm were obtained for the tensile tests and 160x13x3 for the flexural tests.

After the laboratory tests a database was obtained that is composed of the following information: Tensile tests: There are five observations and three properties analyzed, maximum tensile strength, tensile modulus of elasticity and strain, giving a total of 70 processed data by property. Flexural tests: There are five observations and four analyzed properties, deflection, maximum flexural strength, flexural secant modulus of elasticity and flexural strain, giving a total of 70 processed data by property. These data are necessary for the development of the design of experimental.

2.3. *Experimental Design*

The design of experiments aims to answer the question posed of comparing the combinations tested (configuration and type drying) in order to choose the best. For this study it is required measuring the global quality of the adjustment of the regression models of three different experimental designs by means of the determination coefficients (R^2 and $R^2{}_{aj}$), to choose the model that best fits the data. It was proposed to use a 7x2 factorial design, understanding that the effect of the factors, such as configuration and type of drying, is significant; as well as the effect of interaction of these factors on the answers. The factors and levels are described in table 3.

Table 3. Experiment factors and levels.

	Factors	
Levels	Configuration	Drying Type
Level 1	Case 1	Ambient
Level 2	Case 2	Oven
Level 3	Case 3	
Level 4	Case 4	
Level 5	Case 5	
Level 6	Case 6	
Level 7	Case 7	

2.4. *Desirability function analysis*

The desirability function analysis used in this study is based on transforming the responses predicted by the model $\hat{y}$ of each mechanical properties of the tested hybrid for different combinations in dimensionless values within the interval [0,1] denominated as individual desirabilities d_i.

Individual desirability when you want to maximize the output response.

$$d_i = \begin{cases} 0, & \hat{y} \leq y_{min} \\ \left(\dfrac{\hat{y} - y_{min}}{y_{máx} - y_{min}}\right)^r, & y_{min} \leq \hat{y} \leq y_{máx}, r \geq 0 \\ 1, & \hat{y} \geq y_{máx} \end{cases} \quad (1)$$

Individual desirability when you want to minimize the output response.

$$d_i = \begin{cases} 1, & \hat{y} \leq y_{min} \\ \left(\dfrac{\hat{y} - y_{máx}}{y_{min} - y_{max}}\right)^r, & y_{min} \leq \hat{y} \leq y_{máx}, r \geq 0 \\ 0, & \hat{y} \geq y_{máx} \end{cases} \quad (2)$$

Where, $y_{máx}$ is the highest specification of response, y_{min} is the lowest specification response and r is a constant that defines the shape of the desirability function for each response.

After calculating the individual desirability, we proceed to establish a value that measures the average convenience in each combination, which converts the problem of optimizing multiple responses into a problem of a single, objectively analyzable response. Equation 3 is used to calculate global convenience.

$$DG = (d_1{}^{w_1} * d_2{}^{w_2} * \dots * d_i{}^{w_i})^{1/w} \quad (3)$$

Where, w_i is the relative importance of each response and w is the sum of the relative importance.

3. Results and discussions

3.1. Analysis of variance of the mechanical properties

It is determined that the regression model that best fits the data is the model obtained from the complete factorial design. This model presents values of the coefficient of determination closer to 100% compared to other designs of experiments such as the One-way ANOVA (configuration) and the Two-factor ANOVA (configuration, drying). The coefficients of determination are described in table 4.

The analysis of variance identifies the importance of each of the factors as well as their interrelation with the response variables. The p-value of the model, of each factor and of the interrelationships for all the mechanical properties analyzed have a value less than 0.05. Therefore, all of them are defined as statistically significant, there being representative differences between the combinations. The applied experiment combines all input factors, obtaining a total of 14 combinations as dictated by the complete 7x2 factorial design. Five replicas were made for each analyzed property.

Table 4. Coefficients of determination for each model analyzed

	Experimental designs					
	One-way ANOVA		Two-factor ANOVA		Full factorial	
Mechanical properties	R^2	$R^2{}_{aj}$	R^2	$R^2{}_{aj}$	R^2	$R^2{}_{aj}$
Maximum tensile strength	69.08%	66.14%	75.23%	72.43%	92.31%	90.52%
Tensile modulus of elasticity	34.67%	28.45%	67.98%	64.36%	80.69%	76.21%
Strain	47.23%	42.20%	69.39%	65.93%	81.62%	77.35%
Deflection	57.67%	53.64%	76.68%	74.05%	86.37%	83.20%
Maximum flexural strength	58.85%	54.93%	74.18%	71.26%	88.37%	85.71%
Flexural secant modulus of elasticity	62.79%	59.24%	83.33%	81.45%	96.12%	95.22%
Flexural strain	61.19%	57.49%	74.59%	71.72%	84.33%	80.69%

3.2. Desirability Function Analysis

In the individual desirability, the criterion of maximizing the output response is used by equation 1 for the properties: Maximum Traction Effort, Elasticity Module, Elongation Percentage, Maximum Flexion Effort and Bending Sequence Elasticity Module. On the other hand, the criterion of minimizing the output response was used using Equation 2 for: Deflection and Maximum Deformation. For all cases, the value of 1 was assigned to the constant that defines the form of the function, while for the values of the specifications the values described in Table 5 were used.

Table 5. Response specifications.

	Specifications		
Mechanical properties	y_{min}	$y_{máx}$	Objective
Maximum tensile strength [MPa]	19.380	77.150	Maximize
Tensile modulus of elasticity [MPa]	458.480	2459.880	Maximize
Strain [%]	1.890	4.940	Maximize
Deflection [mm]	4.752	17.952	Minimize
Maximum flexural strength [MPa]	11.250	174.928	Maximize
Flexural secant modulus of elasticity [MPa]	495.370	17547.240	Maximize
Flexural strain [%]	0.934	3.180	Minimize

The relative importance was established, these constants being equal to 1. Applying this data to equation 3, a point value was generated that is comparative for each combination. The values of individual desirability and global desirability are detailed in Table 6.

Table 6. individual desirability and overall desirability for answers

	Combinations		Individual desirability								
			d_1 [a]	d_2 [b]	d_3 [c]	d_4 [d]	d_5 [e]	d_6 [f]	d_7 [g]	DG	Rank
1	Case 1	Ambient	1.0000	0.7361	0.6911	0.5190	0.5553	0.2706	0.5227	0.5748	5
2	Case 2	Ambient	0.6865	0.7000	0.4380	0.2077	0.4011	0.1417	0.2342	0.3450	10
3	Case 3	Ambient	0.6160	0.6704	0.4007	0.6845	0.5118	0.2788	0.6073	0.5166	6
4	Case 4	Ambient	0.4364	0.6118	0.2643	0.1997	0.3673	0.1632	0.1941	0.2879	12
5	Case 5	Ambient	0.6510	0.6182	0.4852	0.1849	0.1934	0.0569	0.1149	0.2399	13
6	Case 6	Ambient	0.9013	1.0000	0.3489	0.5098	0.4586	0.2140	0.4746	0.4968	7
7	Case 7	Ambient	0.2785	0.5201	0.1587	1.0000	0.1028	0.1458	1.0000	0.3201	11
8	Case 1	Oven	0.8366	0.6024	0.7154	0.9623	0.7065	0.8584	0.9457	0.7940	2
9	Case 2	Oven	0.7889	0.4281	1.0000	0.4633	0.7200	0.4157	0.4630	0.5785	4
10	Case 3	Oven	0.8234	0.6606	0.6197	0.8639	1.0000	1.0000	0.8620	0.8208	1
11	Case 4	Oven	0.1939	0.3340	0.3036	0.6035	0.6426	0.4730	0.6589	0.4218	9
12	Case 5	Oven	0.0478	0.0510	0.6839	0.4810	0.0272	0.0124	0.1923	0.0910	14
13	Case 6	Oven	0.6195	0.5173	0.5902	0.8618	0.7180	0.6506	0.7952	0.6699	3
14	Case 7	Oven	0.2110	0.2625	0.4334	0.8573	0.4808	0.4790	0.8468	0.4546	8

[a] Desirability of maximum tensile strength
[b] Desirability of tensile modulus of elasticity
[c] Desirability of strain
[d] Desirability of deflection
[e] Desirability of maximum flexural strength
[f] Desirability of flexural secant modulus of elasticity
[g] Desirability of flexural strain

Figure 1 shows a visual interpretation of the overall desirability for each combination. Consequently, the combination 10 that corresponds to Case 3 (Baked Drying), with an overall desirability of 0.8208, is established as optimum.

It was determined that the regression model of the complete factorial design explains from 80.69% to 96.12% the variability of the analyzed mechanical properties, which is why the model is considered as the best fit to the analyzed data. These values are comparable with similar investigations; as for example, E. H. Agung applied an experimental design with response surface methodology in which values of the coefficient of determination were obtained between 96.72 and 98.17%. [18]

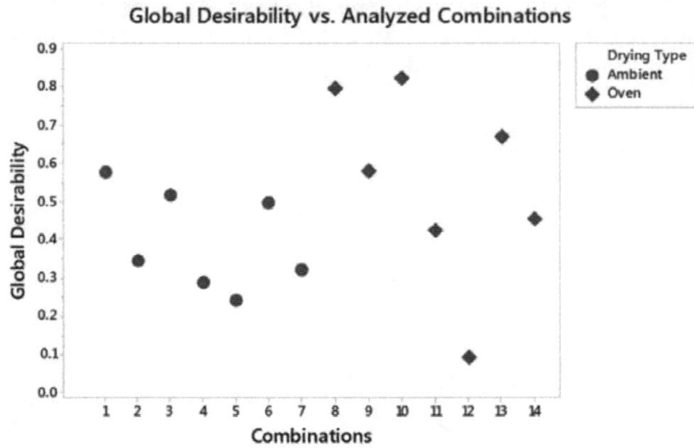

Figure 1. Global desirability of the analyzed combinations.

4. Conclusions

This experimental study uses desirability function based on a full factorial design to establish the optimal combination of input factors for the tensile and flexural mechanical properties of the hybrid composite made of abaca (AbF) and cotton (CF) plus glass (FG), combined with a polymeric polyester resin matrix. From the results of the research you can take of the following conclusions.

1. The configuration and type of drying factors, together with the interaction between these factors explain a high percentage of variability observed in the mechanical properties of the hybrid material analyzed, this percentage is 96.12%.
2. Desirability function with the full factorial design convert a subjective analysis of the mechanical properties, in an objective analysis using a comparable value in each combination.
3. The configuration of hybrid composite is the most significant factor for all response variables or mechanical properties.
4. The best configuration of hybrid material is when use: 20% glass fiber, 7.75% abaca fiber arranged at 45 ° woven with 2.25% of cotton fiber arranged at 135 °, in a polymer matrix polyester and drying oven.

References

[1] Ponce S and Guerrero V 2014 *Rev. EPN* **33(2)** 11.
[2] Pachacama N V P 2015 *Diseño, cyaracterización aplicación de un material compuesto con base de resina epoxi y refuerzo de fibras orgánicas para la aplicación en un prototipo de capot de la camioneta Mazda BT50 2010* 43.
[3] Giglio F and Savoja G 2017 *Vegetable Fibers for Composite Materials In Constructive Sector* IOP Conf. Ser. Mater. Sci. Eng. **225** 012259.
[4] Lucena M P Suarez A and Zamudio I 2009 *Rev. Latinoam. Metal. y Mater.* **1(3)** 1107–14.
[5] Dávila J *et al.*, 2011 *Nuevos Materiales: Aplicaciones Estructurales E Industriales.*
[6] Pickering S J, Liu Z, Turner T A, and Wong K H 2016 *Applications for carbon fibre recovered from composites,* IOP Conf. Ser. Mater. Sci. Eng. **139(1)** 2016.
[7] Hallad S A, Banapurmath N R, Dhage V, Ajarekar V S, Godi M T and Shettar A S 2018 *Kevlar Reinforced Polymer Matrix Composite for Structural Application* IOP Conf. Ser. Mater. Sci. Eng. **376** 012074.
[8] Köhler T, Brüll R, Pursche F, Langgartner J, Seide G and Gries T 2017 *High strength and low weight hollow carbon fibres.*
[9] Matta V, Suresh Kumar J, Venkataraviteja D and Reddy G B K 2017 *Flexural Behavior of Aluminum Honeycomb Core Sandwich Structure*, IOP Conf. Ser. Mater. Sci. Eng. **197(1)**.
[10] Al-Oqla F M and Sapuan S M 2014 *J. Clean. Prod.* **66** 347–54.

[11] Fajardo J I, Valarezo L E and López L M 2013 *Experiencies in obtaining polymeric composites reinforced with natural fiber from Ecuador*, 28–35.

[12] Lewin M 2006 *Handbook of Fiber Chemistry.*

[13] Yang H, Okada Y, Takeuchi K, Jesthi D K and Nayak A 2017 *Aging life evaluation of a new carbon fiber composite core wire Aging life evaluation of a new carbon fiber composite core wire.*

[14] Salleh Z, Ser C and Sci M 2018 IOP Conference Series : Materials Science and Engineering Tensile and Flexural Test on Kenaf Hybrid Composites Tensile and Flexural Test on Kenaf Hybrid Composites.

[15] Gutiérrez Pulido H and De la Vara Salazar R 2012 *Análisis y diseño de experimentos.*

[16] Nurulhuda A *et al.* 2017 *Analysis on Flexural Strength of A36 Mild Steel by Design of Experiment (DOE)* IOP Conf. Ser. Mater. Sci. Eng. **226(1)** 0–8.

[17] Barari F, Morgan R and Barnard P 2014 *A design of experiments (DOE) approach to optimise temperature measurement accuracy in Solid Oxide Fuel Cell (SOFC)* J. Phys. Conf. Ser. **547(1)** 0–10.

[18] Agung E H, Sapuan S M, Hamdan M M, Zaman H M D K and Mustofa U 2011 *Polym. Polym. Compos.* **19(8)** 697–710.

[19] Suardi, Homma H and Abubakar 2018 *Identification of tensile strength properties of abaca fiber by weakest-linkage approach-statistic property of fiber diameter* IOP Conf. Ser. Mater. Sci. Eng. **308** 012037.

[20] Rajmohan T, Yogeswar K, Gayatri S, Dinesh G M, Raghavendra G V and Charan Reddy K R 2018 *Multi response optimization of drilling performance of MWCNT filled banana-glass fibre reinforced composite* IOP Conf. Ser. Mater. Sci. Eng. **390** 012023.

[21] Majumder H and Maity K 2018 *Multi-Response Optimization of WEDM Process Parameters Using Taguchi Based Desirability Function Analysis* IOP Conf. Ser. Mater. Sci. Eng. **338(1)** 0–6.

IOP Conf. Series: Materials Science and Engineering **473** (2019) 012021 doi:10.1088/1757-899X/473/1/012021

DC Bias Risk Assessment of Power Grid Based on Extreme Value Estimation

WU Weili, Jiang Lin

YiLi Normal University, YiNing XinJiang, 835000

wwllxm@163.com

Abstract. The HVDC single-pole operation mode gives the Hami grid be in the risk of DC bias. Grasp the DC bias level of AC power in extreme cases can provide a basis for the defense system failure risk. In this paper, the DC bias risk assessment method of Hami power grid based on extreme value theory is put forward. Firstly, the method of continental plate and soil is analyzed. Secondly, the DC bias risk evaluation index for the power grid is constructed. Finally, the risk level of DC bias of Hami power grid is assessed. The results can provide a theoretical reference for defusing the DC bias risk caused by DC transmission.

1. Introduction

UHVDC transmission has caused a large number of DC bias events in AC system transformers. In order to better prevent and mitigate grid operation accidents caused by DC bias of transformers, it is necessary to evaluate the DC bias of transformers in the most extreme conditions.

At present, there are few researches on the risk of DC bias of transformers caused by HVDC transmission. However, the assessment of grid accident risk caused by geomagnetic storm, another cause of DC bias of transformers, has been made. Literature [1] established Based on the generalized Pareto distribution model of the geomagnetic horizontal field rate of change, the geomagnetic revision rate (nT / min) in the 52 ° N-63 ° N region is estimated at 200 years from 28 geomagnetic observational data in Europe. [2] proposed the geoelectric field conditional probability model for Dst index. Reference [3] combined with the plane wave theory (PWM), using the geomagnetic statistics in the high latitudes to estimate the geoelectric field extremum, quantified Points out the differences of the electric field in the high and low resistivity regions under the same geomagnetic disturbance level. The above study shows that the electrical structure of the earth is also another important factor that determines the electric field [4]. In addition, literature [5] proposed the power system fault risk index considering the cause of geomagnetic storms and quantified the severity of power system faults under extreme magnetic storms. These results provide some references for studying the risk of DC bias in AC / DC hybrid systems.

Therefore, a risk assessment model of DC bias for power system considering the cause of HVDC transmission in Hami power grid is proposed. Based on the bias current model under extreme conditions, a risk assessment index is put forward and the specific grid parameters are used Verify in order to provide a theoretical reference for resisting and preventing the risk of DC bias.

2. DC bias current magnetic bias model

2.1. Earth resistivity model

The earth resistivity information mainly comes from geological prospecting literature [6] and geodetic literature [7, 8], covering 15 areas such as Xinjiang, Tibet, Jiangsu, Luxi and Yangtze areas. The geological components are igneous, Sedimentary rocks and metamorphic rocks.

First of all, according to the geological components of the grid area along the transformer substation distribution and transmission line to the horizontal surface of the regional division.

Second, the vertical stratification of each partition is calculated. By analyzing the geological structure and the electrical structure of the earth, a three-dimensional earth resistivity model can be used. Taking the earth's surface as the xy plane and the z-axis pointing to the geocentric center. The conductivities of the layers corresponding to the n-layer earth resistivity model are respectively for, σ_1, σ_2, $\cdots$, σ_n and h_1, h_2, $\cdots$, h_{n-1}, $h_n \rightarrow \infty$, The method of literature [9] can be used to calculate the equivalent surface resistance of the earth.

2.2. Biasing current impact factor analysis

Factors determining the level of bias current include, the distribution of the electric field around the grounding current, and the structure and parameters of the AC power grid. Using the bias voltage (U_{im}) between substations i and m, we can get,

$$U_{im} = U_i - U_m = E_i L_i - E_m L_m \tag{1}$$

where, L_i and L_m are respectively the straight line distances between substations i and m, and the ground, E_i and E_m are respectively the ground electric field intensities between substations i and m.

Assuming that the ground current is I_g and injected into the ground electrode. There is the following relationship with the subsurface current field consisting of n substation grounding grids and one grounding pole [10]:

$$U_n = I_n R_{nn} + I_g R_{ng} \tag{2}$$

where, R_{nn} is the impedance matrix of the ground network of n substations, R_{ng} is the mutual impedance vector between the grounding pole and the substation. In general, when the substation DC bias voltage U_n is greater, the greater the substation bias current I_n.

2.3. Bias current extreme value model

When the DC grounding electrode is fixed, Resistivity model, for an exchange system to establish bias current model extreme steps are as follows.

Step 1 is to enter the information of the Earth resistivity and power grid.

The second step is to calculate the transmission line and neutral bias current level.

Step 3, adjust the operating mode, for all possible modes of bias bias current level of the estimate, select one of the bias level generally higher than the operating mode stand-by.

3. DC bias risk assessment model

3.1. bias current time product

We define the bias current time product as follows:

$$IT = \int_{t \in T} I_{DCi} \cdot dt \tag{3}$$

Equation (3) shows the consequences of the bias current, which can characterize the heating effect of the DC bias of the transformer.

3.2. Transformer DC bias distribution coefficient

Given n consecutive intervals [0, uu3), [u3, 2u3),···, [(n-1) u3, nu3), the bias current bias can be defined as

$$H_{I_{DC}} = -C_{\text{onst}} \sum_{k=1}^{n} \frac{I_{DC}(k)}{T} \lg \frac{I_{DC}(k)}{T} \tag{4}$$

where, I_{DC} (k) is the number of transformers in the kth continuous range where the neutral point bias DC of the substation falls. T is the total number of transformers. C_{onst} is a constant. The minimum value of H_{IDC} is 0 and the maximum value is C_{onst}lgn. The larger the value is, the more inhomogeneous distribution of the DC over-limit state of the neutral point in the substation is.

3.3. DC bias risk assessment process

The specific methods for assessing the risk of DC bias in the power grid are as follows:

Step 1: According to the geodetic sounding data, establish the earth resistivity model, input the information of the grounding pole position and parameters, the location of the substation of the AC system and the parameters of the power grid.

Step 2: Adjust the power system operation mode in turn and calculate the DC bias risk assessment indexes of equations (3) and (4) on the basis of the first step data.

Step 3, comparing and analyzing the risk assessment indexes under different operation modes, selecting the worst case of grid fault caused by DC bias and outputting the grid parameters.

4. Case analysis

Figure 1 shows the location of Harbin Hazheng ± 800kV DC transmission grounding pole and its surrounding substations. Table 1 shows the high resistivity model of the Hami power network.

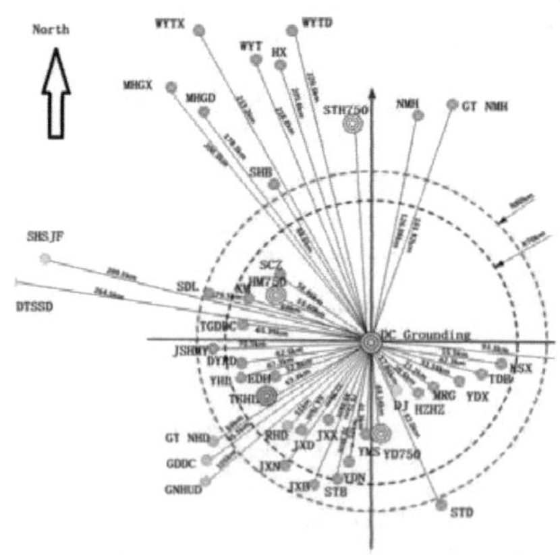

Figure 1. grounding pole and Xinjiang.

Table 1. Layered Earth resistivity parameters of Hami.

Thickness km	Resistivity (ohm-m)
0-15	18000
15-25	200
25-150	1000
150-350	100

Power Grid substation location diagram

Take DC injection I0 = -5000A. Hami grid DC bias risk assessment process is as follows:

Firstly, we estimate the distribution of electric field in the Hami grid coverage area, the results shown in figure 2:

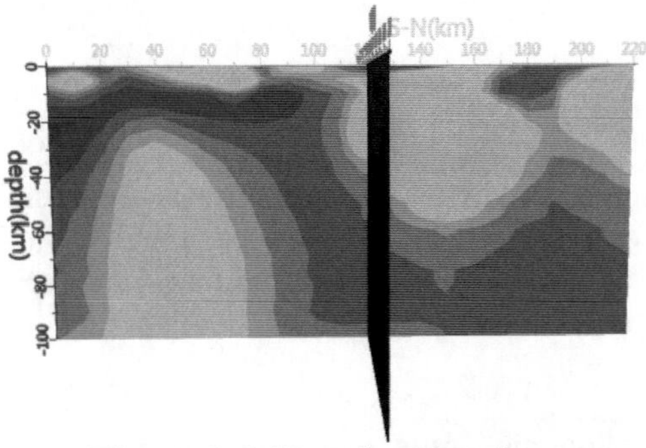

Figure 2. S-N earth resistivity.

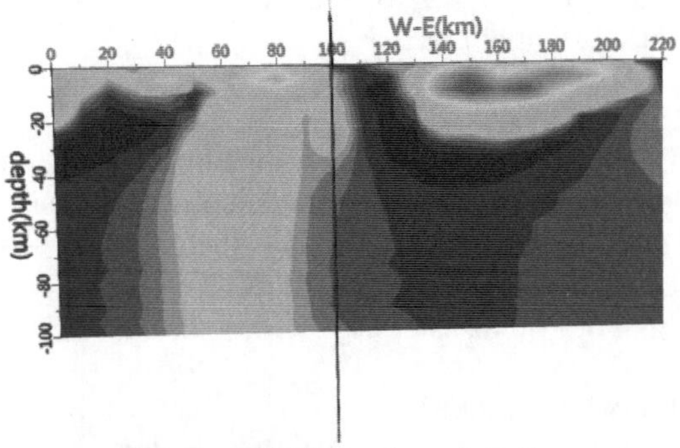

Figure 3. W-E earth resistivity.

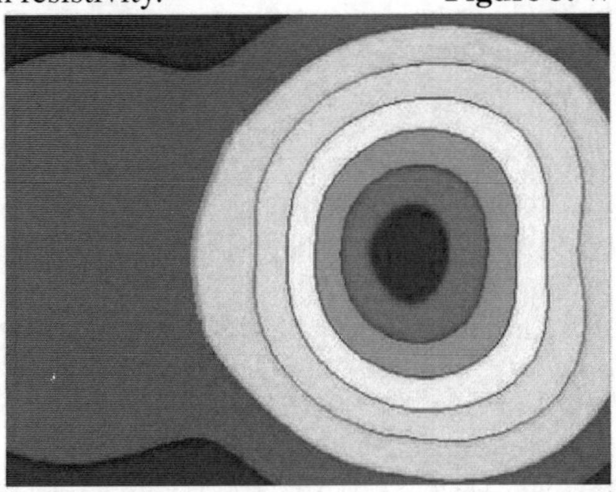

Figure 4. Hami power grid (V) contour distribution.

Secondly, based on the calculation results of geoelectric field distribution, the bias current levels are calculated according to equations. (1) and (2).

Finally, adjust the operation mode of power system, calculate the risk assessment index, and evaluate the DC bias risk of HAMC under different operation modes (the IT index takes unit time).

Table 2. Hami grid DC bias risk assessment results

Index	IT	$H_{I_{DC}}$
Method 1	0.5687	0.1345
Method 2	0.6143	0.4355
Method 3	0.6532	0.7345
Method 4	0.1998	0.6231
Method 5	0.2176	0.0345

IT and $H_{I_{DC}}$ are scalar indicators in table 2, the value interval is [0,1]; Method 4, Method 5 is based on risk assessment after the installation of risk management device risk indicators. It can be seen from table 2 that the IT indexes of Method 3 are the highest in Modes 1, 2, and 3, indicating that the level of bias current is high and the index $H_{I_{DC}}$ of Mode 3 is high, which indicates that the distribution of bias current in substations is extremely uneven. The high level of magnetic current and the risk of burning down. After installation of management methods, the indicators 4 and 5 can be seen that the bias current level is significantly reduced, but the indicator

$H_{I_{DC}}$ in mode 4 is still high, indicating that after the installation of the control device, the whole network of substation bias Current distribution is uneven, there may be individual substation there is still a substantial DC risk.

5. Conclusion

A DC bias risk assessment method for AC system caused by HVDC transmission project was proposed. Based on the Earth data, the high resistivity model of HAMC was established. According to the transformer DC bias effect, the DC bias Risk assessment index are proposed. Finally, the risk assessment method is carried out. The results of the evaluation can provide a reference for avoiding the risk of grid disaster triggered by DC bias in extreme conditions.

Acknowledgement

This research was financially supported by the Xinjiang autonomous region natural fund project(2017D01C417), and the Yili Normal University doctoral start-up fund(2016YSBS01) and the Key projects of Yili Normal University(2015YSZD03).

References

[1] Thomson A W P, Dawson E B and Reay S J, 2011, *Quantifying extreme behavior in geomagnetic activity[J]* Space Weather, **9(10)** 1-12.

[2] Pulkkinen A, Pirjola R and Viljanen A, 2008, *Statistics of extreme geomagnetically induced current events[J]* Space Weather, **6(7)** 1-10.

[3] Pulkkinen A, Bernabeu E, Eichner J et al. 2012, *Generation of 100-year geomagnetically induced current scenarios* Space Weather, **10(4)** S0001-S00013.

[4] Pulkkinen A, Pirjola R and Viljanen A, 2007, *Determination of ground conductivity and system parameters for optimal modeling of geomagnetically induced current flow in technological systems[J]* Earth, Planets, and Space, **59(9)** 999-1006.

[5] Wu Weili, Liu Lianguang and Wang Kairang, 2015, *Risk Assessment Methods and Models of Power System Fault Due to Geomagnetic Disturbance[J]* Proceedings of the CSEE, **35(4)** 830-839.

[6] Lu XinBian and Shi Yan, 1995, *Southwestern Tarim Basin - the Xi Kun regional electrical characteristics is constructed with A-type subduction mode [J]* Petroleum Geology and Experiment, **17(3)** 239-248.

[7] Wei W B, Jin S, Ye G F et al. 2006, *Conductivity structure of crust and upper mantle beneath the northern Tibetan Plateau: Results of super-wide band magnetotelluric sounding* Chinese J. Geophys. (in Chinese), **49(4)** 1215～1225.

[8] Yan Y L, Ma X B, Chen Y et al. 2012, *The study of magnetotelluric sounding on CoQen-Xainza profile in Tibet* Chinese J. Geophys. (in Chinese), **55(8)** 2636-2642.

[9] Wu Weili and Liu Lianguang, 2013, *Calculation Method for Geoelectric Field in Mid-and Low-latitude Area Based on Sparse Geomagnetic Data[J]* Chinese Journal of Space Science, **33(6)** 617-623.

[10] Sixia W U, Yaqian X U, Ping L et al. 2017, *Non-thermal Biological Effect of Environmental Electric or Magnetic Fields Exposure Induced by HVDC Transmission Lines[J]* High Voltage Engineering, **43(4)** 1333-1340.

IOP Conf. Series: Materials Science and Engineering **473** (2019) 012022 doi:10.1088/1757-899X/473/1/012022

Durability of Joint Sealing Tapes on the basis of a Pre-Compressed Polyurethane Foam

B Francke[1] and E Sudoł[2]

[1]adjunct, Construction Materials Engineering Department, Building Research Institute, Filtrowa 1, 00-611 Warsaw, Poland
[2]adjunct, Construction Materials Engineering Department, Building Research Institute, Filtrowa 1, 00-611 Warsaw, Poland

Abstract. In the paper there was analyzed a mechanism of loss of performance in properties of joint sealing tapes on the basis of pre-compressed flexible polyurethane foam. The results of tests reflect the operation of the two main functional factors are discussed, i.e. resistance to change in temperatures and resistance to the effects of UV radiation in the presence of moisture. After those tests the water tightness of the samples, appearance and the remaining compression deformation were assessed. The results of laboratory tests showed that the biggest influence on the remaining compression deformation in the tapes has UV radiation and the main reason for the loss of water tightness are changing temperatures.

1. Introduction

Polyurethane expansion tapes [1, 2] are one of the solution used to seal the joints of the frame with the jamb. Their task is to ensure tightness against water penetration as well as the migration of air. They are gaining popularity as an alternative method to polyurethane foams that are foamed in situ. They are a simple product in usage that does not require processing after application. Expanding tapes are made of polyurethane foam impregnated with an acrylic suspension. They are usually offered in several types, varying in density which results from a diverse level of impregnation saturation. The degree of impregnation and the degree of expansion of the tape after being placed in the joint determine the product's performance. The tapes are delivered in a compressed form, wound on a roll. With free expansion, their thickness increases by 5 times. They are fixed to the frame with a self-adhesive layer. They are offered in a wide range of dimensions. Their effectiveness depends on the appropriate selection of the tape dimensions and the width of the joint. In order for these products to properly perform the function of sealing, they cannot be damaged as a result of usage conditions in the considered period of use. According to well-established technical knowledge, the basic factors that can cause degradation of the products are temperature changes and UV radiation in the presence of moisture.

2. Experiment

2.1. Scope

The Building Research Institute in Warsaw (Poland) has undertaken research aimed at assessing the durability of selected performance properties of expandable polyurethane tapes impregnated with an acrylate suspension. The results of the work, described in this article, do not exhaust the topic, presenting only selected aspects of the problem. The research was carried out in two parallel paths, i.e.

➢ assessment of the resistance to change in temperature,

➢ assessment of the resistance to the effects of actions of UV radiation in the presence of moisture.

Assessment of changes in the appearance and water tightness of the tested products as well as their remaining compression deformation are taken as a measure of durability of tapes after the above-mentioned exposures.

2.2. Materials

After initial elimination tests performed on different products in the same group, we chose seven joint sealing tapes on the basis of a pre-compressed flexible polyurethane foam for sealing around windows and joints in building facades. Basic characteristics of these tapes and basic dimensions that have a significant impact on the performance of the product under the conditions of use in figure 1 were presented in the table 1.

Table 1. Basic characteristics of tapes used in the tests.

Specimen number	Dimensions of the tape, mm		Joint width		Degree of decompression a, %		Density of impregnated sealing tapes, kg/m^3
	width t_F	thickness of the uncompressed sealing tape b_0	min b_{min}	max b_{max}	min a_{min}	max a_{max}	
T1	20	46	8	16	17	35	87
T2	25	70	12	24	17	34	83
T3	20	54	9	18	17	33	84
T4	20	36	6	12	17	33	86
T5	20	46	8	15	17	33	130
T6	20	35	6	12	17	34	147
T7	20	41	7	18	17	44	163

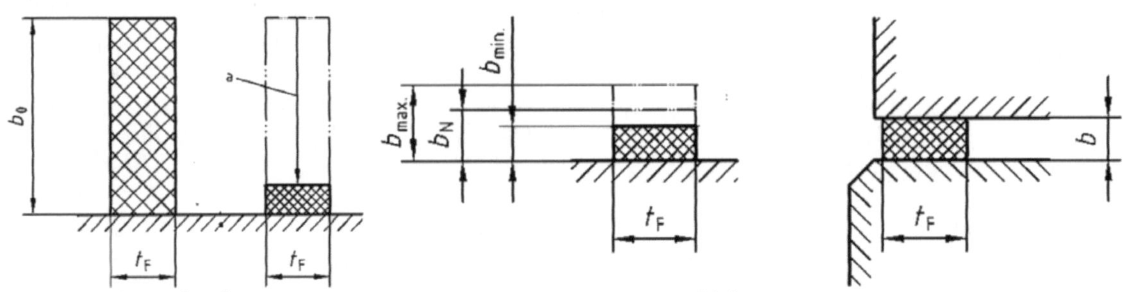

a) production. b) in use.

Figure 1. Geometry of joint and the sealing tape [3].

Nominal width b_N of the joint according to declaration of the manufacturer have to be:

$$b_{min} + (b_N \times 25\ \%) \leq b_N \leq b_{max} - (b_N \times 25\ \%)$$

2.3. Methods of tests

For the needs of the work, 9 samples were prepared for each type of tape. The tapes were placed in metal, rigid frames simulating a joint with a width corresponding to the minimum degree of tape expansion (a_{min} = 17%). The depth of filling of the joints with the tape corresponded to their width t_f (figure 1). Samples have been glued with the self-adhesive side to one side of metal supports. The second surface of the tape was free. The same sample preparation method was used for tested tapes listed in table number 1, i.e. three samples have been prepared with anti-adhesive spacers between the tape and the frame to prevent it from sticking to the frame permanently and other six samples without spacers (figure 2). For all tested tapes mentioned samples have been divided in such way (see figure 3):

- two samples with anti-adhesive spaces and one without them, has been conditioned in standard climate for 24 hours,
- three samples mentioned above were subjected to the tests related to the assessment of tape's resistance to the effects ~~of actions~~ of UV radiation in the presence of moisture,
- three samples mentioned above were subjected to the tests related to the assessment of tape's resistance to change in temperature.

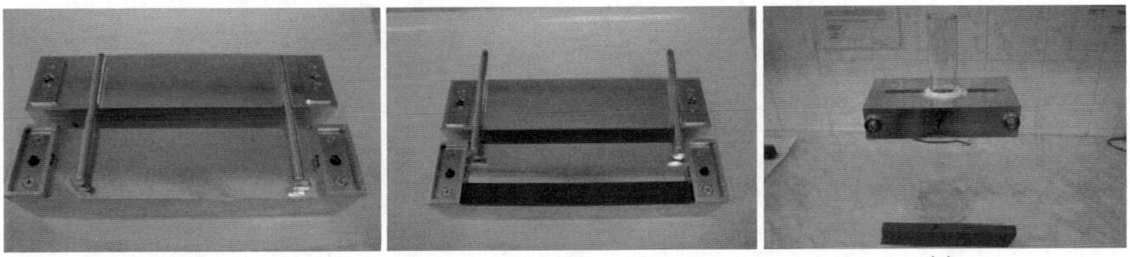

(a) (b) (c)

Figure 2. Rectangular tubes made of aluminum for mounting the tapes for ageing tests: alone (a), with the sample (b) and with the pipe used for water tightness test (c).

for laboratory conditions	for UV and moisture effect	for temperature effect
• degree of decompression (samples with anti-adhesive spacer) • appearance (samples without anti-adhesive spacer) • watertightness (samples without anti-adhesive spacer)	• degree of decompression (samples with anti-adhesive spacer) • appearance (samples without anti-adhesive spacer) • watertightness (samples without anti-adhesive spacer)	• degree of decompression (samples with anti-adhesive spacer) • appearance (samples without anti-adhesive spacer) • watertightness (samples without anti-adhesive spacer)

Figure 3. Types of samples used in the research.

Specimens were subjected to the test of resistance to the effects of action of light and moisture [3]. They were hung up in a Xenon test apparatus and were weathered for a total of 3 months in the test chamber at a light/dark alternation with cycles according to EN ISO 4892-2, table 3, Method A, cycle 1. The sum of the radiation during this period per test area amounts to 240 $MWs/m^2 \pm 5$ %. Conditions in the test chamber are shown on the figure 4. Afterwards the samples have been stored in the drying oven at temperature of 50 °C for 1 h and a further 24 h at standard climate [3].

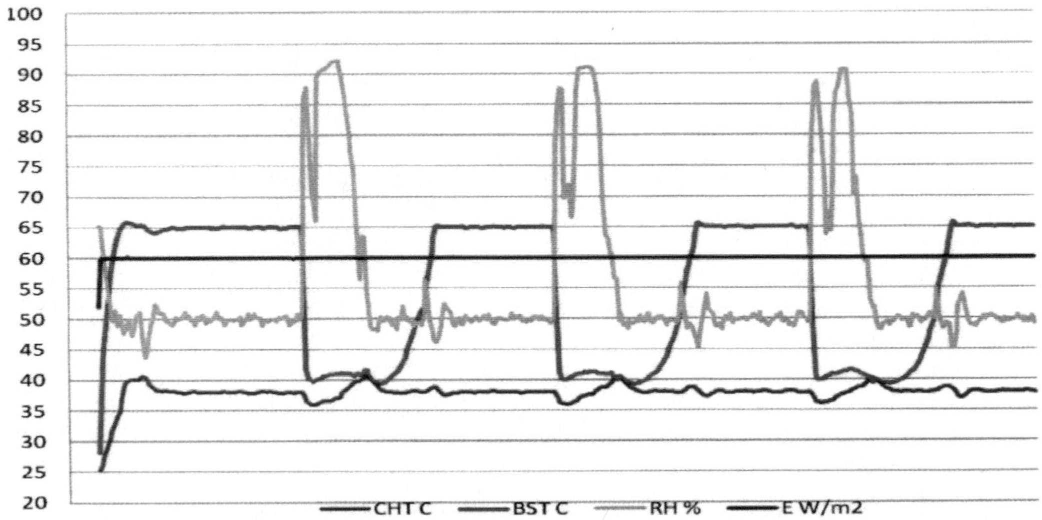

Figure 4. Conditions in the chamber during the action of UV radiation in presence of moisture.

To test the resistance to change in temperature specimens were pre-stored for 7 days at standard climate and after that have been subjected to three times the following cycle of storing (figure 5):

- exposure in an oven at 50 ± 2 °C for 22 h, followed by 2 h in an oven at 80 ± 2 °C (this process have been repeated twice),
- immersion in distilled water at 23 ± 2 °C for 1 day,
- exposure in a freezer at -20 ± 2 °C for 3 days,

Subsequently the test specimen has been stored one more day at standard conditions.

(a) (b) (c)

Figure 5. Samples during the test : in the oven(a), in distilled water (b), in freezer(c).

After each of the aging test, the change in the appearance and water tightness of the samples as well as remaining compression deformation were assessed. Remaining compression deformation was determined after 6h after finishing aging action, in accordance with EN ISO 1856. Water tightness has been tested with a head of water of 60 mm in a tube test during 180 second at standard climate conditions.

2.4. Results and discussion

For the first batch of samples, in none of the tested cases water leaks were found after the test was completed. But after two ageing tests only 4 tapes fulfill requirements of water tightness. Results of these tests are presented on the figure 6.

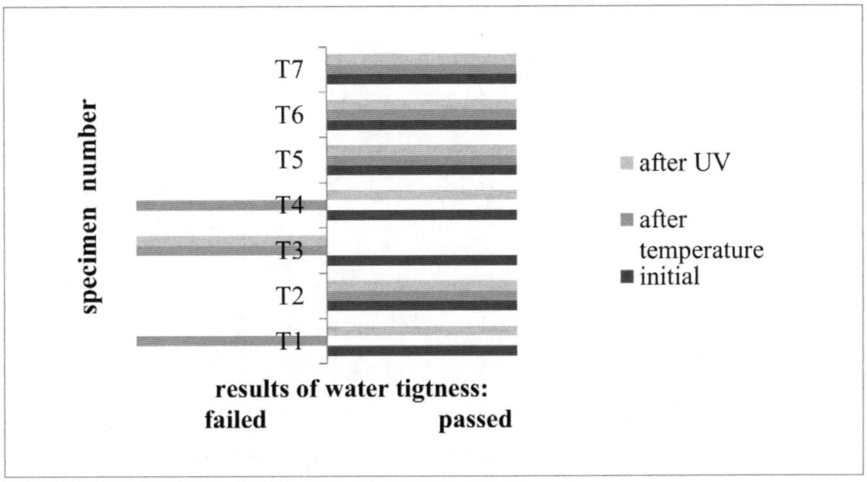

Figure 6. Results of water tightness.

There is no visible leakage of impregnating agent. Comparative combination of the results of remaining compression deformation after actions of UV radiation in the presence of moisture and after temperature aging conditions are shown on figure 7.

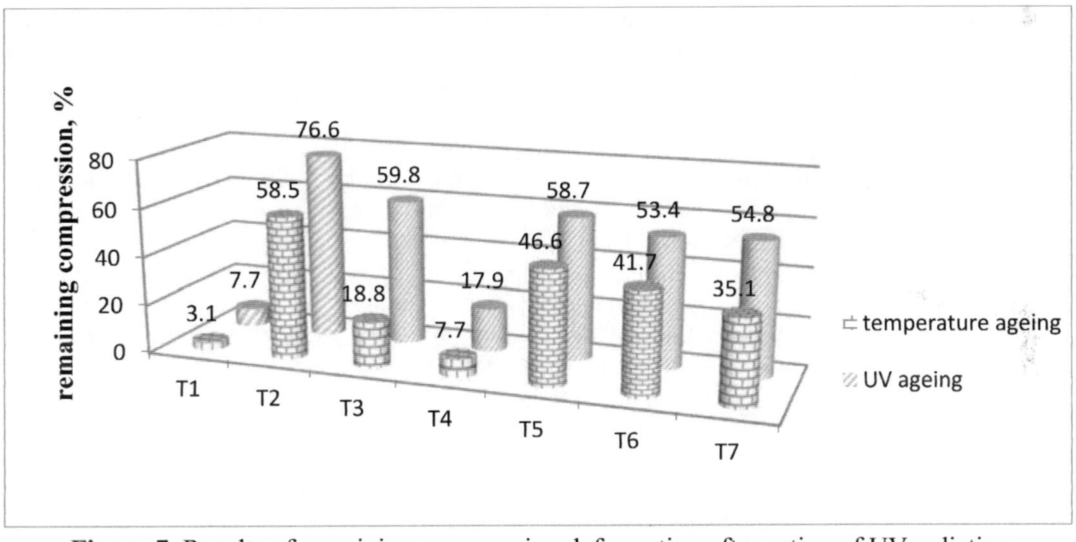

Figure 7. Results of remaining compression deformation after action of UV radiation in the presence of moisture and after temperature ageing conditions.

The values of remaining compression deformation of tapes as a result of simulation of working conditions are higher in the case of UV radiation in the presence of moisture than in the effect of variable temperatures, which is graphically shown in figure 8. There is a correlation between effects of these two influences as an ageing coefficient submitted as a quotient residual deformation after UV treatment in the presence of moisture and residual deformation after changing temperatures. This coefficient ranges from 1.3 to 2.5. If the coefficient varies between 1.3 and 1.6, tapes fulfill requirements of water tightness after temperature ageing. In the case, when it rises above 2, the water resistance of the tapes is lost. There is no relation between discussed factor and the density of the tapes.

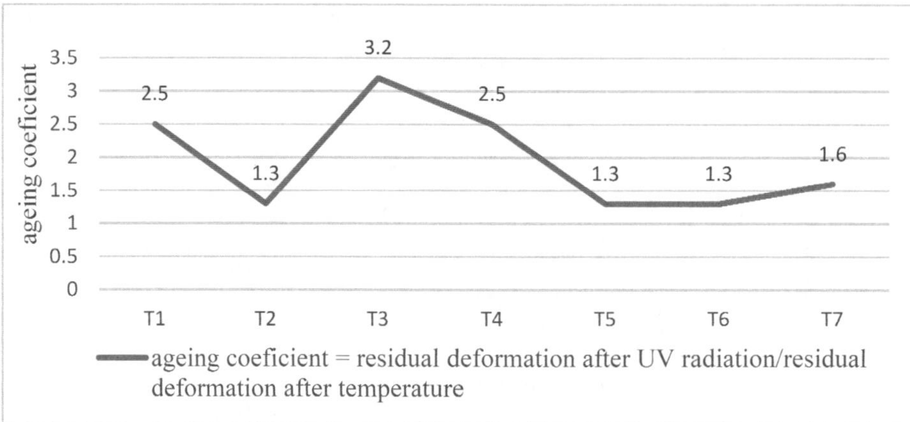

Figure 8. Influence of aging factors on the rate of tape's deformatiom.

3. Conclusion

Test results discussed in this article suggest conclusions described below regarding the durability of pre-compressed flexible polyurethane joint sealing tapes:

- there are no visible changes in external appearance of all tapes after aging tests, i.e there is no visible leakage of impregnating agent,
- working conditions have a significant impact on the successive loss of functional properties of the tapes, identified as remaining compression deformation of the cross-sectional area after squeezing the tape in the test joint,
- the influence of UV radiation in the presence of moisture has the greatest influence on the remaining compression deformation of the tapes. In the majority of cases studied, the remaining compression deformation is from 1.3 to 2.5 times higher after UV treatment in the presence of moisture than in the effect of changing temperatures. The exception is one sample showing even greater disproportions in this case,
- observation mentioned above is not reflected in the results of the watertight test. The waterproofness of the joint is adversely affected mainly by the influence of variable temperatures and not by the action of UV radiation in the presence of moisture. The deviation from this statement is visible only on one tested tape, in which also the deformation caused by the UV radiation was deviated from the average.

Acknowlegments

The paper was prepared within the statutory research work at ITB (Warsaw- Poland) NZM-051/2017 *„Wdrożenie do praktyki laboratoryjnej metod badań rozprężnych taśm poliuretanowych, przeznaczonych do uszczelniania połączeń okien ze ścianą"* and with the technical support of Artur Kupisz i Marcin Kupisz

References

[1] Bracke W, Van Den Bossche N and Janssens A 2014. *Airtightness of Building Penetrations: Air Sealing Solutions, Durability Effects and Measurement Uncertainty* 35th Conference: Ventilation and Airtightness in Transforming the Building Stock to High Performance, Proceedings AIVC 488–500

[2] Langmans J, Desta T, Alderweireldt L and Roels S 2017 *Int. J. Ventilat.* **16(1)** 30-41.

[3] *Joint sealing tape on the basis of a pre-compressed flexible polyurethane foam for sealing window and joints in building facades* European Organization for Technical Approvals, European Assessment Document 320001-00-0605 2014.

IOP Conf. Series: Materials Science and Engineering **473** (2019) 012023 doi:10.1088/1757-899X/473/1/012023

Comparative Study of DAM and ELM to One-Storey Eccentrically Braced Frames Subjected to Seismic Load in Indonesia

M Orientilize[1], H Purnomo[1], S A Rahim[1] and A Y Priyadi[2]

[1] Lecturer, Civil Engineering Department, Universitas Indonesia, Kampus UI Depok, Indonesia

[2] Student, Civil Engineering Department, Universitas Indonesia, Kampus UI Depok, Indonesia

Abstract. Indonesia is late in updating its structural steel building design code to AISC 360-10. Revision was done in 2015 replacing the 2002 code. The new version changes the method of steel stability analysis from effective length method (ELM) to direct analysis method (DAM). The 2015 code implements DAM, which is based on second-order elastic analysis which directly calculates P-delta effect in analysis. DAM takes into account geometric imperfection and strength reduction. Notional load as 0.002 of gravity load should be applied horizontally to represent geometric imperfection. It is allowed to adjust the notional load coefficient. Numerical study has been conducted to compare the use of DAM and ELM on eccentrically-braced frame subjected to Indonesian seismic load. Different values of notional load coefficient are considered. Experimental results conducted by other researchers were used as reference. The frames were reanalysed using four different stability methods: ELM first-order analysis, ELM second-order analysis and DAM with different notional load coefficients: 0.002 and 0.003

1. Introduction

The design code of steel structural building in Indonesia, SNI 03-1729-2015, is a revision of the former code of SNI 03-1729-2002. It is a translated version of AISC 2010. Revision was made in 2015, five years after AISC released AISC 360-10 in 2010 and replaced the former code AISC 2005. One significant change from AISC 2005 to 2010 was structure stability analysis. AICS 2010 uses direct analysis method (DAM) as the main method, as mentioned in Chapter C: Design for Stability [1]. As stated, the direct analysis method allows a more accurate determination of the load effects in the structure through the inclusion of the effects of geometric imperfections and stiffness reductions directly within the structural analysis. This also allows the use of K = 1.0 in calculating the in-plane column strength. The former code, AISC 2005, used effective length method (ELM) as the main method. As mentioned in Appendix 7, ELM is still permissible to be used if the ratio of maximum second-order drift to maximum first-order drift in all storeys is equal to or less than 1.5. ELM can be approached in two ways, first is second-order analysis, similar to DAM, but without stiffness reduction, and, secondly, first-order analysis where P-delta effect is counted through moment amplification.

Notional loads to represent geometric imperfection are described in section C2.2b AISC 2010. The loads shall be applied as lateral loads at all structural levels. The notional load coefficient of 0.002 is based on a nominal initial storey out-of-plumbness ratio of 1/500 of column length and it is permissible to adjust the notional load coefficient proportionally.

Laid on seismic zones where earthquakes frequently occur, hence, structural building in Indonesia should be designed according to the SNI 1726:2012 design code for earthquake resistance. Eccentrically Braced Frame (EBF) is an earthquake resistant structural system for steel structure. It is a hybrid system, offering lateral stiffness from its concentric brace, and ductility from moment frame system. In addition, EBF has a link that serves as an element of energy dissipation as well as the place of plastic joints. The plastic joints occur away from the column so that, in the planning, EBF structure has a high endurance and security [2].

Several studies comparing ELM and DAM as regard structural stability have been conducted by other researchers [3, 4], but most have been carried out on moment resisting frames and none on varied notional loads coefficient. Therefore, a numerical study of different notional load coefficients has been conducted on a one-storey concentric steel braced frame. The study was performed using different stability methods, ELM and DAM. One-storey EBF was simulated against three different seismic loads in the Indonesian seismic zone. Experimental results conducted by other researchers were used as reference. The experiments were reanalysed with four methods: ELM first-order analysis, ELM second-order analysis and DAM with two different notional load coefficients: 0.002 and 0.003.

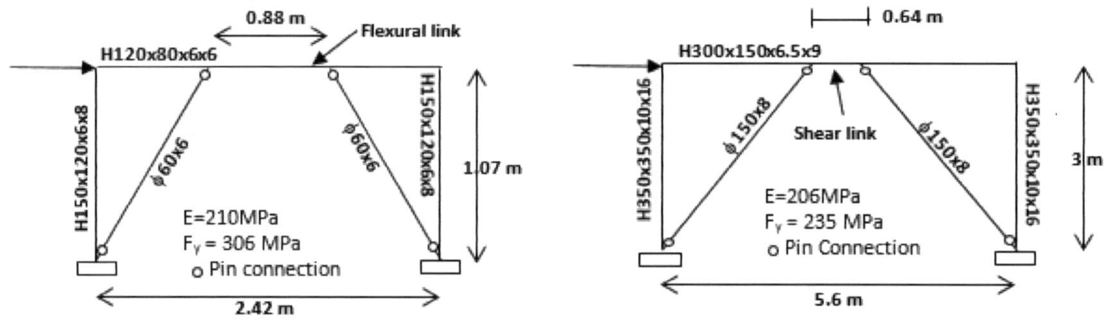

Figure 1. Two Eccentrically Braced Frames as a reference for numerical studies.

2. Numerical Study

2.1. Numerical Model of EBF with flexural and shear link

An experimental studies reported by Shi [5] and Shujun [6], as shown in figure 1, are used as a reference in modelling EBF using SAP software [7]. The two frames have different link characteristic, flexural and shear link. As described in AISC 341 [8], flexural link is known as long link whereas shear link is known as short link. Frame elements are assigned for all elements, column, beam and bracing. Nonlinearity of material and p-delta effect plus large displacement were conducted in the analysis. Rigid connections were applied on beam column joint whereas pin connections were used for brace element according to the experimental set-up. Pushover analysis was carried out to represent the experimental test where horizontal load was applied incrementally until the frames reached maximum capacity. Characteristics of links were modelled according to moment curvature as described in FEMA 356 [10]. Plastic hinges on both beam and column end, as well as on link and bracing, were assigned.

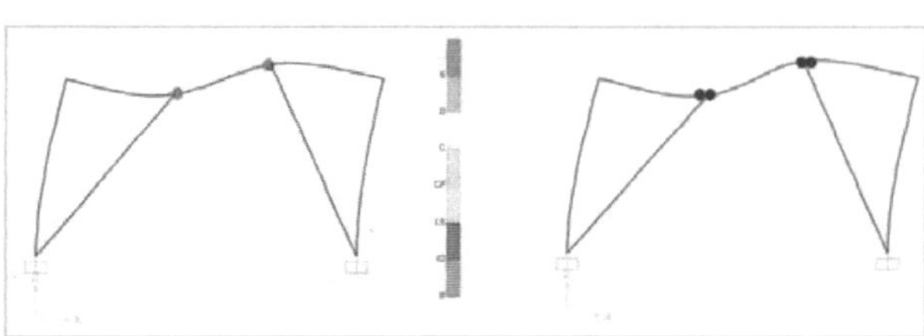

Figure 2. The first plastic hinge and the final stage of EBF with Flexural Link.

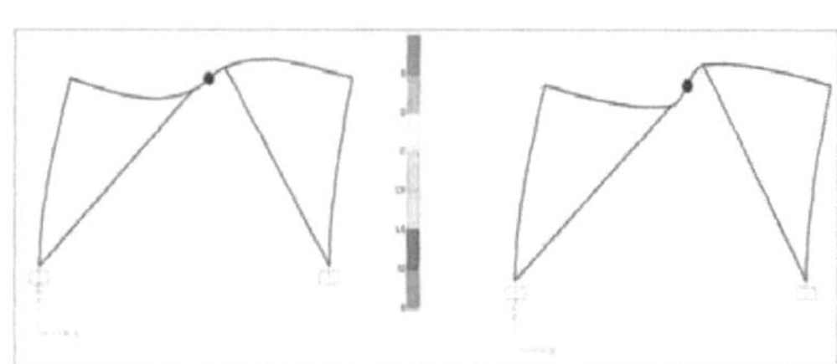

Figure 3. The first plastic hinge and the final stage of EBF with Shear Link.

Numerical simulations detect different collapse mechanism between the two frames, as can be seen on figures 2 and 3. The first plastic hinge at EBF with flexural link is formed at edge link and a subsequent plastic hinge is formed at beam segment outside of the link, as can be seen in figure 2. The SAP result agrees with numerical results obtained by Shi et al. [7]. For EBF with shear link, the shear yielding occurs in mid-link and is the only plastic hinge which is formed, as shown in figure 3.

Figure 4 and 5 show load – displacement curve based on SAP and and experimental results. As can be seen, in the linear stage, the test results are slightly above the numerical results. After the first plastic hinge is formed, results of SAP have close agreement with the test result. Moreover, the numerical model can accurately predict the maximum capacity of EBF. It can be concluded that SAP can represent the test well and, hence, the model can be used for further numerical study.

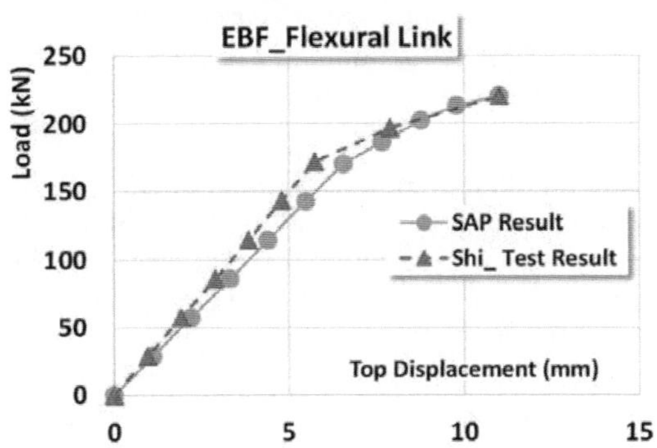

Figure 4. Load – Displacement Curves of EBF with Flexural Link.

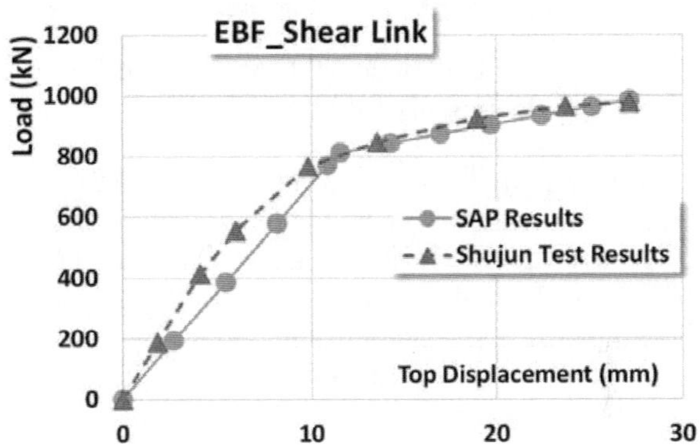

Figure 5. Load – Displacement Curves of EBF with Shear Link.

2.2. Results and Discussion

The effect of initial imperfections should be taken into account, either by direct modelling of imperfections in the analysis or by the application of notional loads, as specified in Section C2.2b [1]. The magnitude of the notional loads shall be:

$$Ni = 0.002\alpha Yi \tag{1}$$

where α =1.0 (LRFD); α = 1.6 (ASD)

Ni = notional load applied at level i, kips (N)

Yi = gravity load applied at level i from the LRFD/ASD load combination, kips (N)

The notional load coefficient of 0.002 in equation (1) is based on a nominal initial storey out-of-plumbness ratio of 1/500L and it is permissible to adjust the notional load coefficient proportionally [1]. ELM can be referred to as a first-order elastic analysis where moment due to P-delta effect is calculated based on amplification factor [2]. ELM can also be analyzed as second-order analysis if the P-delta and P- δ effect are included in the analysis. In order to see the difference in the ELM approaches of first-order and second-order analysis, the EBF frames were reanalyzed by both method. Meanwhile, different notional coefficients were used for the direct analysis method, which are DAM_0.002 and DAM_0.003.

Pushover analysis were carried out on both eccentrically braced frames. The maximum force is based on the limitation of ELM, where the ratio of second-order to first-order analysis is less or equal to 1.5. Hence the maximum pushover forces are: 1940.09kN for EBF with flexural links and 414.27kN for EBF with shear link. Results of pushover analysis are plotted on figure 6 and 7. Load-displacements curves resulting from different stability analysis methods are compared to experimental tests performed by Shi et al. and Shujun et al. The curves of the four methods are almost similar for EBF with flexural link, whereas slightly different results are observed on EBF with shear link. It can be concluded that using ELM or DAM as stability analysis of EBF with single storey leads to similar structural responses in term of force – displacement value.

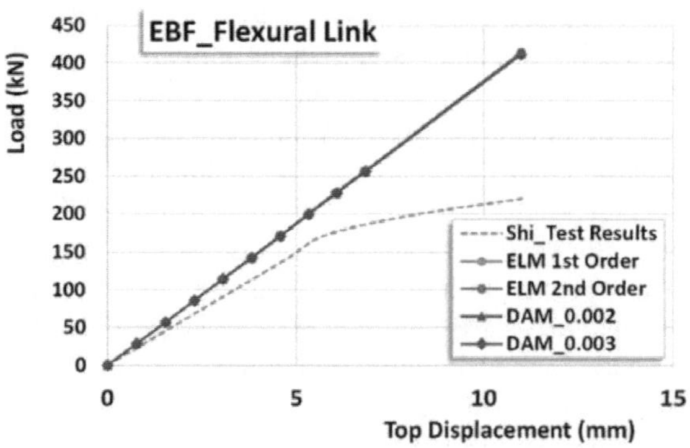

Figure 6. Load-Displacement Curves of EBF with Flexural Link analyzed by ELM and DAM.

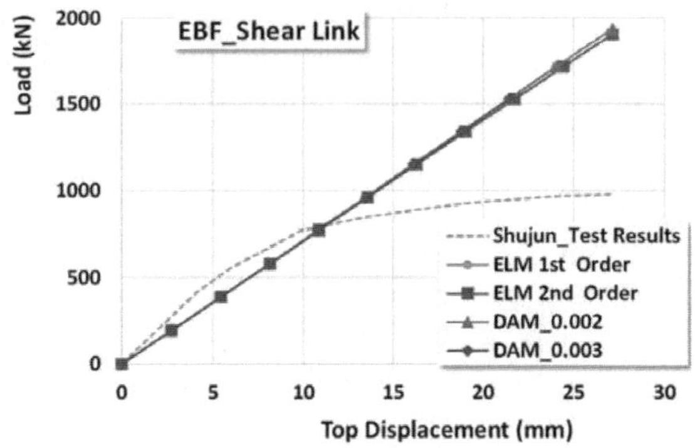

Figure 7. Load-Displacement Curves of EBF with Shear Link analyzed by ELM and DAM.

Further study was carried out by taking into account seismic load. The EBF were simulated in three different seismic zones in Indonesia, from the lightest and medium to the strongest seismic load, which are Samarinda, Jakarta and Padang, respectively. Seismic simulation also considered three different soil conditions: hard, medium and soft soil, which are named as SC, SD and SE, respectively. Horizontal forces presenting earthquake load are determined based on SNI 1726:2012 [13] using linier elastic seismic analysis, as follows:

$$F_x = C_{vx}V \tag{2}$$

where: F_x : Seismic load for x-storey
C_{vx} : Distribution factor for each floor
V : Base shear of structure, where $V = C_s.\ W$
W : Effective Weight of the structure, and Cs : Seismic response factor

Table 1. PM Interaction of EBF.

Seismic Zone	Soil Type	Earthquake Loads (kN)	PM Interaction of EBF with Flexural Link					PM Interaction of EBF with Shear Link				
			Advanced Analysis	ELM 1st Order	ELM 2nd Order	DAM _0.002	DAM _0.003	Advanced Analysis	ELM 1st Order	ELM 2nd Order	DAM _0.002	DAM _0.003
Samarinda	SC	12.54	0.016	0.016	0.016	0.016	0.016	0.007	0.007	0.007	0.007	0.007
	SD	16.73	0.022	0.022	0.022	0.022	0.022	0.009	0.009	0.009	0.009	0.009
	SE	26.13	0.034	0.034	0.034	0.034	0.034	0.014	0.014	0.014	0.015	0.015
Jakarta	SC	72.74	0.095	0.094	0.094	0.094	0.094	0.040	0.040	0.040	0.040	0.040
	SD	82.77	0.108	0.107	0.107	0.107	0.107	0.046	0.046	0.046	0.046	0.046
	SE	94.06	0.122	0.122	0.122	0.122	0.122	0.052	0.052	0.052	0.052	0.052
Padang	SC	145.68	0.190	0.188	0.189	0.189	0.189	0.081	0.081	0.081	0.081	0.081
	SD	145.68	0.190	0.188	0.189	0.189	0.189	0.081	0.081	0.081	0.081	0.081
	SE	131.30	0.171	0.170	0.170	0.170	0.170	0.073	0.073	0.073	0.073	0.073

Results of the analysis are presented on PM interaction and drift. PM is determined based on interaction formula where factored axial force Pu and bending moment Mu are compared to each's nominal strength, ϕPn and ϕMn, multiplied by their resistance factors. Drift is top lateral displacement of the frames. Table 1 and 2 present PM interaction results and drift. Advanced analysis based on numerical simulation as mentioned in Section 2 is used as a reference. As presented, similar PM values are found for both EBF with flexural or shear links, which means that stability analysis using ELM and DAM gives the same results. Moreover, there is no effect of different value of notional load coefficient to the PM ratio.

Different results are found at drift, as shown in table 2. Although ELM and DAM do not give significant results, ELM second-order analysis gives closer results to advanced analysis than other methods for EBF with flexural links. DAM with notional load coefficients as 0.002 or 0.003 has similar drift. Meanwhile, similar drift is also detected on EBF with shear link when it is analyzed with ELM second-order and DAM. Again, changing notional load coefficient from 0.002 to 0.003 does not change the drift. In addition, the value of PM interaction and drift is far below the allowable value. The allowed drift of EBF with flexural links and shear links is 2.675mm and 7.5mm, respectively.

Table 2. Drift of EBF.

Seismic Zone	Soil Type	Earthquake Loads (kN)	Drift of EBF with Flexural Link					Drift of EBF with Shear Link				
			Advanced Analysis	ELM 1st Order	ELM 2nd Order	DAM _0.002	DAM _0.003	Advanced Analysis	ELM 1st Order	ELM 2nd Order	DAM _0.002	DAM _0.003
Samarinda	SC	12.54	0.39	0.31	0.40	0.52	0.52	0.18	0.18	0.18	0.18	0.18
	SD	16.73	0.51	0.42	0.53	0.69	0.69	0.23	0.24	0.24	0.24	0.24
	SE	26.13	0.80	0.65	0.83	1.08	1.08	0.37	0.37	0.37	0.37	0.37
Jakarta	SC	72.74	2.32	1.82	2.31	3.02	3.01	1.02	1.04	1.02	1.03	1.02
	SD	82.77	2.54	2.07	2.63	3.43	3.42	1.16	1.18	1.17	1.17	1.17
	SE	94.06	2.89	2.35	3.99	3.90	3.89	1.32	1.34	1.32	1.33	1.32
Padang	SC	145.68	4.47	3.64	4.63	6.04	6.03	2.04	2.07	2.05	2.05	2.05
	SD	145.68	4.47	3.64	4.63	6.04	6.03	2.04	2.07	2.05	2.05	2.05
	SE	131.30	4.03	3.28	4.17	5.45	5.43	1.84	1.87	1.85	1.85	1.85

3. Conclusion

The effect of different methods of steel stability analysis, ELM and DAM, to one-storey EBF is not significant. This is because of the study is only limited to one-storey structure where the instability do not appear significantly. Hence the buckling factor "K" between ELM and DAM seems similar. Two different notional load coefficients also do not significantly change the EBF response. This is due to less gravity load worked on structure. However, notional load is a function of gravity load. The finding is confirmed by PM interaction and drift results when the frames are loaded by seismic load according to the seismic design code in Indonesia. The EBF has enough strength against seismic loads in all seismic zones in Indonesia. Further study should be conducted on multi-storey EBF with gravity load.

Acknowledgment

This works is supported by Hibah PITTA 2018 funded by DRPM Universitas Indonesia No. 2451/UN2.R3.1/ HKP.05.00/2018.

References

[1] AISC 2010 *Specification for Steel Structural Building.*

[2] Prajzner J 2006 *Evaluation of the Effective Length Method and the Direct Analysis Method for the Design of Steel Columns in Frame Structures* (University of Maryland).

[3] Surovek A E and Ziemian R D 2005 *The Direct Analysis Method: Bridging the Gap from Linear Elastic Analysis to Advanced Analysis in Steel Frame Design* (ASCE library).

[4] SAP 2000 Adv. V.15 2011 *Analysis Reference Manual, Computers and Structures Inc.* (Berkeley California).

[5] Shujun H 2015 *Adv. Steel Constr.* **11(1)** 95-110.

[6] AISC 341-16 2016 *Seismic Provisions for Structural Steel Buildings* (Chicago: American Institute of Steel Construction).

[7] Shi Y J, Xiong J and Wang Y Q 2010 *J. Build. Struct.* **31(2)** 29-34.

[8] FEMA 451 2006 *NEHRP Recommended Provisions: Design Examples* (California: Federal Emergency Management Agency).

[9] Badan Standardisasi Nasional 2012, *SNI 1726 Tata Cara Perencanaan Ketahanan Gempa untuk Struktur Bangunan Gedung dan Non Gedung* (Jakarta: Badan Standardisasi Nasional).

[10] AISC and Structural Steel Educational Council 2006 *AISC Seismic Design Manual first edition* (AISC inc.).

Fundamental Comparative Study on Band-Gap Properties of 1-D Conventional Photonic Crystal and 1-D Function Photonic Crystal

Yungao Cai[a1], *, Yinghong Dong[1]

[1]College of Engineering Technology, Baoshan University, Baoshan Yunnan, 678000, China

Abstract. Photonic crystal is a kind of artificially material composed of two or more dielectrics arranged periodically, which can control the flow of photons due to the photonic band-gap. In this article, the band gap properties of the one-dimensional (1-D) conventional photonic crystal and the 1-D function photonic crystals are comparativly analyzed according to dispersion relation, periodicity and incident angle systematically. The results have some guiding significance for the 1-D photonic crystal and photonic crystal device fabrication.

1. Introduction

Since the concept of photonic crystal was proposed in 1987 by Yablonovitch [1] and John [2] respectively, it has many applications and has become one of the most important theses of the world because it can control the flow of photons due to the photonic band-gap. The author project group are concentrating on the 1-D conventional photonic crystal, and made some achievements, such as references [3-5]. In recent years, a kind of photonic crstal has been proposed, named function photonic crstal, which the refractive index is a function of the space position, this concept stems from the X.Y. Wu project group of Jilin University, China in 2012 [6-10].

In this article, the band gap properties of the 1-D conventional photonic crystal and the 1-D function photonic crystals are comparativly analyzed based on the preliminary work of the author project group and part referencs mainly from the X.Y. Wu project group systematically in order to guide the photonic crystal research and photonic crystal device fabrication better.

2. Comparative analysis and discussion

2.1. Difference in transfer matrix

As shown in figure 1, it's schematic diagram of the 1-D function photonic crystal with the propagation of light; the author project group are concentrating on the 1-D conventional photonic crystal, the transfer matrix method had been used for the research process. According to references [3-5], the transfer matrix for the 1-D conventional photonic crystal could written as follow,

$$\vec{M} = \begin{bmatrix} \cos\beta_2\cos\beta_3 - \frac{p_3}{p_2}\sin\beta_2\sin\beta_3 & -\frac{i}{p_3}\cos\beta_2\sin\beta_3 - \frac{i}{p_2}\sin\beta_2\cos\beta_3 \\ -ip_2\sin\beta_2\cos\beta_3 - ip_3\cos\beta_2\sin\beta_3 & \cos\beta_2\cos\beta_3 - \frac{p_2}{p_3}\sin\beta_2\sin\beta_3 \end{bmatrix} \quad (1)$$

Here, $\beta_1 = 2\pi n_1 b / \lambda \cos\theta_1$, $\beta_2 = 2\pi n_2 b / \lambda \cos\theta_2$, $p_1 = n_1 \cos\theta_1$, $p_2 = n_2 \cos\theta_2$, θ_1 and θ_2 are the angles between the incident normal and the stratified direction of the two kinds of media with different refractive index in the basic periodic structure respectively.

According to the research achievements of the X.Y. Wu project group of Jilin University, China, the transfer matrix for the 1-D function photonic crystal could written as follow [6-10],

$$M = M_1 M_2 = \begin{bmatrix} \cos\delta_1 & -\dfrac{i\sin\delta_1}{\sqrt{\dfrac{\varepsilon_0}{\mu_0}} n_1(b)\cos\theta_i^I} \\ -n_b(0)\sqrt{\dfrac{\varepsilon_0}{\mu_0}}\cos\theta_t^I \sin\delta_1 & \dfrac{n_b(0)\cos\theta_t^I \cos\delta_1}{n_1(b)\cos\theta_i^I} \end{bmatrix} \begin{bmatrix} \cos\delta_2 & -\dfrac{i\sin\delta_2}{\sqrt{\dfrac{\varepsilon_0}{\mu_0}} n_2(a)\cos\theta_i^{II}} \\ -n_a(0)\sqrt{\dfrac{\varepsilon_0}{\mu_0}}\cos\theta_t^{II} \sin\delta_2 & \dfrac{n_a(0)\cos\theta_t^{II} \cos\delta_2}{n_a(a)\cos\theta_i^{II}} \end{bmatrix} \quad (2)$$

Where,

$$\delta_b = \frac{w}{c} n_1(b)(\cos\theta_i^I b + \sin\theta_i^I \int_0^b \frac{dz}{\sqrt{(1+k_0^2)(\frac{n(z)}{n(0)})^2 - 1}}) \quad (3)$$

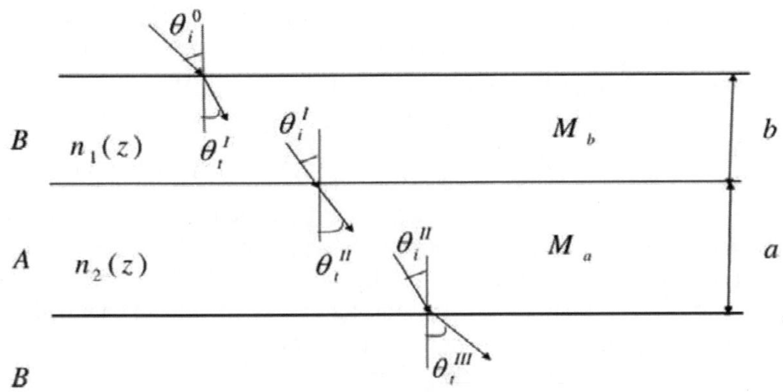

Figure 1. Schematic diagram of the 1-D function photonic crystal.

2.2. Dispersion relation

The expression of the dispersion relation for the 1-D conventional photonic crystal are written as follow [3],

$$\cos kd = \cos\beta_1 \cos\beta_2 - \frac{1}{2}(\frac{n_2}{n_1} + \frac{n_1}{n_2})\sin\beta_1 \sin\beta_2 \quad (4)$$

For 1-D function photonic crystal, it's written as follow [6]

$$\cos kd = \cos\delta_1 \cos\delta_2 - \frac{1}{2}(\frac{\eta_b}{\eta_a} + \frac{\eta_a}{\eta_b})\sin\delta_1 \sin\delta_2 \quad (5)$$

Here, $d = b + a$, $\beta = (2\pi \frac{n_a n_b}{n_a + n_b})(\frac{wd}{2\pi c})$, $\eta_x = \sqrt{\frac{\varepsilon_0}{\mu_0}} n_x \sqrt{1 - \frac{n_0^2}{n_x^2}\sin^2\theta_i^0}$ ($x = a$ or b).1-D conventional photonic crystal is a special case of the 1-D function photonic crystal. Since 2012, various types 1-D function photonic crystal has been studied by the X.Y. Wu project group [6], the type of refractivity are mainly include sinusoidal function, polygonal function and so on, the functional expressions are written as follow [6].

For sinusoidal function,

$$\begin{cases} n_1(z) = n_1(0) + A_1 \sin(\frac{z}{b}\pi), & 0 \le z \le b \\ n_2(z) = n_2(b) + A_2 \sin(\frac{z-b}{a}\pi), & b \le z \le b+a \end{cases} \tag{6}$$

For polygonal function,

$$n_1(z) = \begin{cases} n_1(0) + \frac{2(m-1)n_1(0)z}{b}, & 0 \le z \le \frac{b}{2} \\ n_2(b) + \frac{2(m-1)n_1(0)}{b}(b-z), & \frac{b}{2} \le z \le b \end{cases} \tag{7}$$

$$n_2(z) = \begin{cases} n_2(0) + \frac{2(m-1)n_2(0)(z-b)}{a}, & b \le z \le b + \frac{a}{2} \\ n_2(0) + \frac{2(m-1)n_2(0)}{a}(a+b-z), & b + \frac{a}{2} \le z \le b+a \end{cases} \tag{8}$$

The results shown that there will be bandgap in the disconnected region on the dispersion curve; and with the increase of frequency, the bandgap width decreases when $kd = 3$, but the bandgap increases when $kd = 0$ for the 1-D function photonic crstal [6]. But the bandgap width is constant with the increase of frequency for 1-D conventional photonic crstal [3].

2.3. Effect of periodicity on the band-gap properity

Take the sinusoidal function as an example, as described in reference [7], the number of the oeriod cycle were 4, 6,and 8, the relationship between transmittance and frequency were anlyzed, figure 2 [4] and figure 3 [7] are the transmission spectra of 1-D conventional photonic crystal and 1-D function photonic crystal with different periodicity, as described in figure 2 and figure 3 in turn, no matter 1-D conventional photonic crstal or 1-D function photonic crstal, it shown that with the increase of the periodicity, the transmittance gradually decreases, and the transmittance in the bandgap will be kept unchangeable when the periodicity reach critical point [4-5].

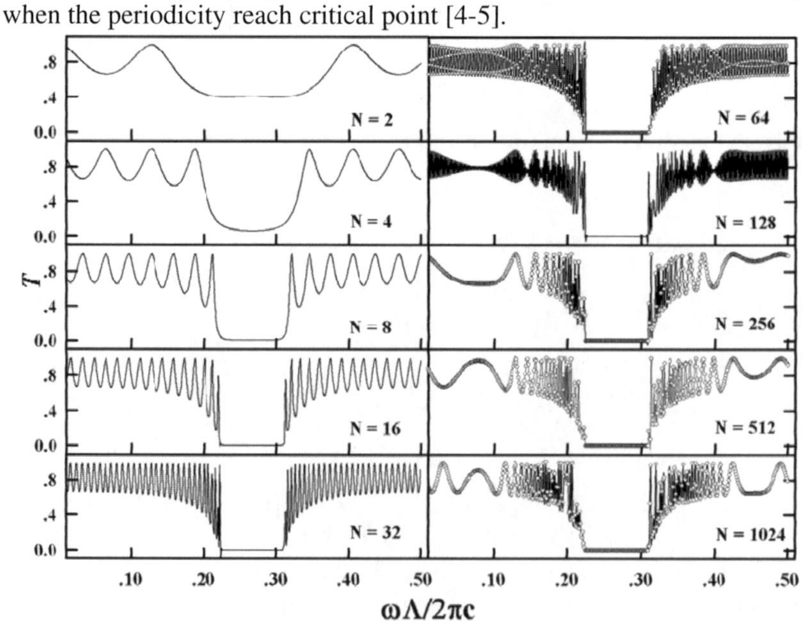

Figure 2. Transmission spectra of the 1-D conventional photonic crystal with different periodicity.

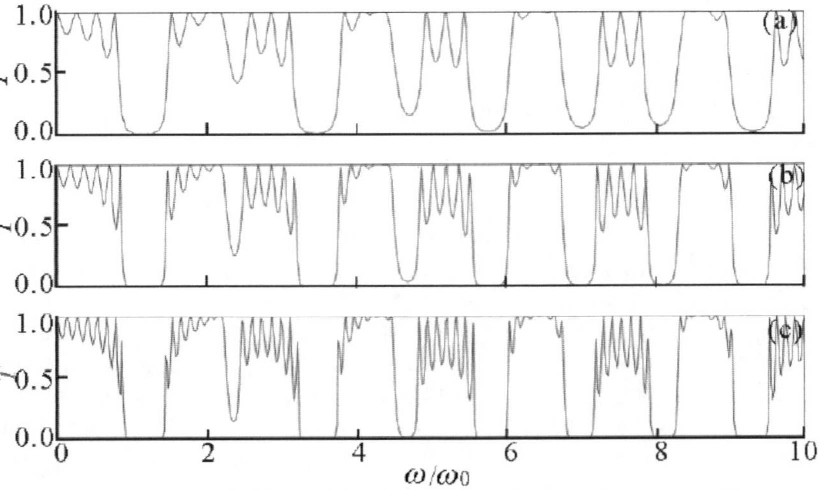

Figure 3. Transmission spectra of the 1-D function photonic crystal with different periodicity.

2.4. *Effect of incident angle on the band-gap properity*

In reference [4], when other parameters kept unchanged, and the incident angle were 0, $\frac{\pi}{18}$, $\frac{\pi}{6}$, $\frac{\pi}{3}$, and $\frac{4\pi}{9}$ for the 1-D conventional photonic crystal, the result shown that with the increase of the incidence angle, the bandgap width of the photonic crystal also increase as shown in figure 4, but the center postion and the transmittance of the bangap remains unchanged. Acoording to references [6-10], for the 1-D function photonic crystal, with the increase of the incidence angle, the bandgap width of the photonic crystal increase, this conclusion is in agreement with the 1-D conventional photonic crystal, but the transmittance of the bangap is getting lower, which is obviously different from the 1-D conventional photonic crystal.

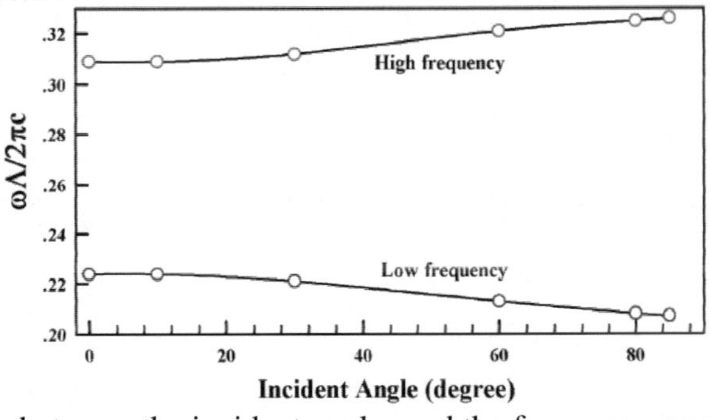

Figure 4. Relationship between the incident angles and the frequency range of the band gap [4].

3. Summary

Generally, fundamental stduy on band-gap properities of 1-D conventional photonic crystal and 1-D fuction photonic crystal1-D are contrasted based on part achievements of author project group and part referencs mainly from the X.Y. Wu project group according to expression of transfer matrix, dispersion relation, periodicity and incident angle. The conventional photonic crystal is a special case of the function photonic crystal, and the 1-D function photonic crystal has more, wider or narrower bandgap structures than the 1-D conventional photonic crystal. The results could have some guiding significance for the 1-D photonic crystal and specific photonic crystal device fabrication.

Acknowledgements
This work are supported by the Yunnan Applied Basic Research Project (No. 2017FD165), the Sixth Group Young and Middle-Aged Academic and Technical Leaders Project of Baoshan (No. bszqnxshjsdtr201705)

References

[1] Yablonoviteh E 1987 *Phys. Rev. Lett.* **58(20)** 2059-62.
[2] John S 1987 *Phys. Rev. Lett.* **58(20)** 2486-9.
[3] Cai Y G., Dong Y H 2017 *Appl. Mech. Mater.* **872** 3-7.
[4] Cai Y G, Dong Y H, Wang X M 2013 Crystal Micronanoelectronic Technol. **50(2)** 90-4.
[5] Cai Y G, Fang X and Dong Y H 2015 *Micronanoelectronic Technol.* **52(1)** 37-41/68.
[6] Wang G H, Wang Q C, Wu X Y et al. 2012 *Acta Phys. Sin.* **61(13)** 134208-11.
[7] Wang Q C, Wang Y and Wang G H 2012 *J. Jishou Univ. (Nat. Sci. Ed.)* **33(1)** 36-40.
[8] Wu X Y, Zhang S Q, Zhang B J et al. 2013 *Physica E Low Dimens. Syst. Nanostruct.* **53** 1-6.
[9] Ba N, Wang Q C, Liu X J et al. 2012 *J. Jilin Univ. (Nat. Sci. Ed.)* **50(1)** 34-7.
[10] Liu X J, Li N, Zhang S Q et al. 2012 *Acta Photonic Sinica,* **41(10)** 1193-9.

IOP Conf. Series: Materials Science and Engineering **473** (2019) 012025 doi:10.1088/1757-899X/473/1/012025

The Effect of Pier and Deck Connection on the Seismic Response of U-Turn Curved Bridge

Hansel Loshaless[1], Yuskar Lase[2], and Josia Irwan Rastandi[2]

[1]Student, Civil Engineering Department, Universitas Indonesia, Kampus UI Depok, Indonesia
[2]Lecturer, Civil Engineering Department, Universitas Indonesia, Kampus UI Depok, Indonesia

Abstract. The maximum seismic response of irregular U-turn curved bridge is significantly related to the pier and deck connection. An irregular U-turn curved bridge with radius of 45m, 8.5m total width and 154m total length was analyzed in this paper. This highly horizontally curved bridge was evaluated by performing nonlinear time history analysis on the representative bridge model with 3 types of connection between pier and deck: (i) rigid connection, (ii) hinge connection, and (iii) lead rubber bearing connection. Nonlinear hinge was modelled at the base and the top of the pier by using fiber hinge. Moreover, the effect of the number of connection between pier and deck was evaluated. The results indicate that bridges with one connection give better performance level than those with two connections; however, bridges with only one connection is more critical to the torsion issue.

1. Introduction

The long distance on the freeway is a major issue for drivers who need to make a U-turn. To solve this major issue, drivers need a U-turn curved bridge on the freeway, so they can make a U-turn. From a civil engineering point of view, this bridge presents many challenges in its planning process. Past studies investigated that the maximum response of each bridge structure component rose due to earthquake loading in different direction [1]. Past studies also showed that curved bridges are more susceptible to earthquakes than straight bridges, and the seismic response of bridges will be more dangerous as the curvature of the bridge increases [2]. Due to the nature of this U-turn curved bridge, the bridge should not be analysed like any other straight bridge as it is classified as an irregular bridge under the American Association of State Highway and Transportation Officials (AASHTO) rules [3].

2. Sample structures

The bridge is a ten-span, solid slab superstructure (depth = 950mm), supported on reinforced concrete single pier bents with the rectangular section (1400mm x 1400mm). This bridge has 8.5m total width and 154m total length. The bridge radius is 45m. The super elevation of the bridge is 10% and the slope of the road in longitudinal direction is 5%. The design speed of this U-turn curved bridge is 40 km/h. See figure 1 for Plan View. Figure 2 shows the section of bridge's superstructure. The location of this study is Jakarta, Indonesia, with soft soil classification, and the structural design follows AASHTO LFRD Bridge Design Specifications using the modification factor, R, as 3.

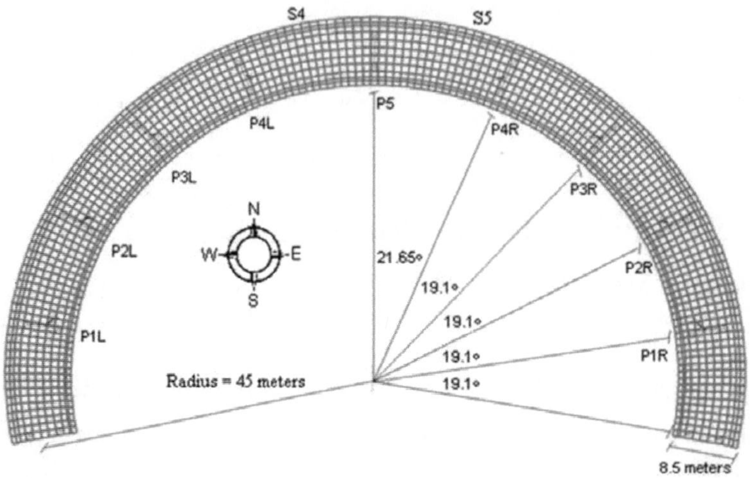

Figure 1. Plan View.

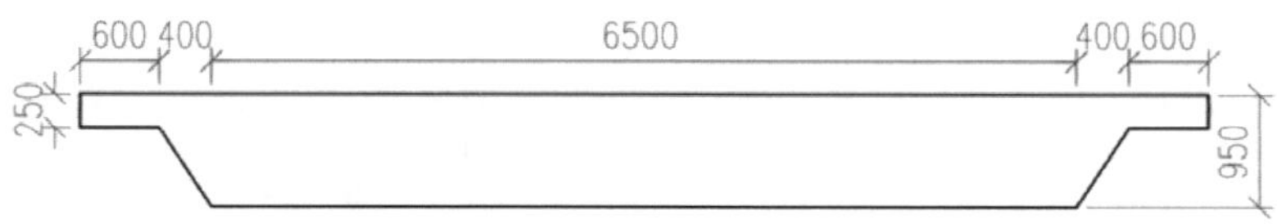

Figure 2. Section of bridge's superstructure.

The investigation and analysis include total six bridge structures, namely Rigid-1, Rigid-2, Hinge-1, Hinge-2, LRB-1, and LRB-2. All bridges have the same geometry as described before. At Rigid-1 and Rigid-2, rigid connections are used between the pier and the deck. At Hinge-1, one hinge connection is used between the pier and the deck. At Hinge-2, two hinge connections are used between the pier and the deck. At LRB-1 and LRB-2, lead rubber bearings (LRB) are used as seismic isolators and placed between the pier and the deck. The dimensions and properties of isolators used in LRB-1 and LRB-2 are shown in table 1. The connection between superstructure and the abutment in all analyzed bridges is lead rubber bearing. The abutments would not be included in earthquake resisting systems since the connection between superstructure and abutment is LRB.

Table 1. LRB isolators properties

Parameters	Abutment	LRB-1	LRB-2
Stiffness of elastomer, K_r (kN/mm)	4.52	4.52	2.36
Stiffness of lead core, K_p (kN/mm)	45.16	45.16	23.64
Yield strength, F_y (kN)	561.78	561.78	328.51
Effective Stiffness (kN/mm)	7.52	7.52	3.86
Effective Damping (%)	29.75	29.75	29.22

3. Bridge modelling

The demand of the bridges was analyzed using Acceleration Response Spectrum Analysis (ARSA) and the capacity is analyzed using nonlinear pushover analysis for ultimate displacement. The demand and capacity were then compared to ensure the safety of the bridge. Then, Nonlinear Time History Analysis (NLTHA) was performed to bridges that meet the demand and capacity criteria that were analyzed before. Three groups of time history for NLTHA were selected and matched to represent high seismicity of Jakarta. They were chi-chi (TCU120), landers (MEL), and sitka earthquake record (212V50). Jakarta is classified as a high seismic performance zones (seismic zone 4) in accordance to values of Sds (0.837 g) and Sd1 (0.573 g). Each time history record then shall be modified to be compatible with target response spectrum (7% probability of exceedance in 75 years) using the time-domain procedure. Then, each record was rotated in 3 directions (0°-90°, 45°-135°, 90°-0°) due to the structure configuration of the bridge. For each suite of ground motion, 100% of the input ground motion in each horizontal directions are given simultaneously to structure. Direct Hilber-Hughes-Taylor was used. NLTHA was performed after nonlinear dead loads analysis. Nominal material properties were used in ARSA, while expected material properties were used in NLTHA. Table 2 shows those properties used in the analysis.

The pier was modeled as a single frame element that starts from the top of the foundation and ends at the bottom of the connection. Rigid links were used to correctly model the structural behavior between the connection and the deck. For the purpose of NLTHA, a plastic hinge was modelled at the base and the top of the pier. The plastic hinge was assigned using fiber hinge. Mander's uniaxial nonlinear concrete model was used for modelling the stress-strain behavior of unconfined and confined concrete [4]. The hysteretic behavior for concrete nonlinear model used Takeda hysteresis model, while the reinforced steel nonlinear model used the kinematic hysteresis model. The length of the hinge at pier was calculated according to AASHTO Guide Specifications for LFRD Seismic Bridge Design and the location of the plastic hinge was assumed at mid-height of the hinge zone [5].

The deck structure was modelled with elastic shell element. The elastic element was selected because nonlinear behavior was not expected in deck structure during earthquake. The super elevation (10%) and the slope of the road in longitudinal direction (5%) was considered in structure modelling. So that the height of each pier varies following the elevation of the bridge deck. The bridge model with 2 connections can be seen in figure 3 below.

Table 2. Properties of materials.

Conrete Properties			
Unconfined compressive strength	f'_{co}	35	Mpa
Expected strength of f'_{co}	f'_{ce}	45.5	Mpa
Reinforcing Steel Properties			
Specified yield strength	f_y	400	Mpa
Specified tensile strength	f_u	550	Mpa
Expected yield strength	f_{ye}	465	Mpa
Expected tensile strength	f_{ue}	655	Mpa

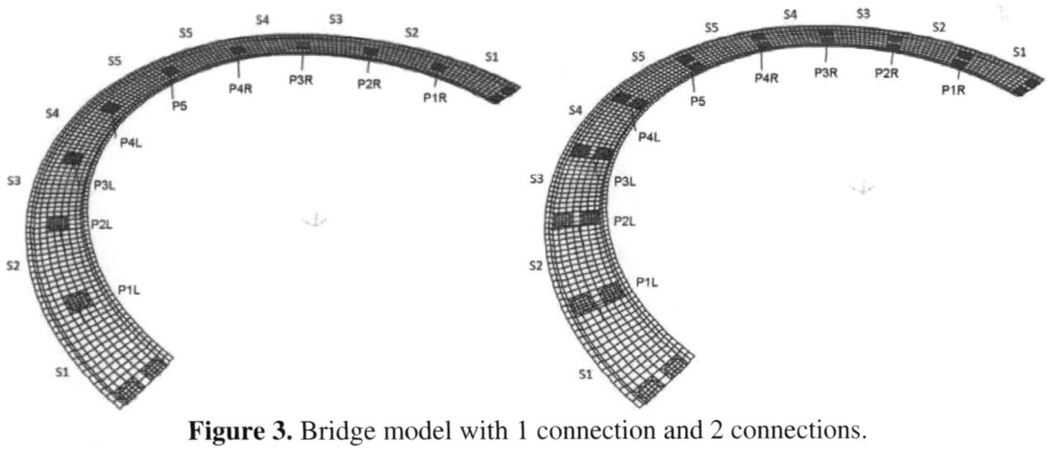

Figure 3. Bridge model with 1 connection and 2 connections.

4. Results and discussion

4.1. Modal analysis

Modal analysis was conducted and periods and shapes of vibration were obtained for all three types of connection. Values of the fundamental mode of vibration in transverse and longitudinal direction are presented in table 3. For all of analyzed bridge models, the first fundamental mode was in E-W direction. This result was expected because the bridge was symmetrical in N-S direction, but asymmetrical in E W direction. The influence of the number of connections between column and deck was not very important regarding to their periods.

Table 3. Periods of the first fundamental mode (seconds).

	1 Connection			2 Connections		
First Fundamental Mode	Rigid	Hinge	LRB	Rigid	Hinge	LRB
E-W	0.39	0.37	0.71	0.38	0.41	0.69
N-S	0.36	0.35	0.60	0.35	0.38	0.61
Torsional	0.28	0.30	0.51	0.27	0.31	0.52

4.2. Top pier displacement

A brief summary of the maximum seismic response of the pier of all bridge models is shown in figures 4 and 5. It can be seen that the displacement curvature curves in figures 4 and 5 are almost similar to the shape curvature of the bridge. The reason is that the pier on the edge (P1L and P1R) is stiffer than the pier in the middle of the bridge (P5) due to its height. It also can be seen that the maximum displacement occurs in the bridge model with two connections. However, the top pier displacement of Rigid-1 and Rigid-2 model are almost the same. From figure 4 and figure 5, it can be seen that the lowest displacement at the top of pier P5 occurs at the bridge with LRB and rigid connection. The reason is that the bridge with rigid connection tend to have more redundancy than the one with hinge or LRB connection.

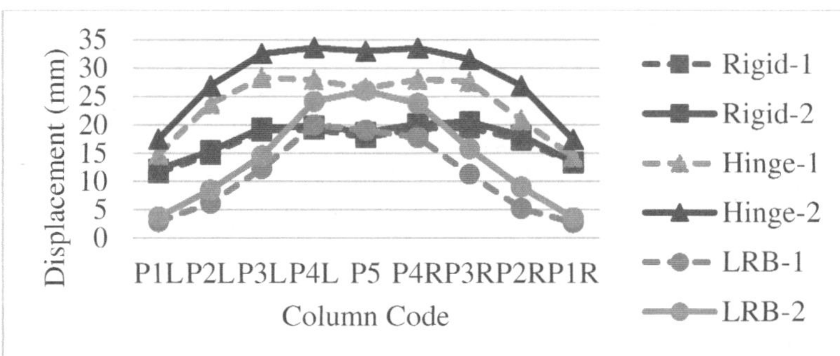

Figure 4. Top pier displacement (E-W direction).

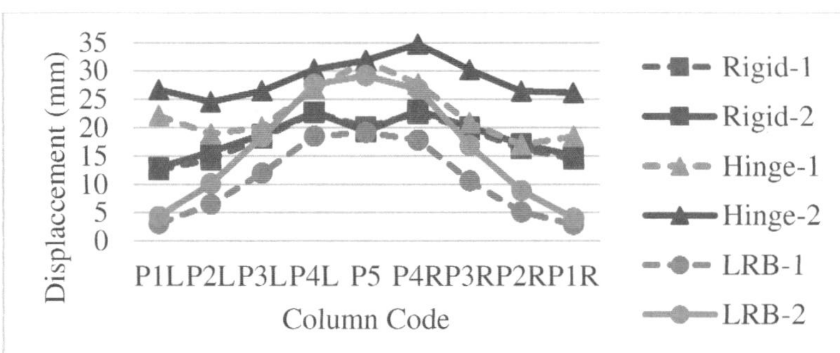

Figure 5. Top pier displacement (N-S direction).

4.3. Pier base shear

From figure 6 and figure 7, it can be seen that the number of connections of U-turn curved bridge is not important regarding to their seismic response, except the one with the hinge model in figure 6. It can be seen that the Hinge-1 and Hinge-2 curves in figures 6 and 7 are almost similar to the shape of the curved bridge. Therefore, the seismic response of the bridge with hinge connection depends on the pier stiffness, while the seismic response of the bridge with LRB connection is governed by LRB stiffness. The seismic response of the bridge with rigid connection in E-W direction does not depend on the pier stiffness. Its base shear curve in figure 6 is almost similar to the response of the straight bridge, while its base shear curve in N-S direction connection similar to the shape of the curved bridge.

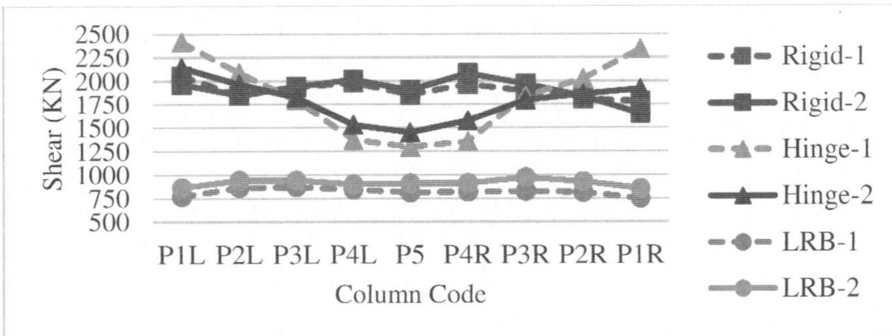

Figure 6. Base shear in each pier (E-W direction).

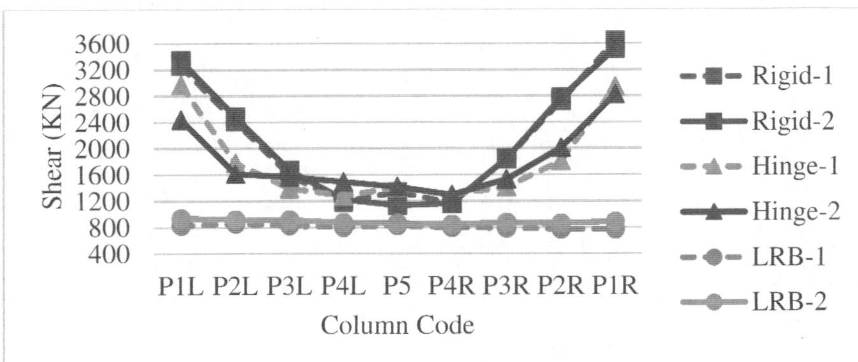

Figure 7. Base shear in each pier (N-S direction).

4.4. Deck longitudinal rotation

Figure 8 shows the longitudinal rotation at deck S4 (shown in figure 1). The "B" in figure 8 Shows the width of the bridge deck which is 8500 mm. It can be seen that the longitudinal rotation of bridge deck with one connection is two times larger than the bridge with two connections, except the bridge with one rigid connection. This observation may alert design engineers that when U-turn curved bridge is designed with only one connection, rigid connection should be used to connect the pier and the deck.

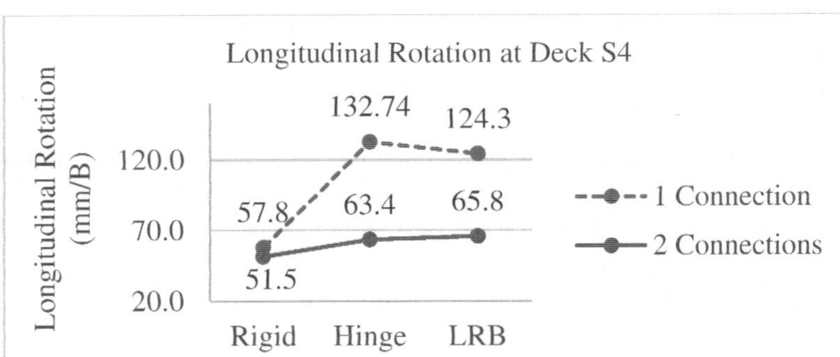

Figure 8. Longitudinal Rotation at Deck S4.

4.5. Performance level
Table 4 shows the maximum rotation at plastic hinge on each pier. The performance level target on preliminary design is life safety according to AASHTO LFRD Seismic Bridge Design. It can be seen that the maximum rotation occurs at Hinge-2 and the bridge with two connections tend to have bigger plastic hinge rotation than the bridge with one connection. The performance level of each pier is calculated based on ASCE 41-13 [6]. The result shows that although the bridge was designed for life safety performance, the actual performance of all the bridge was below immediate occupancy except for Hinge-2 bridge which is immediate occupancy to life safety.

Table 4. Maximum rotation at plastic hinge.

Rotation (x10^{-3} rad)	P1L	P2L	P3L	P4L	P5	P4R	P3R	P2R	P1R	Status
Rigid-1	-2.46	-2.26	-2.29	-2.14	-1.83	-1.97	-2.69	-2.55	3.12	<IO
Rigid-2	-2.49	-2.29	-2.29	-2.09	-1.89	-2.07	-2.71	-2.58	3.08	<IO
Hinge-1	-4.42	-3.61	-3.34	-2.94	-3.34	-2.92	-2.86	2.93	3.53	<IO
Hinge-2	-5.56	-4.92	-4.33	3.43	-3.52	3.93	4.09	3.85	5.32	IO-LS
LRB-1	0.26	-0.64	-1.39	-2.20	2.03	2.00	1.25	0.44	0.26	<IO
LRB-2	0.35	-1.00	-1.89	-3.02	3.29	3.33	2.33	1.18	-0.31	<IO

Table 5. Time history record associated to table 4.

	P1L	P2L	P3L	P4L	P5	P4R	P3R	P2R	P1R
Rigid-1	LD 90	CC 90	CC 45	LD 45	LD 0	LD 90	LD 90	SK 0	CC 90
Rigid-2	LD 90	CC 0	CC 45	LD 45	LD 0	SK 0	LD 90	SK 45	CC 90
Hinge-1	CC 45	CC 90	CC 90	CC 90	SK 0	LD 0	LD 90	LD 90	CC 90
Hinge-2	CC 45	LD 90	LD 90	CC 90	SK 90	CC 45	CC 0	CC 45	CC 45
LRB-1	LD 45	LD 45	LD 0	LD 0	LD 45	LD 90	LD 90	CC 45	LD 45
LRB-2	LD 45	LD 45	LD 90	LD 0	LD 45	LD 90	LD 90	LD 90	LD 90

Table 5 shows the time history record (chi-chi/CC, landers/LD, or sitka/SK) and the directions (0°-90°, 45°-135°, 90°-0°) associated with table 4. It can be seen that the maximum response of each bridge structure component arises due to earthquake loading in different direction. This observation may alert design engineers that when U-turn curved bridge was designed, the time history record or earthquake loading should be rotated to produce maximum response of each pier.

5. Conclusion
The comparative analysis of the seismic response of six curved bridge models by applying the nonlinear time history analysis was carried out with the aim of studying the influence of connection type variations. The conclusions are:

1. Bridges with the hinge connection have the lowest performance level and have the highest pier displacement.
2. Bridge with only one connection give better response and performance than bridge with two connections. When U-turn curved bridge was designed with only one connection, the rigid connection should be used to prevent the higher longitudinal rotation in bridge deck.
3. Even though bridges with rigid connection have the highest pier base shear than bridges with hinge connection, bridges with rigid connection have better performance than bridges with hinge connection.

4. The result showed that although the bridge was designed for life safety performance, the actual performance of almost all bridges were below immediate occupancy.

Acknowledgment
This works is supported by Hibah PITTA 2018 funded by DRPM Universitas Indonesia No. 7185/UN2.R3.1/ PPM.00.01/2018.

References
[1] Ni Y, Chen J, Teng H and Jiang H 2015 *J. Traffic Transp. Eng. (English Ed.)* **2** 233-41.
[2] Soleimani F, Yang C and DesRoches R 2017 *Structures Congress* 469-80.
[3] *AASHTO LFRD Bridge Design Specifications* 2017 American Association of State Highway and Transportation Officials.
[4] Mander J B, Priestley M J N and Park R 1988 *J. Struct. Eng.* **114(8)** 1804-26.
[5] Aviram A, Mackie K R and Stojadinovic B 2008 *Guidelines for Nonlinear Analysis of Bridge Structures in California* PEER Center.
[6] ASCE 2014 *Seismic Evaluation and Retrofit of Existing Buildings (ASCE 41-13)* American Soc. Civil Eng.

Some Reflections On Styrene Suspension Polymerization Experiment Teaching

Dongbo Guan[1], Jinliang Li[1], Guoen Sun[1], Xiaojie Zhai[1*]

[1]College of Materials Science and Engineering, Key Laboratory of Automotive Materials, Ministry of Education, Jilin University, Changchun 130025, China

Abstract. The results of suspension polymerization of styrene were studied by changing the dosage of initiator, stirring speed, size of stirring paddle, content of dispersant and reaction temperature. The results showed that when using 500 ml four flasks, styrene 40 ml and polyvinyl alcohol solution (3 g polyvinyl alcohol dissolved in 1000 ml deionized water) 200 ml, initiator 0.5 g, stirring speed 300-450 r/min, temperature range 86-95°C, using the stepped heating method and using the impeller with a spreading blade diameter of 43 mm can ensure the success of the experiment, and the yield is relatively high.

1. Introduction

Suspension polymerization is one of the important methods to prepare synthetic resin. It is a suspension polymerization in which the initiator monomer is dispersed in a medium (usually water) which is not soluble with the monomer under the action of strong mechanical stirring force and dispersant. Polystyrene is one of the five general plastics in the world today and has a wide range of USES. Suspension polymerization is the main method of synthesis of polystyrene. Zhang Jianli et al. [1] studied the factors affecting particle size in styrene suspension polymerization. Wang qiao et al. [2] studied the effects of initiator dosage, stirring speed, temperature, dispersant dosage and ratio of water to monomer on suspension polymerization of styrene. Tang Xiaohong et al. [3] studied the effects of initiator dosage, suspension agent dosage, stirring speed and reaction temperature on the suspension polymerization of styrene. Ge Jianfang et al. [4] have studied the addition of hexagonal boron nitride to suspension polymerization system to change the size uniformity of styrene microspheres. Meanwhile, through the combination of classroom teaching and extracurricular experiments, students can cultivate individualized and innovative thinking and improve the teaching quality and effect. Zhao Dongmei et al. [5] has seven different uses. The color polystyrene microspheres were prepared with styrene as monomer, benzoyl peroxide as catalyst, polyvinyl alcohol as dispersant and divinylbenzene as crosslinking agent. Arouse students' interest. Li Qiuxia et al. [6] used polyvinyl alcohol as suspension dispersant, benzoyl peroxide as initiator, tertiary dodecyl mercaptan as chain transfer agent to prepare white granular polystyrene with controllable molecular weight by suspension polymerization. The author found that the suspension polymerization experiment was easily influenced by the reaction conditions in the process of the teaching of styrene suspension polymerization experiment.

In this paper, the success rate of styrene suspension polymerization was discussed by changing the dosage of initiator, stirring speed, the size of impeller blade, temperature range and dispersant content.

2. Experimental

2.1. Reagent and instrument

Styrene (chemical purity), guangdong shantou xilong chemical company; Sodium hydroxide (analytical purity); Pva (chemical pure), jinan hengrui chemical co., LTD. Benzoyl peroxide (BPO, analytically pure).

1 standard 4-mouth bottle (500ml);1 condensate tube (300mm); One thermocouple; One separating funnel (125ml);Beaker (100ml) two;1 electric heating package; One set of electric mixer.

2.2. Experimental processes

(1) In the beaker, 40ml styrene washed by sodium hydroxide solution and 0.5g peroxybenzoyl, the initiator, were added and stirred by hand to dissolve.

(2) Add 0.6g polyvinyl alcohol and 200ml deionized water in another beaker, heat it to boiling and dissolve it into a uniform transparent solution for later use (note: it has been dissolved in advance).

(3) Add 200ml polyvinyl alcohol solution to 500ml four-necked flask with thermometer, mixer and reflux condenser. Stir and heat up. After being up to 80 ℃, molten have joined the initiator of the styrene monomer. Careful adjustment of stirring speed is generally controlled at 300-450r/min, so that the monomer is dispersed into droplets of appropriate size.

(4) The droplet size after adjust well, up to 90 ℃, heat preservation after 3 h extracted several particles, observe whether the cooled into bead solids. If so, will raise the temperature to 95 ℃, insulation 1 h, the reaction to an end.

(5) Immediately pour the reactant into the beaker, wash with room temperature deionized water for three times, and then filter. Bead polymers in petri dishes, in 50 ℃ drum wind drying oven drying to constant weight, weighing, calculating the yield.

3. Results and discussion

3.1. Effect of initiator content on polymerization reaction

Experimental conditions: after alkali washing styrene 40ml, polyvinyl alcohol solution 200ml, stirring speed 400r/min, reaction temperature of small stirrer (blade diameter: width when blade is fully expanded, 43mm) was described in the heating step of the experimental steps. The influence of initiator BPO dosage on the polymerization reaction was examined, and the results were shown in table 1.

It can be seen from table 1 that in the case of using a small stirring propeller (blade diameter: 43mm), the yield gradually increased with the increase of initiator. When the content of initiator was 0.6g, the polystyrene balls were larger and distributed uniformly, the experiment was successful. This is because when the content of initiator increases, the concentration of free radical is higher, the reaction speed of chain growth is faster, and the resulting polystyrene balls are larger. On the other hand, as the concentration of the initiator increases, the number of active species in the reaction system increases, the probability of chain growth increases, and the yield increases. The distribution of the diameter of polystyrene is uniform. Taking into account the size of the diameter of polystyrene and the yield of polystyrene, the content of BPO is set as 0.5g in the experimental teaching.

Table 1. Influence of amount of initiator on polymerization reaction.

BPO dosage (g)	Particle size and distribution	Experiment is successful or not	Productivity (%)
0.4	Smaller, nonuniform	Yes	96.2
0.5	Smaller, uniform	Yes	96.8
0.6	Larger, uniform	Yes	97.4
0.7	Larger, nonuniform	Yes	97.6

3.2. Effect of stirring speed on polymerization reaction

Experimental conditions: after alkali washing styrene (40ml), polyvinyl alcohol (200ml) and BPO (0.5g), the reaction temperature of the small stirrer (blade diameter: 43mm) is described in the heating step of the experimental procedure. The effect of stirring speed on the polymerization reaction is examined, and the results are shown in table 2.

As can be seen from table 2, with the increase of mixing rate, the yield gradually increased, and all the experiments in the four groups were successful. However, when the mixing speed reached 400r/min, the diameter of the polystyrene ball decreased. This is because when the mixing speed is increased, the shear force is enhanced. When the droplet formed in the early stage of mixing is too large, small droplets are separated from the initial droplet, making the dispersed droplet smaller. The size and yield of the polystyrene ball diameter are taken into account comprehensively, and the mixing speed is set as 400r/min in experimental teaching.

Table 2. Influence of mixing speed on polymerization reaction.

Stirring rate(r/min)	Particle size and distribution	Experiment is successful or not	Productivity (%)
300	Larger, nonuniform	Yes	97.2
350	Larger, uniform	Yes	97.3
400	Smaller, uniform	Yes	98.5
450	Smaller, nonuniform	Yes	98.7

3.3. The influence of the size of stirrer on the polymerization reaction

Experimental conditions: 500ml four-necked flask, 40ml of styrene after alkaline washing, 200ml of polyvinyl alcohol solution, and the dosage of BPO were 0.5g, stirring speed was 400r/min. The reaction temperature was described in the heating step of the experimental steps.

It can be seen from table 3 that when the blade diameter of the agitator was 55mm and 62mm, the experimental results were all unsuccessful. That is to say, instead of forming spherical polystyrene, an agglomerate formed around the impeller, resulting in the scrapping of four-necked flask and the experiment failed. Owing to 500ml of four-necked flask of solvent volume, the use of lamina of 43mm diameter of impeller, under the regulation of speed, formed a relatively stable state of turbulence, the probability of collision between dispersed droplets is reduced, but when using a leaf of the diameter of the impeller, under the regulation of rotation speed, the form is relatively severe turbulence state, dispersed droplets collision in such a state of easy adhesive into pieces each other cause the failure of the experiment.

Table 3. Influence of the size of impeller on the polymerization reaction.

Blade diameter of stirrer(mm)	Experiment is successful or not
43	Yes
55	No
62	No

3.4. Effect of dispersant content on polymerization

Experimental conditions: after alkali washing, styrene 40ml, BPO 0.5g, small impeller (unfolding blade diameter: 43mm), stirring speed 400 r/min, reaction temperature as described in the step of heating up, the effect of dispersant content on the polymerization reaction was investigated, as shown in table 4.

From table 4, it can be seen that with the increase of dispersant content, the yield of polymerization increases at first and then decreases, but the diameter of polystyrene spheres decreases. This is because when the dispersant content is low, the whole dispersing system is unstable, and droplets collide with each other easily to form larger droplets, so the droplets appear at this time. When the content of dispersant decreases, the concentration of dispersant around the polymer droplets is higher. At this time, the existence of a certain concentration of dispersant will hinder the merging of small droplets to form large droplets. In this case, the probability of forming small droplets is higher, so the final polystyrene sphere has smaller diameter. However, the concentration of dispersant should not be too large, otherwise the growth of polymers will be affected.

Table 4. Effect of dispersant content on polymerization.

Dispersant content (ml)	Particle size and distribution	Experiment is successful or not	Productivity (%)
180	Larger, nonuniform	Yes	97.6
200	Smaller, uniform	Yes	98.2
220	Smaller, uniform	Yes	98.6
240	Smaller, uniform	Yes	97.5

3.5. Effect of reaction temperature on polymerization

Experimental conditions: after alkali washing, styrene 40 ml, BPO 0.5 g, small impeller (unfolding blade diameter: 43 mm), stirring speed 400 r/min, the effect of dispersant content on the polymerization reaction was investigated, polyvinyl alcohol solution 200 ml, the results are shown in table 5.

As can be seen from Table 5, when the reaction temperature is lower, the reaction speed is slower. At this time, there is no polymer in the system. With the increase of the system temperature, the reaction speed accelerates, and the spherical polymer begins to form. In this experiment, the spherical polystyrene microspheres with uniform spherical diameter are obtained by stepped heating method.

Table 5. Effect of reaction temperature on polymerization.

Reaction temperature (°C)	Particle size and distribution	Experiment is successful or not	Productivity (%)
68-74	No particle formation	No	-
74-80	Larger, nonuniform	Yes	97.2
80-86	Larger, nonuniform	Yes	97.9
86-95	Smaller, uniform	Yes	98.8

3. Conclusions

The effects of initiator content, stirring speed, agitator size, dispersant content and reaction temperature on suspension polymerization of styrene were studied. It was found that for the specific volume of four-necked flask, using a certain size of the impeller play a key role in the success or failure of the experiment. The content of initiator, stirring speed and dispersant had little effect on the success or failure of the experiment, but the temperature ranges and the way of heating had great influence on the suspension polymerization of styrene. Therefore, in the styrene suspension polymerization of macromolecular chemistry experiments, it is recommended to use lamina of 43mm diameter of impeller when using 500 ml of four-necked flask, the temperature range is selected at 86-95 °C, and the heating method is stepped by heating.

References

[1] Zhang J L, Chi C L 2008 *J. Henan Inst. Eng. Nat. Sci. Ed.* **1** 57-8.
[2] Wang Q, Zhang Y Q and Liu Y 2010 *Chem. Biol. Eng.* **4** 71-3.
[3] Tang X H, Wang J and Zheng H G et al 2012 *Chem. Biol. Eng.* **2** 44-6.
[4] Ge J F, Ge X and Pu X et al 2018 *Guangdong Chem.* **16** 212-213+211.
[5] Zhao D M, Sun L G, Du Y H et al 2014 *Lab. Sci.* **3** 57-60.
[6] Li Q X, Yan X H, Liu X F et al 2014 *Guangdong Chem.* **18** 64-5.

Comparative Analysis of Fixed base and Isolated Structure in "L" Shaped Plan with Time History Analysis based on ASCE 7-16

Rastandi Josia Irwan[1], Rahim Sjahril A.[1], Lase Yuskar[1], and Yan Hendro[2,*]

[1]Senior lecturer, Civil Engineering Department, Universitas Indonesia, Kampus UI Depok, Indonesia
[2]Graduate student, Civil Engineering Department, Universitas Indonesia, Kampus UI Depok, Indonesia

Abstract. An "L" shaped plan is often being chosen for residential, office, or hospital, whereas this configuration plan does not meet structural torsion requirement. This selection is done by the consideration of limited area and architectural needs, e.g. hospital needs for ventilation. Since the development of technology is very rapid this era, an innovation emerges beyond conventional solution, in terms of base isolation. In this paper, the selected research object is known as lead rubber bearing (LRB) with damping ratio 27%. To fulfill the research of L-shaped, the variation of length of the wings are proposed. Six models are functioned as office buildings in 6-story tall; three fixed base models are designed with dual system and another three isolated models are design using linear distribution lateral forces according to ASCE 7-16 code. Three-dimensional nonlinear time history analysis for isolated models is performed and will involve seven pairs of ground motions, which are matched to MCER target spectra of Jakarta in soft soil condition. In the end, the non-linear dynamic main responses of isolated structure may provide better and optimal results. In addition, estimated cost for design phase of pre-construction can be done by the assessment of rebar ratio and equivalent thickness of concrete, known from the results of this study.

1. Introduction

Indonesia is one of countries with a larger population growth, majority based in the capital city of Jakarta. One the major problems that comes up in Jakarta is the gap between land demand and the growth of local population or incoming migrants, causing land to be very expensive and the availability of symmetrical land to be scarce. Apart from these aspects, architectural design for natural ventilation in some building is also one of crucial reason of having L-shaped plan. L-shaped plan is one of the typical asymmetric floor plans surround besides T, H and + configuration. By several studies that have been conducted, L-shaped has strong torsion responses due to inconsistency between the center of mass and stiffness [1]. Therefore, it should be avoided since it does not meet the dynamic design principle of a structure, which states torsion response was not being expected to be occurred in the fundamental mode when the building is given any earthquake excitation energy [2]. Another problem caused by this shape is variations of rigidity, resulting in a local stress concentration at the "notch" of the reentrant corner. Both problems, the stress concentration and torsion response are interrelated. To sustain of using the L-shaped plan in design, several solutions are provided; separating buildings into each other, stiffen the ends, and using collector beams or walls [3]. By using these conventional solutions, material costs and improvements at the notch will certainly need to be considered.

As rapid technological development and science in the field of seismic engineering, an innovation was emerged to be solution of seismic problems occurred in fixed base structure, called base isolation. Several studies related to base isolation was mostly conducted; one of them is using FPS (friction pendulum system) as its research object under near-fault excitations [4]. By assigning an appropriate variation of rigidity of isolation system to produce a small eccentricity is a good strategy in rehabilitate low-rise asymmetric structures [5]. Besides reducing torsion, lateral forces occurred in superstructure could be minimized by using base isolation [6]. Considering for these positive impacts, this paper is likely to be a recommendation for choosing fixed base or isolated structure in the real case, by giving the global response using time history analysis and the quantity of rebar ratio and equivalent thickness of concrete. The results of time history analysis will only be done on isolated model due to the spend computation time on fixed base models.

2. Base isolation – lead rubber bearing

Base isolation is one of the most important devices in the last decade for earthquake engineering, which is used for decoupling the superstructure from accelerated foundation. To minimize damage to buildings, the superstructure is needed to be design stiffer enough to provide rigid body motion. There are two main principle performances of base isolation; to extend the natural period of the whole structure and provide higher damping through its lead component. Using base isolation system will not generate any amplification of shear forces on each floor above, resulting a significant reduction in floor accelerations and inter-story drifts compared to fixed base structure, as shown in figure 1 and figure 2. There are several types of base isolation; one of them is lead rubber bearing (LRB). Unlike the others, LRB has the ability of attenuation of large scale earthquake energy, because it equipped with lead material in the center inside.

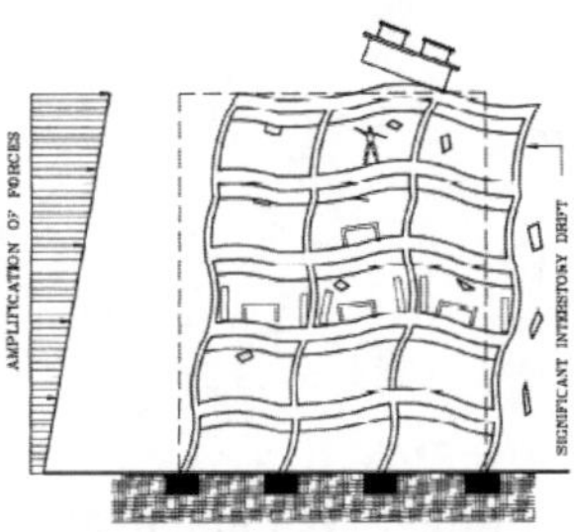

Figure 1. Fixed base (without base isolation).

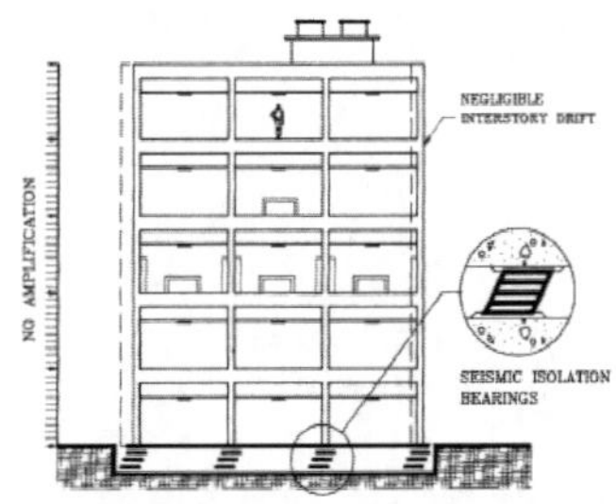

Figure 2. Base-isolated structure (using base isolation).

The characteristic of LRB was in nonlinear condition when earthquake excitation was applied on it. Therefore, the force-displacement behaviour of the LRB is modelled as bilinear model or hysteresis curve. To determine the lateral stiffness and damping ratio, base isolation could be tested dynamically to plot hysteresis curve. The isolator parameters that describe the bilinear law are initial elastic stiffness K_u, the post-yield stiffness K_d, characteristic strength Q_d, and force yield F_y. These all parameters are acquired from manufacturer of isolator, matching the criteria of the hysteresis loop of the analytical model and testing experimental [7].

3. Description of study model

This all study models are office buildings using 6-story tall height, with typical floor height of 3.3 meters and 5 meters for the ground floor height. All the models are made asymmetrically L-shaped with variation of the wings length ratio (shown in table 1). The buildings fully used reinforced concrete with concrete quality fc' 30 MPa (for beam, column, shear wall, and slab); for rebar "U40" is used, with specific quality as stated; fy= 400 MPa, fu= 570 MPa, fye = 468 MPa, and fue = 655 MPa. To make it into the real comparison, the L-shaped plan and code provisions are becoming the main reference for designing all models. The configuration and types of beams and columns in isolated models are made slightly different from the fixed base models, following the dimensional requirement.

The most types of beams using in both models are B36 (300x600 mm) as primary beams and B5A25 (250x500 mm) as secondary beams, but there is also additional collector beams B58 (500x800 mm) only in fixed models around the "notch" area. For columns, in isolated models are using C66 (600x600 mm) whereas in fixed base models are using bigger dimension; C88 (800x800 mm) and C75 (750x750 mm). To modelling the damping on the superstructure of isolated models, the stiffness-proportional damping is being applied, as recommended by [8].

Table 1. List of model variations.

Model	Ratio of bays (L/B)	Note
L1-FB	9:5	Fixed base
L1-BI		Base-isolated
L2-FB	9:7	Fixed base
L2-BI		Base-isolated
L3-FB	9:9	Fixed base
L3-BI		Base-isolated

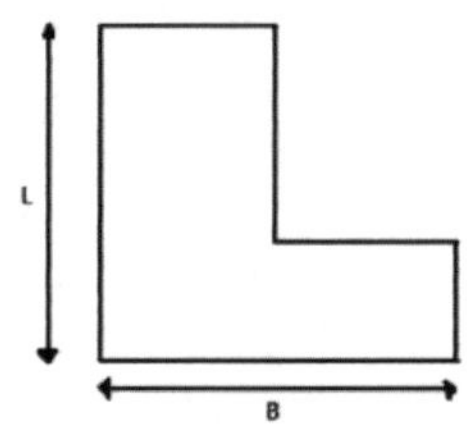

Fixed base models are then modified by inserting link elements on the base of the structure to generate isolated models (details can be seen in figure 3). To avoid torsion effect, the specifications in the layout of base isolation used in each column are designed to proportional to the gravity load resisted in each supports. In order to determine nonlinear parameters of isolators, an approach for the global and each of effective stiffness could be calculated by using equation (1) and equation (2). The specification of each base isolation based on the previous equations can be improved by the empirical formula and gave values as shown in table 2.

$$(K_{eff})_{total} = \frac{4\pi^2 W_{eff}}{T_M^2 g} \quad (1)$$

$$K_{effective-i} = \frac{W_i}{\sum W_i} * (K_{eff})_{total} \quad (2)$$

In which,

$(K_{eff})_{total}$ = total effective stiffness of the isolation system at the maximum displacement [kN/m]

$K_{effective-i}$ = effective stiffness of i-isolation system at the maximum displacement [kN/m]

W_{eff} = effective seismic weight [kN]

T_M = effective period of seismically isolated structure at the maximum displacement [s]

g = acceleration caused by gravity [m/s^2]

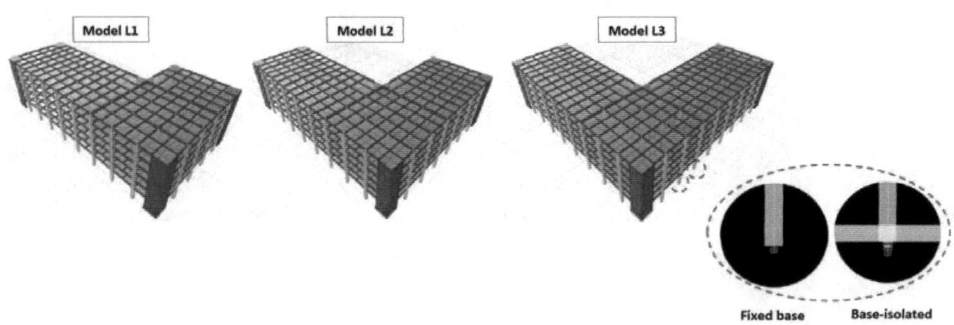

Figure 3. 3-Dimensional model on each variation and the difference between fixed base and base-isolated.

Table 2. The specifications of each base isolation are used in each L plans.

Model	L1-BI					L2-BI					L3-BI				
Parameter	Tipe isolator					Tipe isolator					Tipe isolator				
	LRB 1	LRB 2	LRB 3	LRB 4	LRB 5	LRB 1	LRB 2	LRB 3	LRB 4	LRB 5	LRB 1	LRB 2	LRB 3	LRB 4	LRB 5
Initial stiffness (kN/m)	2881	2842	5265	4802	9166	2907	2894	5298	4848	9125	3011	2973	5385	4909	9135
Stiffness post-yield (kN/m)	288	284	526	480	917	291	289	530	485	912	301	297	538	491	913
Characteristic strength, Qd (kN)	77	76	141	129	245	80	77	142	130	244	81	80	144	131	244
Force yield (kN)	86	84	157	143	273	88	86	158	144	271	90	88	160	146	272
Effective stiffness (kN/m)	492	486	900	821	1567	501	495	906	829	1560	515	508	920	839	1561
Effective damping ratio (%)	27	27	27	27	27	27	27	27	27	27	27	27	27	27	27

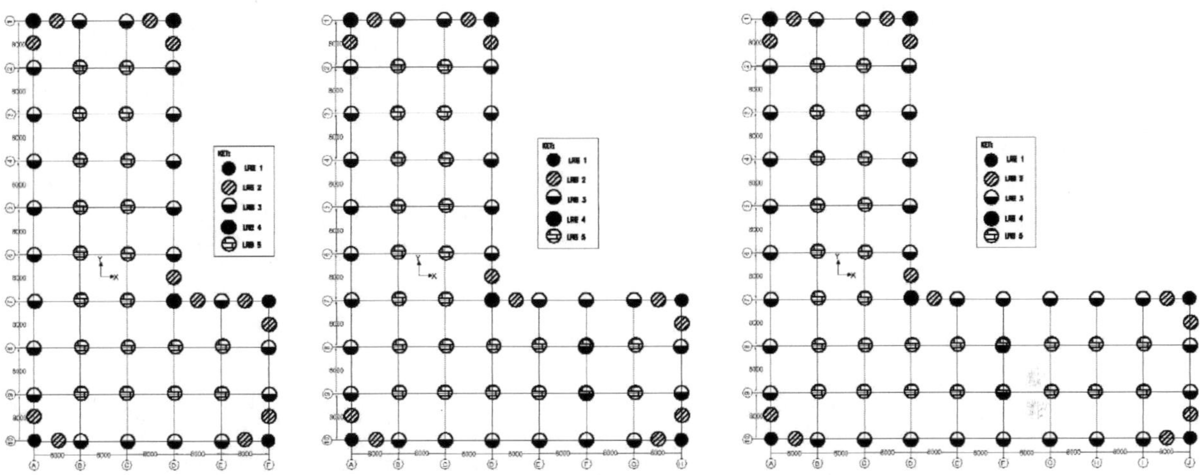

Figure 4. Layout of base isolation on L1, L2, and L3 models.

4. Loading assumption for design

In general, there are two types of loading to be included: gravity load and lateral load (limited to earthquake load). Gravitational loads are consisted of self-weight load, superimposed dead load, and live load. Superimposed dead load will be assigned 250 kg/m^2 on typical floor and 600 kg/m^2 on the roof. Live load will fulfill the SNI 1727:2013 (equivalent as [9]) code provision used for office: 240 kg/m^2 on typical floor (not including ground floor), 500 kg/m^2 on the ground floor, and 100 kg/m^2 on the rooftop. Meanwhile, earthquake load is using DBE spectrum response of Jakarta type SE site (soft soil) for the design of structural elements and using MCER spectrum of Jakarta for the time history analysis to determine the real responses. The comparison of both spectra are shown in figure 5.

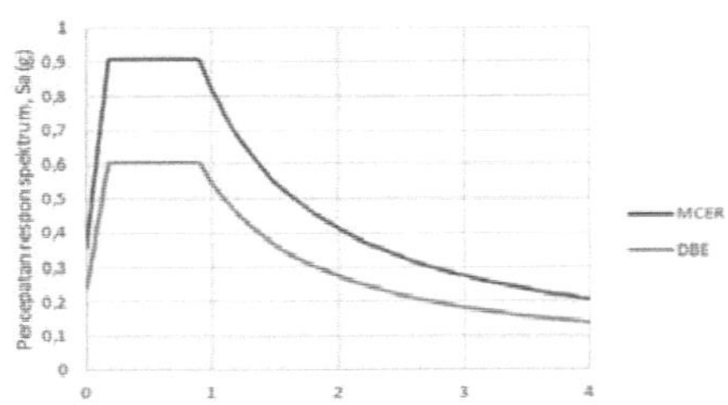

Figure 5. The comparison of DBE and MCER spectrum.

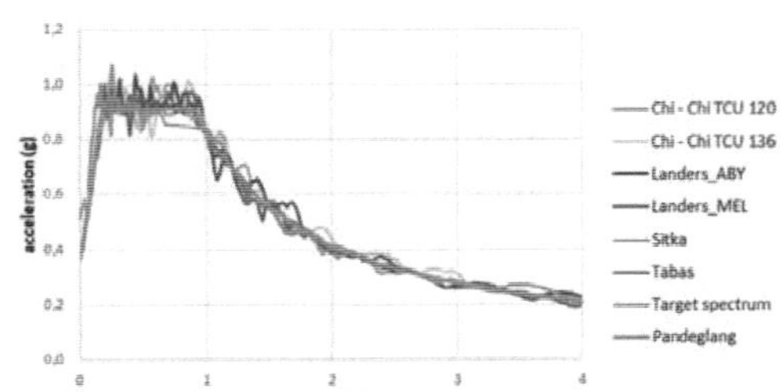

Figure 6. Seven selected ground motions are matching to target spectrum MCER.

5. Nonlinear time history analysis (NLTHA)

In base-isolated models, nonlinear modelling is only applied to the LRB isolator element (for U2 and U3 direction). Since the inner forces generated in superstructure are insignificant, all the structural elements are not necessary for nonlinear modelling. According to [9], nonlinear time history analysis (NLTHA) will involve not less than seven pairs of horizontal scaling ground motion records, either by scaling or spectral matching to the target spectrum MCER Jakarta of soft soil. The scaling process will not change the frequency content from the original earthquake but only changes in the amplitude in each period with different scale factor, the scaling result is shown in figure 6.

6. Results

6.1. Base shear

To acquire the result of response spectrum analysis (RSA) to be compared, this analysis will consider 100% in critical direction and 30% in the perpendicular, to include any additional torsion. While the results of time history analysis are the average responses of seven selected ground motion records. In figure 7, the base shear obtained by nonlinear time history analysis (NLTHA) is consistent to be higher than RSA method in inelastic response of isolated modes by the concept of equal displacement rule. The base shear of NLTHA in isolated models was not clearly define a significant reduction, since it is slightly larger than fixed base models using modification factor (R) of 7. The common value of R for designing dual system structure is 7, however, when nonlinear time history analysis is performed, the obtained R value may not exactly the same as initial design assumption. It depends on the number of hinge yielding occurs during earthquake; when the number is increasing, the reduction factor becomes greater. The difference of both values can be noted 34.23%, 33.31%, and 36.75%, respectively for L1, L2, L3 models.

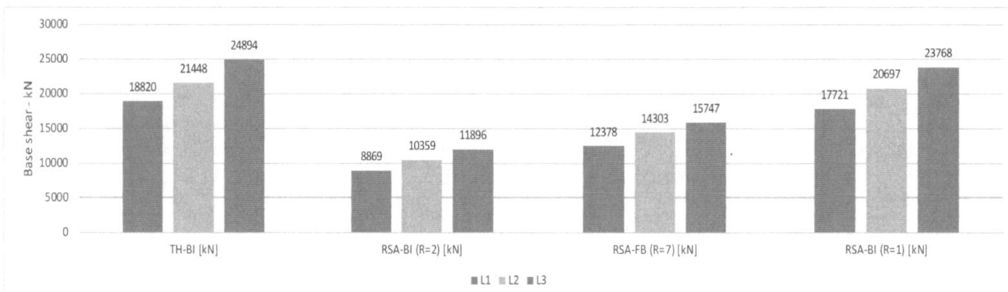

Figure 7. The comparison base shear of fixed base and base-isolated.

Furthermore, the results of time history analysis are giving closely to the unreduced response spectra analysis, with percentage difference of less than 6%. It does happen because the mass participation over 90% had occurred in the first mode. Therefore, by using RSA method to analyze isolated structure can be a reasonable alternative to acquire dynamics response.

6.2. Drift story

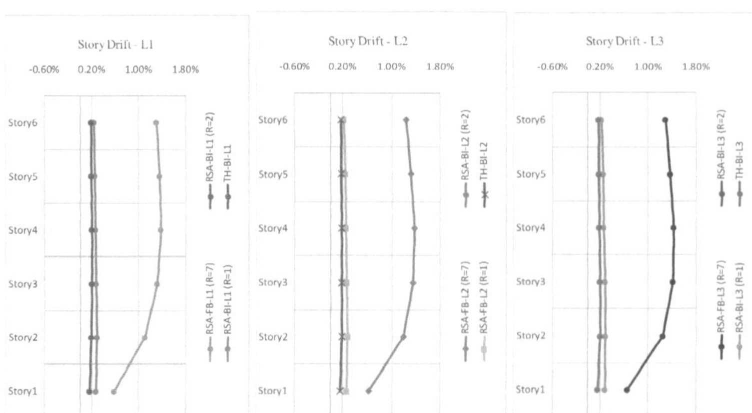

Figure 8. The comparison drift story of fixed base and base-isolated on each model.

According to the results in figure 8, it shows that the isolated structure can significantly reduce the rate of drift story at the MCER condition. By comparing the performance of base isolation in NLTHA to fixed base in RSA, will be obtained a reduction range of 72.15 – 86.57% in all variation models.

6.3. Rebar ratio and equivalent thickness of concrete

Before comparing to each other, it is important to note that all the isolated models are designed using reduction factor of 2 and following the linear vertical distribution of lateral forces approach on the superstructure based on MCE parameter; while the other models of fixed base are designed using dual system in accordance with the current earthquake design requirement. In this study, RSA is chosen as the basic of design process for all models since RSA method was often used in mid-rise building. The results obtained is an estimation on several structural elements, including beams, columns, slabs, and shear walls on the superstructure, but the detail of rebar calculation is not including the length of anchorage at the end and lap splicing. In order to determine the spacing of shear rebar in each element, ACI code was used for guidance. Overall the results will be acquired and calculated ideally by program, so that the value might be smaller than the reality.

In figure 9, the rebar ratio of fixed base obtained consistently in the range of 136 – 140 kg/m^3. While in isolated model, the rebar ratio is increasing along the various asymmetric L-shaped plan – generating 15%, 7%, 7% differences on model L1, L2, L3, respectively. Besides rebar, the total quantity of concrete known, as equivalent thickness will have slightly differenced i.e. the 239-242 mm range for model fixed base and the 229-234 mm range for isolated models (shown in figure 10). The usage of base isolation is more likely to save steel rebar more than concrete, since the reduction in concrete is not significant. This is following to the rules of condition that allow the design of superstructure over base isolated models with low ductility, in this way, the details of reinforcement between joints and larger local stress element in "notch" area can be designed in simple way.

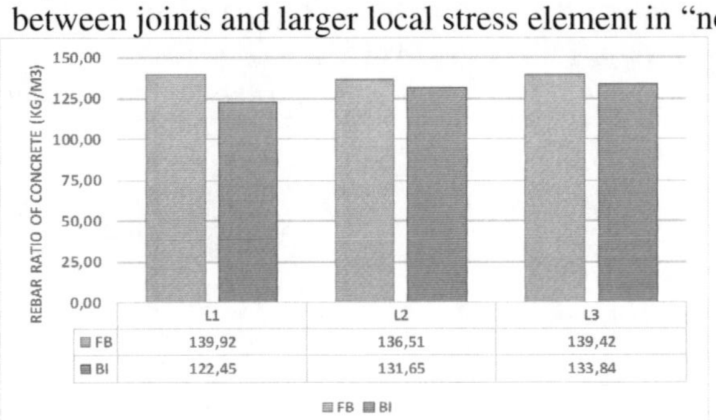

Figure 9. The comparison of rebar ratio in a cubic concrete (kg/m^3) of fixed base and isolated base.

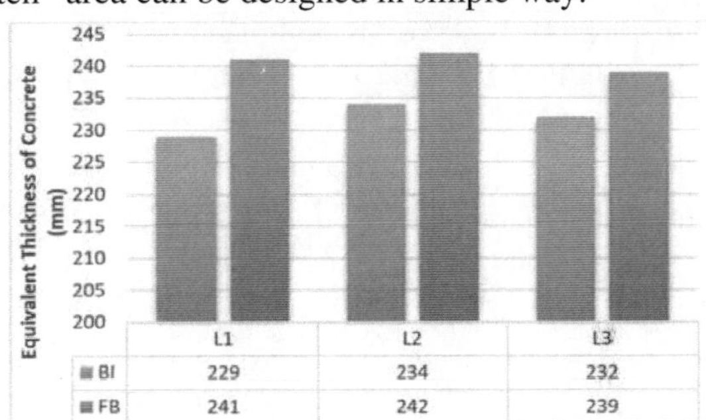

Figure 10. The comparison of equivalent thickness concrete of fixed base and isolated base.

Table 3. Detail of material saving of concrete and rebar on each variation.

Variation	Component	Model		
		FB	BI	% Saving
L1	Rebar (ton)	506,24	428,38	15%
	Concrete (m3)	3617,96	3498,46	3%
L2	Rebar (ton)	576,30	538,47	7%
	Concrete (m3)	4221,65	4090,28	3%
L3	Rebar (ton)	672,51	626,06	7%
	Concrete (m3)	4823,46	4677,85	3%

7. Conclusion

From the study, the base isolation is proven to be effective in dissipate seismic energy and more efficient in rebar usage. Here are the details of some conclusion remarks:

1. The torsion effect could be avoided in fundamental modes with the proper types and isolator layout
2. By using base isolation, mass participation factor in translation and torsion over 90 percent will be more easily achieved, so the analysis is enough to represent the real vibrated mass
3. The base shear calculated from the NLTHA is consistent higher than RSA on fixed base. The difference of both values is 34.23%% for L1 model; 33.31% for L2 model; and 36.75% for L3 model
4. The isolated structure is giving the smallest drift story rate of not more than 0.2% for each story, avoiding any damages to nonstructural elements inside the building
5. Using base isolation can lead to the saving effort; i.e. rebar by 7-15% and concrete by 3%.

Acknowledgements

This research is financially supported by the Directorate of Research and Community Engagement of Universitas Indonesia (DRPM UI) through Hibah Publikasi Internasional Terindeks Untuk Tugas Akhir Mahasiswa (PITTA) UI 2018 No.7185/UN2.R3.1/HKP.05.00/2018

References

[1] Deng P, Li X, Dong W and Tai L 2014 *Appl. Mech. Mater.* **501–504** 1547–50.
[2] Chopra A K 2012 *Dynamics of Structures* Prentice Hall.
[3] Naeim F 2001 *The Seismic Design Handbook (second edition)* Springer Science & Business Media New York.
[4] Mazza F and Mazza M 2016 *Soil Dyn. Earthq. Eng.* **90** 299–312.
[5] Etedali S and Sohrabi M R 2011 *Int. J. Civil Environ. Eng.* **2.**
[6] Cancellara D and De Angelis F 2017 *Comp. Struct.* **180** 74–88.
[7] Hameed A, Koo M S, Thang D D and Jeong J H 2008 *KSCE J. Civil Eng.* **12(3)** 187-96.
[8] Ryan K. L, Asce M and Polanco J 2008 *J. Struct. Eng.* **134(11)** 1780–4.
[9] ASCE/SEI 7-16 2017. *Minimum Design Loads for Buildings and Other Structures* American Soc. Civil Eng.

IOP Conf. Series: Materials Science and Engineering **473** (2019) 012028 doi:10.1088/1757-899X/473/1/012028

Postbuckling Analysis of Functionally Graded Beams

K Soncco[1], X Jorge[1] and R Arciniega[1]

[1]Department of Civil Engineering, Universidad Peruana de Ciencias Aplicadas (UPC), Surco, Lima, Peru

Abstract. This paper studies the geometrically non-linear bending behavior of functionally graded beams subjected to buckling loads using the finite element method. The computational model is based on an improved first-order shear deformation theory for beams with five independent variables. The abstract finite element formulation is derived by means of the principle of virtual work. High-order nodal-spectral interpolation functions were utilized to approximate the field variables which minimizes the locking problem. The incremental/iterative solution technique of Newton's type is implemented to solve the nonlinear equations. The model is verified with benchmark problems available in the literature. The objective is to investigate the effect of volume fraction variation in the response of functionally graded beams made of ceramics and metals. As expected, the results show that transverse deflections vary significantly depending on the ceramic and metal combination.

1. Introduction

Functionally graded materials (FGMs) have attracted the attention of the scientific community since their discovery by material scientists in Sendai, Japan [1]. FGMs are manufactured by grading the desired properties of two or more materials, usually ceramics and metals, in the selected direction. They offer great advantages from the materials science point of view because their continuous variation prevents stress concentrations, thus preventing the occurrence of plastic deformations and cracking [2]. Owing to these advantages, a large number of publications based on the analysis of FGM structures exist in the literature. The most relevant contributions to the topic are discussed below.

Kapania and Raciti [3] wrote a detailed review regarding the shear theories used for static cases, vibrations, and buckling for beams and thin shells. Sankar and Tzeng [4] obtained an elastic solution for FGM beams subjected to transverse loads. Kang and Li [5] investigated large deformations in functionally graded beams subjected to transverse loads. Ngyuen et al. [6] studied the geometrically non-linear behavior of functionally graded beams and structural frames. Likewise, Arciniega and Reddy [7] studied the geometrically non-linear behavior of shells under different types of loads; their model was validated using different benchmark problems found in the literature.

The behavior of structural members is of equal importance in research and in the industrial field. FGMs were designed to be used under sudden changes in temperature. However, structural members must also be analyzed under any type of conditions, since temperature variations only occur in a short period of time. Therefore, the static behavior of FGM beams must be evaluated in terms of transverse deflection [8]. For example, Reddy [9] proposed a super-convergent finite element Timoshenko model for static problems. Sankar [10] provided an elasticity solution based on the Euler–Bernoulli theory for functionally graded beams subjected to static transverse loads. The static behavior of functionally graded beams using the high-order shear deformation theory has been also studied by Kadoli et al. [11].

This work presents a computational model for the non-linear analysis of functionally graded beams subjected to buckling loads. A theoretical formulation based on the improved first-order shear deformation theory (FSDT) has been developed; the computational model incorporates finite deformations under the Lagrangian framework. An abstract finite element formulation with spectral high-order interpolation functions was then derived. The verification results of the formulation show that the computational model presented is satisfactory.

2. Beam formulation and finite element model

Let $\{x^i\}$ be a set of Cartesian coordinates with orthonormal basis $\{\mathbf{e}_i\}$. The neutral axis of the beam is defined by the coordinate x^1. The displacement vector is assumed to be of the following form

$$\mathbf{v}(x^1, x^3) = \mathbf{u}(x^1) + x^3\,\boldsymbol{\varphi}(x^1) + (x^3)^2\,\boldsymbol{\psi}(x^1) \tag{1}$$

where $\mathbf{u} = u_i\mathbf{e}_i$ denotes the displacement vector of the neutral axis, $\boldsymbol{\varphi} = \varphi_i\mathbf{e}_i$ and $\boldsymbol{\psi} = \psi_3\mathbf{e}_3$ are difference vectors $(i = 1,3)$. Equation (1) contains five independent variables. The quadratic term $\boldsymbol{\psi}$ is included to avoid the Poisson locking, therefore, no enhanced methods are needed.

For the given displacement field, we define the Green-Lagrange strain tensor as

$$\boldsymbol{\varepsilon} = \boldsymbol{\varepsilon}^{(0)} + x^3\boldsymbol{\varepsilon}^{(1)} + \underline{(x^3)^2\boldsymbol{\varepsilon}^{(2)} + (x^3)^3\boldsymbol{\varepsilon}^{(3)} + (x^3)^4\boldsymbol{\varepsilon}^{(4)}} \tag{2}$$

where high order terms are neglected.

The weak form can be easily constructed from the principle of virtual work. The configuration solution of the beam is defined by the triplet $\boldsymbol{\Phi} \equiv (\mathbf{u}, \boldsymbol{\varphi}, \boldsymbol{\psi})$. Thus, we obtain

$$\begin{aligned}\mathcal{G}(\Phi,\delta\Phi) &= \mathcal{G}_{\text{int}}(\Phi,\delta\Phi) - \mathcal{G}_{\text{ext}}(\Phi,\delta\Phi) \\ &= \int_{\mathscr{X}} \sum_{k=0}^{1} (\mathbf{N}^{(k)} \cdot \delta\boldsymbol{\varepsilon}^{(k)})\, dx^1 - \int_{\mathscr{X}} \mathbf{p} \cdot \delta\mathbf{u}\, dx^1 = 0\end{aligned} \tag{3}$$

where $\delta\boldsymbol{\Phi} := \delta\mathbf{u}, \delta\boldsymbol{\varphi}, \delta\boldsymbol{\psi}) \in \upsilon$ and υ is the space of admissible variations. The stress resultants are evaluated from the following equation

$$\mathbf{N}^{(i)} = \sum_{j=0}^{1} \mathbb{B}^{(i+j)}\boldsymbol{\varepsilon}^{(j)}, \qquad i = 0,1 \tag{4}$$

Therefore, the final expression for the weak form will be

$$\mathcal{G}(\Phi,\delta\Phi) = \int_{\mathscr{X}} \sum_{k=0}^{1}\sum_{l=0}^{1} \delta\boldsymbol{\varepsilon}^{(k)} \cdot \mathbb{B}^{(k+l)}\boldsymbol{\varepsilon}^{(l)}\, dx^1 - \int_{\mathscr{X}} \mathbf{p} \cdot \delta\mathbf{u}\, dx^1 = 0 \tag{5}$$

The components of the tensor $\mathbb{B}^{(i)}$ are the material stiffness coefficients and they are given by

$$\mathbb{B}^{(k)} = \int_{-h/2}^{h/2} (x^3)^k\, \mathbb{C}\, dx^3, \qquad k = 0,1,2\ . \tag{6}$$

In this two-phase functionally graded materials the properties are assumed to vary through the thickness of the beam. The materials in the bottom and top surfaces are metal and ceramic respectively. Therefore, the fourth-order elasticity tensor $\mathbb{C}$ is a function of the thickness x^3. Their components are expressed as

$$C_{ijkl}(x^3) = C^{c}_{ijkl} f_c + C^{m}_{ijkl} f_m \tag{7}$$

where f_c, f_m are the volume fractions of the ceramic and metal phases which are computed by mean of the power law. That is

$$f_c = \left(\frac{x^3}{h} + \frac{1}{2}\right)^n \qquad f_m = 1 - f_c \tag{8}$$

Let $\mathscr{X}$ be the domain of the neutral axis parametrization where the finite element domain lies in. Recall that $\hat{\Omega}^e \equiv [-1,1]$ is a parent domain in ξ-space and $x^1(\xi) : \hat{\Omega}^e \in \mathbb{R} \to \mathscr{X}$. The finite element equations are obtained by interpolating the components of the field variables written in terms of the base vectors. Namely

$$\mathbf{u}^{hp}(x^1) = \left(\sum_{j=1}^{m} u_k^{(j)} \phi^{(j)}(\xi)\right)\mathbf{e}_k,\ \boldsymbol{\varphi}^{hp}(x^1) = \left(\sum_{j=1}^{m} \varphi_k^{(j)} \phi^{(j)}(\xi)\right)\mathbf{e}_k,$$
$$\boldsymbol{\psi}^{hp}(x^1) = \left(\sum_{j=1}^{m} \psi_3^{(j)} \phi^{(j)}(\xi)\right)\mathbf{e}_3, \qquad k = 1,3 \tag{9}$$

The adopted basis functions $\phi^{(i)}$ are C^0 interpolant polynomials of Gauss–Lobatto-Legendre quadrature points, which are particularly suitable for high-order expansions. Explicitly, the one-dimensional basis functions of order $p = m - 1$ are expressed using the p-order Legendre polynomial P_{m-1}, as shown

$$\phi^{(i)}(\xi) = \frac{(1-\xi^2)P'_{m-1}(\xi)}{m(m-1)P_{m-1}(\xi_i)(\xi-\xi_i)} \tag{10}$$

High-order spectral elements, in contrast to low-order finite elements, do not exhibit locking problems.

Finally, the linearization of the discrete form of equation (5) is carried out leading to a symmetric tangent operator. The resultant set of algebraic nonlinear equations is solved by the incremental-iterative Newton-Raphson method and the predictor-corrector cylindrical arc-length method.

3. Results and discussions

In this section, some results are presented to study the non-linear behavior of FGM beams with compressive loading applied to the neutral axis. To activate the post-buckling equilibrium path, a small perturbation d is applied transverse to the P load of ratio $d/P = 0.001$.

The geometry of the beam, based on Massin and Al Mikdad [12], is depicted in figure 1: $L = 0.50\,\text{m}$ $b = 0.075\,\text{m}$ $h = 0.0045\,\text{m}$. The values of Young's modulus for aluminum and zirconium are given by:

$$E_m = 70\,\text{GPa (aluminum)},\ \ E_c = 151\,\text{GPa (ceramic)},\ \ \nu = 0.3 \tag{11}$$

Two boundary conditions are analyzed: clamped-free (CF) and simply supported (SS) beams. The example problem is solved using polynomials of fourth order ($p = 4$) with a finite element mesh of ten elements. The shear locking effect was minimized utilizing high-order interpolation functions.

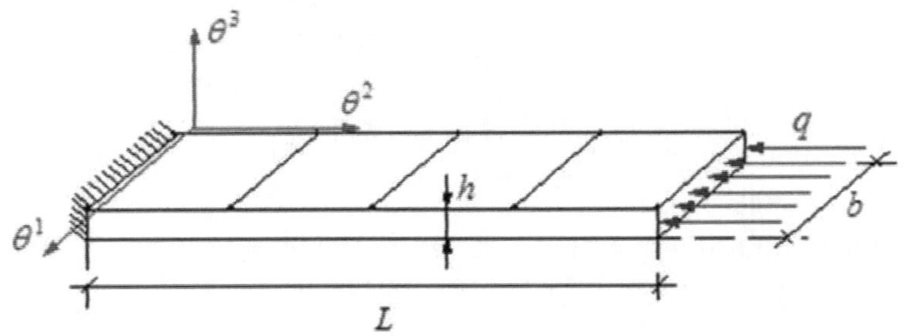

Figure 1. Postbuckling of a plate strip under compressive load.

Figure 2 shows load versus displacement curves for a clamped-free functionally graded beam. Here, we can see how the proposed model describes correctly the non-linear behavior of the beam. Similar results for isotropic cases are shown in reference [13]. The critical load for the example is $P_0 = \pi^2 \mathrm{E_m I}/4\mathrm{L}^2$.

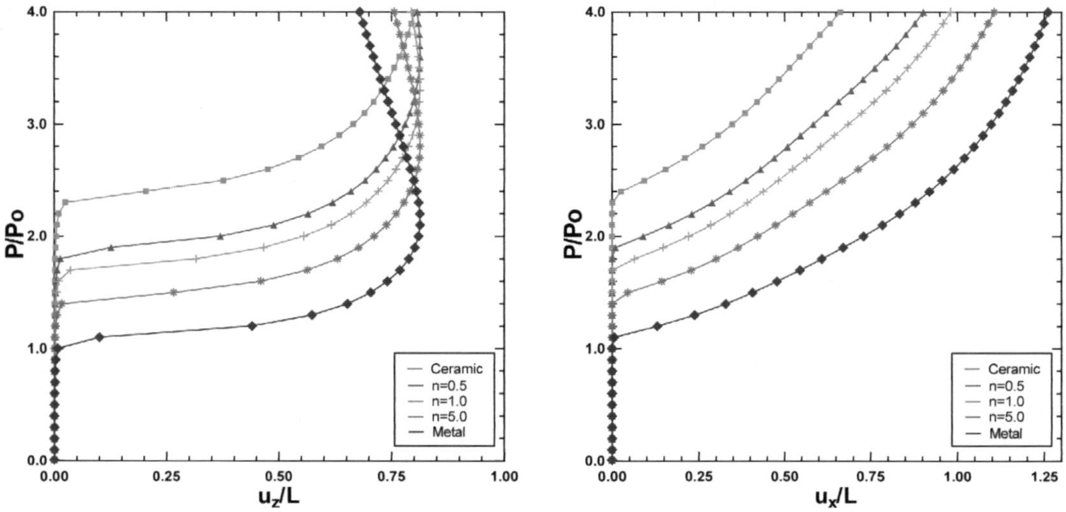

Figure 2. Load-displacement curves for clamped-free beam with different volume fraction exponent *n*.

Likewise, figure 3 shows the load-displacement curves at the center of a simply supported beam subjected to buckling load. The critical load is $P_0 = \pi^2 \mathrm{E_m I}/\mathrm{L}^2$. This exercise shows the effectiveness of the formulation to describe finite deformations of FGM beams, as depicted by Pagani and Carrera [13].

Finally, figure 4 shows several deformed configurations for the simply supported FGM beam under axial compressive loading and a volume fraction equal to 1.

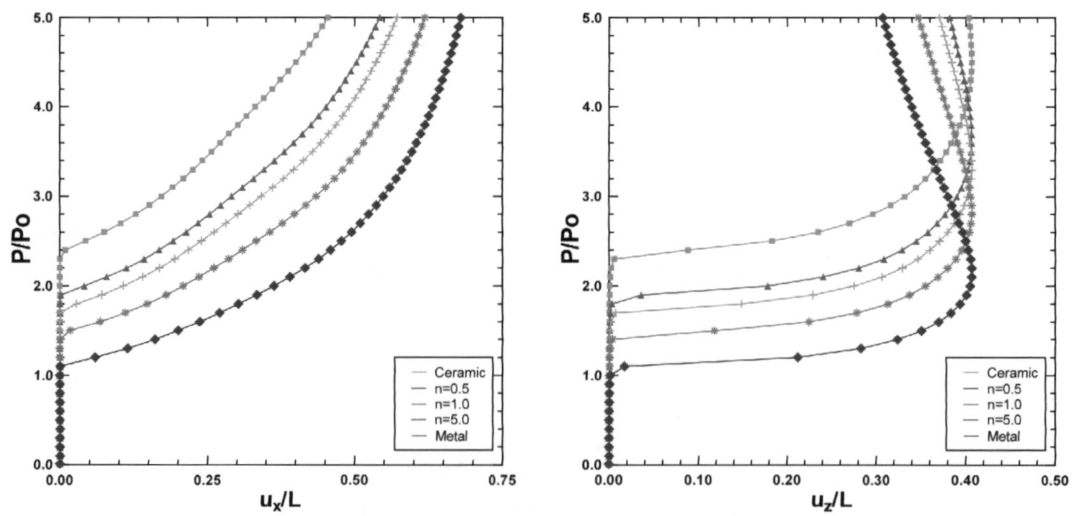

Figure 3. Load-displacement curves for simply supported (SS) beams with different volume fraction exponent *n*.

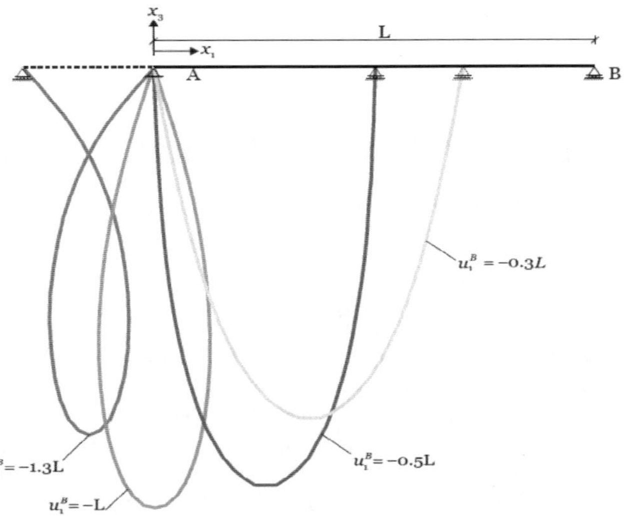

Figure 4. Deformed configurations of the SS functional graded beam under compressive loading.

4. Conclusions

A finite element model has been formulated to study the non-linear behavior of FGM beams. The formulation was based on an improved FSDT with five parameters and derived using the principle of virtual work. To assess the formulation, two benchmark problems were analyzed under CF and SS boundary conditions, subjected to buckling loads. The set of nonlinear equations were resolved through the arc-length method. Results for the displacements at the end (CF) and center (SS) of the beam were presented. The effect of the volume fraction exponent on the non-linear behavior of these beams was evaluated. We obtained the following conclusions:

(a) As expected, differences in the displacements for different volume fraction exponents and both cases (CF and SS) were observed.

(b) The proposed tensor formulation provides a robust model for the study of mechanical phenomena of FGM beams undergoing finite deformations.

References

[1] Koizumi M 1997 *Comp. Part B Eng.* **28B** 1–4.
[2] Finot M and Suresh S 1996 *J. Mech. Phys. Sol.* **44** 683–721.
[3] Kapania M R K and Raciti S 1989 *AIAA J.* **27(7)** 923–34.
[4] Sankar B V and Tzeng J T 2002 *AIAA J.* **40(6)** 1228–32.
[5] Kang Y A and Li X F 2009 *Int. J. Non-Linear Mech.* **44** 696–703.
[6] Nguyen D K, Gan B S and Trinh T H 2014 *Struct. Eng. Mech.* **49** 727-43.
[7] Arciniega R A and Reddy J N 2006 *Int. J. Sol. Struct.* **44(6)** 2036-52.
[8] Kadoli R et al. 2008 *Appl. Math. Model.* **32** 2509–25.
[9] Reddy J N 1997 *Sadhana* **24** 175–98.
[10] Sankar B V 2001 *Compos Sci. Technol.* **61** 689–96.
[11] Kadoli R, Akhtar K and Ganesan N 2008 *Appl. Mathe. Model.* **322** 2509–25.
[12] Massin P and Al Mikdad M 2002 *Comput. Struct.* **80** pp 835-47.
[13] Pagani A and Carrera E 2018 *Mech. Adv. Mater. Struct.* **25** 15-31.

Temperature Changes Effects to Dynamics Performances of a Pinned-Supported Steel-Arch-Bridge

C Christian[1], J I Rastandi[1] and Y Lase[1]

[1]Department of Civil Engineering, Universitas Indonesia, Depok, 16424, Indonesia

Abstract. Seldom is temperature changes effect taken into account in structural design and monitoring. However, in some cases temperature may be the governing load case, especially for a long span bridge. Moreover, statistics have shown that the world's temperature is gradually increasing nowadays and may cause the temperature effect to increase. This may cause an alteration to dynamics parameters of the structure, which will be a highlighted object in this research. These parameters are directly corelated to physical parameters, e.g. mass and stiffness. Hence, as the physical parameters has any kind of change, dynamic parameters will also immediately change. This alteration may modify the health index of the structure. In some regulations, the deviation of the value of natural frequencies, as one of dynamics characteristics, is limited to be 10%. This research is aimed to observe the health index of a determined case study bridge and to observe the effects of temperature changes to dynamics performances of a pinned-supports-arch bridge in the Province of Riau, Indonesia, and it was found, by modeling the structure in Finite Element (FE) Model and performing vibration testing, that temperature difference has caused the natural frequencies of the bridge to vary up to 3.63%.

1. Introduction

Structural failure may occur in a couple seconds for many reasons, one of those is due to strength and stiffness degradation, since it may lead to resonance condition. To avoid this to happen, applying structural monitoring system is considerably a good choice. Some parameters are usually selected and observed to monitor the health index of a structure, such as natural frequencies of the structure. Natural frequencies may be occupied as a governing parameter since it is directly related to the physical parameters, i.e. structural mass and stiffness [1]. Hence, as the physical parameters have any kind of alteration, the natural frequencies of the structure will also definitely be modified.

However, the presence of temperature change may also modify natural frequencies of the structure, since the material of the structure has a thermal coefficient, that may lead the structure to elongate and then change the stiffness of the structure due to elongation and the resulting forces [2,3]. Therefore, temperature effect should be thoroughly observed to justify, if the natural frequencies of the structure are modified due to structural deterioration or temperature change [4].

This research is aimed to observe the health index of a determined case study bridge in the Province of Riau, Indonesia. Furthermore, the effects of temperature changes to the case study bridge will also be analyzed by employing finite element model and vibration testing to the real bridge, and both results will be compared to justify structural health index and the effects of temperature changes to structure.

2. Dynamics parameters of structure

In general, dynamic parameters of structure consist of natural period of vibration, natural frequencies, and mode shapes. Natural period of vibration is the total time needed for a structure to oscillate in a full cycle, and natural frequency is the number of cycle in a range of time of a second. And mode shape is the dynamic characteristic of the structural deflection due to the corresponding modes and frequencies.

In dynamic analysis, structure can be divided into two classifications, single degree of freedom (SDOF) and multi degree of freedom (MDOF). Degree of freedom is the number independent displacements required to define the displaced positions of all the masses relative to their original positions [2]. The number of natural frequencies of the structure depends on the number of modes, which is also the number of the degree of freedom.

Theoretically, dynamic parameters of structure can be obtained by using the dynamic equilibrium equation, which is a second order differential equation. Here it is:

$$[M]\{\ddot{u}\} + [C]\{\dot{u}\} + [K]\{u\} = \{p(t)\} \tag{1}$$

Equation above can be solved by using matrices operation, square matrices with an ordo *n* x *n*, depends on the number of modes. For SDOF structure, this equation can be simply solved by performing an ordinary differential equation analysis.

The natural frequencies of structure can also be experimentally obtained by conducting vibration testing. This kind of testing intend to record the vibrational behavior of the structure as a function of time and then to transform it into frequency domain data by performing Fast Fourier Transform (FFT). FFT is a powerful tool to analyze the behavior of a structure in a frequency domain, since the result may provide the peak response of the structure in the corresponding frequency.

Fourier transform is commonly used to analyze the frequency of a series of signal. Basically, signal is a combination of several sinusoidal functions, hence Fourier Transform is used to decompose the complex function into a simpler function to solve.

If the input domain in Fourier Transform is a time domain function, then the output of the calculation will be a frequency domain function. This transform can be mathematically expressed as follow:

$$F(\omega) = \sum_{-\infty}^{\infty} f(t)e^{-i\omega t}\partial t \tag{2}$$

3. Temperature and heat analysis

Steel bridges responds to changing temperature quickly. Steel structure may react differently during heating and cooling periods. Temperature different between the inside and outside may cause the steel structure to react differently [5].

Several previous researches have been conducted to observe the relationship of temperature and responses of structure, and many of those have concluded that the amplitude of the structural response may increase as the temperature increases. Structural responses can be either static or dynamic, and can be either displacement, velocity, or acceleration.

Here is the governing equation to analyze the heat transfer on structure in case of conduction:

$$Q_x = -KA\frac{\partial T}{\partial x} \tag{3}$$

Where:

Q_x = Rate of heat transfer
K = Thermal conductivity
A = Area
$\frac{\partial T}{\partial x}$ = Temperature gradient

4. Research design

This research is classified as the combination of case study and experimental study research, since it occupies a bridge as a case study object and holds an experimental field testing to fulfill this research.

The case study bridge is located in the district of Bangkinang, Province of Riau, Indonesia. This bridge spans over Kampar River, with three main spans with a total span length of 200 m. The side spans of the bridge are 50 m each, and the middle span 100 m. This bridge is 7 m wide and pinned in every supports it has. It has two lanes of one-way traffic flow.

The first thing to do was to model the structure in FE model by employing MIDAS Civil Software. The objective of conducting a finite-element analysis modelling is to determine the parameters of the structure, in this case are natural frequencies, in the ideal condition, however, the real condition will not show the exactly the same value to the result of the modelling, since the condition in the real structure may not be as ideal as in the finite element model. And here are the real and the modelled structure.

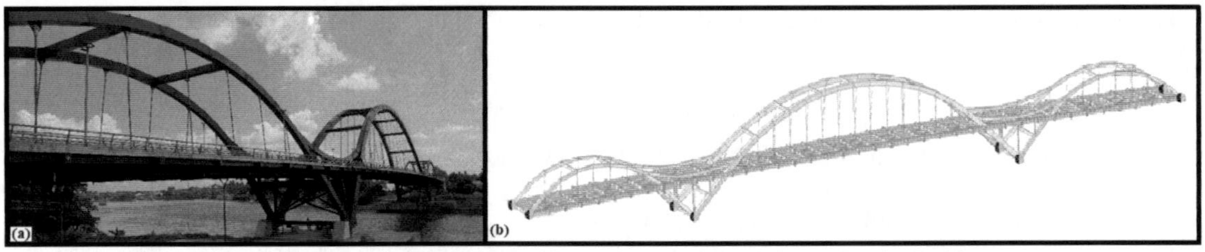

Figure 1. (a) Case Study Bridge; (b) Modelled Case Study Bridge.

After modeling the structure, the next step to do is to conduct vibration testing to extract the real natural frequencies of the bridge. Vibration testing will be conducted by occupying ambient loadings of the bridge to be the source of excitation, since the bridge has been already accessible by the people of the city. Ambient forces here mean the functional loadings of the bridge causing the bridge to vibrate, in this case is dominated by transportation loadings. The transportation loadings will be the source of excitation and vibrate the bridge.

To conduct this vibration testing, several equipment and instrumentations will be occupied, such as vibrating wire strain gauges to measure the strain and temperature of a specified point, a set of vibration sensors, to transmit the mechanics displacement and movement of the structure into electrical or digital signal which will be acquired by operator, in this case is a set of computers. Here are several used instrumentations to conduct this testing.

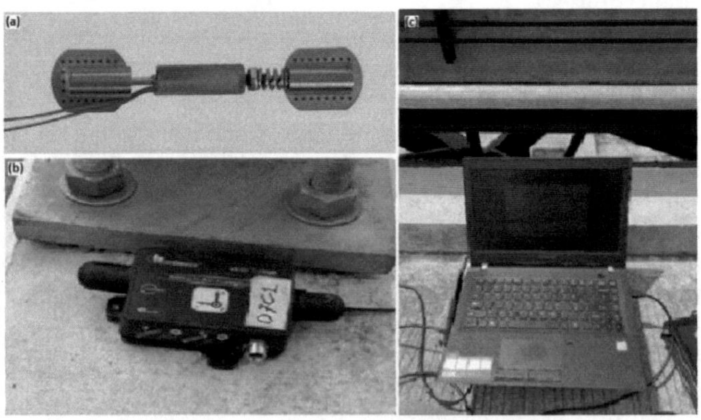

Figure 2. (a) Vibrating Wire Strain Gauge; (b) Vibration Sensor; (c) A set of Laptop.

Vibration testing were conducted in five different times of the day, at which the history showed that the temperature different is the most extreme. The testing was conducted at 5th – 6th of June 2018, at 12 pm, 6 pm, 3 am, 9 am, and 2 pm. At these point of times, the system temperature, showed by the sensor of the installed sensors, varied in a range of 23.4^{o} C – 45.1^{o} C, which has a deviation of 21.7 centigrade.

This research employed four wireless sensors installed in the mid span of the bridge in order to observe the maximum bending response under an exact same excitation. This position was also selected due to the limitation of the sensors, where it can only be installed in a range of around-fifty-meter length. to Figure below shows the locations of the installed vibration sensors.

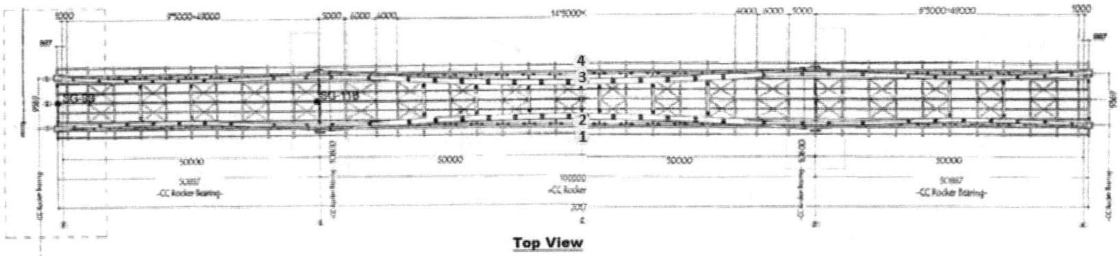

Figure 3. Vibration Sensors Installation Layout.

The structure was excited by employing the travelling vehicle over the bridge as it ambient loading, and the vibrational behavior was recorded before and after the vehicle pass over the bridge. However, the measurable vibration was coming from the travelling truck, and another vehicle might not have enough power to vibrate the whole bridge at once. Sampling frequency of the sensors was set to be 200 Hz (at 12 pm) and the others 600 Hz.

This research is also aimed to observe the health index of the specified case study bridge. Some regulations stated that the allowable difference value of the natural frequencies is limited to be 10%. If the natural frequencies' fluctuation more than 10%, the structure is supposed to be in a fit condition.

5. Results and discussions

5.1. Finite element result

In this section, the result of finite element analysis of the bridge will be provided. The observed output of the MIDAS Civil software is limited only in natural frequencies and mode shapes. The number of mode is also limited up to the first 12 modes. These results will then be employed as a benchmarking value to vibration testing result. This table below sums up the first 12 natural frequencies of the structure.

Table 1. Natural Frequencies and Natural Periods of Vibration from FE Model.

Modes	**Frequency**		**Period**
	(rad/sec)	(cycle/sec)	(Sec)
1	4.329929	0.689130	1.451106
2	6.121680	0.974296	1.026383
3	7.295267	1.161078	0.861269
4	10.278603	1.635890	0.611288
5	11.100667	1.766726	0.566019
6	11.304089	1.799102	0.555833
7	11.659843	1.855722	0.538874
8	11.676054	1.858302	0.538126
9	14.841583	2.362111	0.423350
10	15.856338	2.523615	0.396257
11	16.565074	2.636413	0.379303
12	16.645014	2.649136	0.377482

And figure below shows the corresponding mode shapes from the twelve modes.

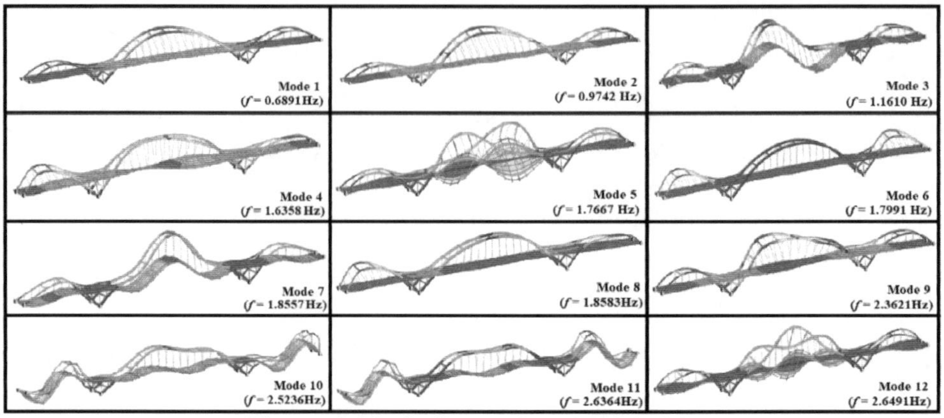

Figure 4. Mode Shapes from FE Model.

Default condition modeling of the structure shows that the structure has the highest response in Z direction (gravity direction) at the local installation point of the sensors, that is in the midspan of the bridge, is in the seventh mode, which has the natural frequency of 1.8557 Hz. Therefore, the first peak response of the FFT operations will then be assumed to be the seventh mode, since the structure was excited in the gravity direction during the vibration testing, so that the dominant response of the structure will also be recorded in the gravity direction.

5.2. Vibration testing result

Vibration testing was conducted by employing the passing vehicle over the bridge as the source of the excitations. However, the structure may adequately vibrate only if a big truck pass by. Private car may also create vibration, however it may be poorly recorded by the sensor, since the vibration induced is not great enough to consider.

To observe the behavior of the structure, vibration signal must be extracted or cut, so that the data only shows the vibrational behavior of the structure. To generate this, the signal must be cut in the range of time, when free vibration took place. Here is the picture of the cut signal, where free vibration took place, and the corresponding FFT result respectively.

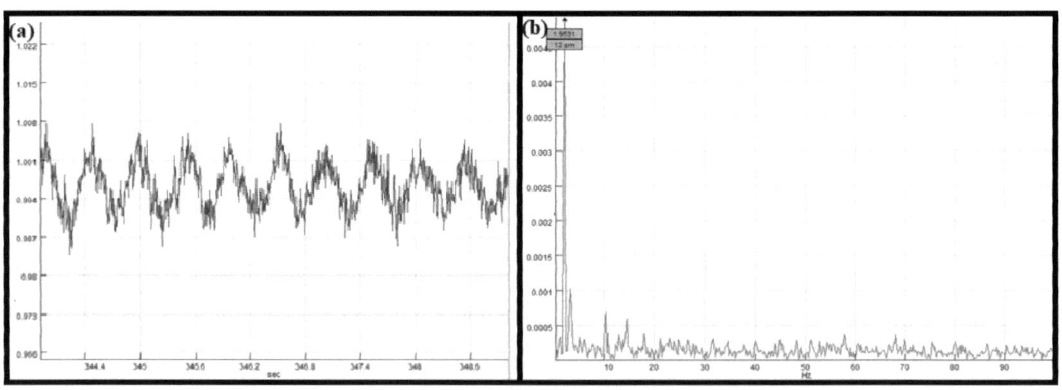

Figure 5. (a) Cut Signal; (b) Corresponding FFT Result.

From this FFT result, it is known that the first peak response is in the frequency of 1.9531 Hz. This may show the dominant mode of the bridge in bending direction. This may not be the first mode of the structure in global, since the first mode of the structure is known to be a lateral movement, and in the frequency of 0.5397 Hz.

This table below sums up the summary of the results of vibration testing, which shows the deviation of the obtained natural frequencies to 1.93 Hz, as a previously-pre-obtained natural frequency of the bridge.

Table 2. Summary of Obtained Frequencies of the Structure.

No.	Time	Temp. (°C)	Natural Freq. (Hz)	Deviation to Previous Vib. Testing (Hz)	Deviation (%)
1	12:14 PM	42.8	1.9531	0.0231	1.20
2	5:57 PM	37	1.92	0.01	0.52
3	3:19 AM	23.4	1.875	0.055	2.85
4	9:41 AM	32.6	1.9097	0.0203	1.05
5	1:52 PM	45.1	2	0.07	3.63

Table above shows that the natural frequency of the real bridge obtained by vibration testing was higher than of that by FE model. This may indicate that the real bridge has higher value of strength and/or stiffness. This may happen, since the stiffness of the concrete slab was neglected in the FE model. This plot below may show the effect of temperature changes to natural frequencies of the case study bridge, comparing the vibration testing results to FE model result.

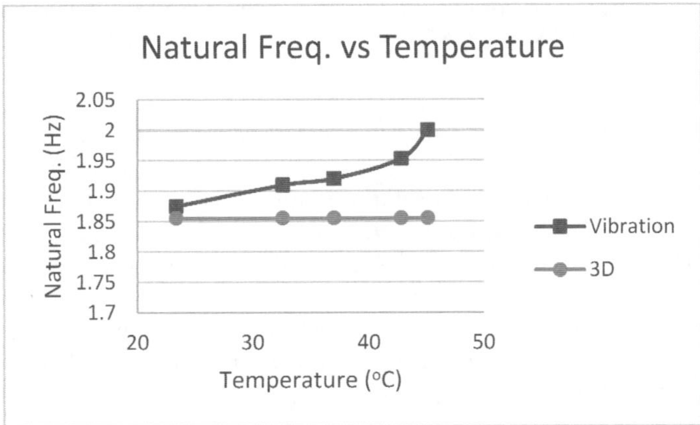

Figure 6. Natural frequency to the change of temperature.

Figure above shows that natural frequencies of the case study bridge rises up as the temperature increases. This may not be as the same as another previous researches about the effect of temperature changes to natural frequencies. Previous researches found that the natural frequency of a simple beam is inversely proportional to the changes of temperature. This may happen because the simple beam structure has pin and roll support, where elongation is allowed. However, in this case study bridge, elongation is prohibited, since the bridge has pin supports in all of the boundary condition, where the point must stay there.

This condition may lead the structure to experience an additional axial force due to the boundary condition. The structure itself wants to elongate, due to the temperature appraisal, however it may not be possible, due to the boundary conditions. Hence, according to the geometry theory, the structure may have additional stiffness in the stiffness matrices components.

Finally, the health index of the structure will be justified by using equation below, by calculating the different value of the natural frequency of the bridge in the daily average temperature of the bridge, that is in 32-33°C, so that the experimental natural frequency is taken to be 1.9097 Hz. Hence, the fitness ratio of the structure can be determined by utilizing this equation, proposed by Thomas (2009) [6].

$$FR = 1 - \left|\frac{1.8557 - 1.9097}{1.8557}\right| = 0.9709 = 97.09\% > 90\%$$

Hence, the structure is still in a health condition, since the fitness ratio is more than 90%.

6. Conclusions

These conclusions may be generated from this research:

1. The observed structure is still in a health condition, since the fitness ratio of the structure is more than 90%.
2. The natural frequencies of the real structure may have higher values than of it in the FE model, which means the real structure is stiffer than the modelled structure. And the observed structure is still in a safe and healthy condition by the time of the vibration testing.
3. Temperature changes may cause the natural frequencies of the case study bridge to change too in directly proportional way, due to the boundary condition of the structure.
4. Temperature changes may cause the natural frequency of this bridge to vary up to 3.63% to the common condition of the case study bridge.
5. Effect of temperature changes should be put into account to conduct structural health monitoring (SHM), since the alteration due to temperature changes itself might reach 3.63% in a range of 10% permission of the structure to change in natural frequencies according to the regulations.

Acknowledgement

This research is financially supported by the Directorate of Research and Community Engagement of Universitas Indonesia (DRPM UI) through Hibah Publikasi Internasional Terindeks Untuk Tugas Akhir Mahasiswa (PITTA) UI 2018 No. 7185/UN2.R3.1/ HKP.05.00/2018.

References

[1] Gillich G R and Mituletu I C 2017 Smart Sensors, Measurement, and Instrumentation 26. 13 – 37.
[2] Paz M and William L. 2004 *Structural Dynamics: Theory and Computation Fifth Edition.* Massachusetts: Kluwer Academic Publisher.
[3] Chopra A K. 2015 *Dynamics of Structure*. Prentice-hall International Series.
[4] Okumatsu T et al. 2015 *Life-Cycle of Struc Sys.* 168–75.
[5] Wenzel H and Dieter P. 2005 *Ambient Vibration Monitoring* England: Wiley.
[6] Terrel T 2009 *Structural Health Monitoring for Damage Detection Using Wired and Wireless Sensors Clusters* The University of Central Florida, Orlando: Master of Science Thesis.

Cracking Behavior of Reinforced Lightweight Concrete Beam Using Hot Water Pre-treated Oil Palm Shell Coarse Aggregate

N Handika[1], F D Sofyani[1], E Tjahjono[1], E Arijoeni[1]

[1]Civil Engineering Department, Faculty of Engineering, Universitas Indonesia, Indonesia

Abstract. As one of the largest palm oil producers in the world, by-products of this commodity are also produced in Indonesia. One of the solid by-products that are obtained from palm oil production is called Oil Palm Shell (OPS). These days, this material is interesting to be studied as it has a lightweight but sufficiently tough and rough outer surface. In previous studies done in laboratory, OPS was used as alternative coarse aggregate in concrete mix proportion. The designed compressive strength of OPS concrete is in the range of 20-23 MPa. In this research, the authors are interested in studying the crack behaviour of reinforced lightweight concrete beams using OPS replacing natural coarse aggregates. First, pre-treatment of OPS is performed by using hot water at 50-°C as based on previous studies. Second, two samples of identical beam with 15 × 25 × 300 cm^3 of dimension were cast and tested under four-point loading using force control hydraulic actuator. Third, crack opening evolution due to flexural test that occurs in OPS lightweight concrete beam is measured, to be precise, on the pure bending area. The crack opening is compared to service limit cracks from building code to find the corresponding loading application.

1. Introduction

Reinforced concrete (RC) has been one of the most commonly used building materials in the world since its development in the 20th century [1, 2]. In most cases, the durability of RC structures becomes the main concern since it is necessary to ensure the operation of the facility. Durability is often related to crack opening of structures. In certain circumstances, cracks can lead to degradation of concrete and further damage the function of an RC building. Therefore, the behaviour of cracking on RC structure is a critical issue. In SNI 2847: 2013 [3], the effect of crack on RC elements, e.g. beam, is interpreted as stiffness reduction. Seventy per cent of the moment of inertia may be taken when beam has passed its first cracks phase. In the calculation of beam's deflection, this cracked moment of inertia is also considered.

Moreover, Eurocode 2 has a design crack opening formulation in addition to crack spacing for RC Element. The maximum crack opening criteria are regulated based on the utilisation of buildings as service limit criterion. It limits the crack opening at 300 μm [4].

On the other hand, the development of new material is escalating along with the increasing demand for housing needs. Some potential substitution materials have been recently studied. One of the materials is called Oil Palm Shell (OPS). These solid shell-formed by-products are obtained from palm oil production [5]. As one of the largest producers of palm oil, OPS from this industry are produced massively in Indonesia. Ineffectiveness of this solid side product waste processing has led some researchers to study the potential OPS waste in the civil engineering domain [6-8]. According to Yew et al. [8], OPS can be used as coarse aggregate material (Biosource material).

Research in OPS concrete conducted in 2016 in Laboratory of Structural and Material, Civil Engineering Department, Universitas Indonesia resulted in the most effective and possible concrete mix design with compressive strength reaching 20-23 MPa [7]. Regarding this condition, the authors decided to study the performance of OPS RC structures.

In this paper, crack behaviour of lightweight RC beam using OPS substituting natural coarse aggregates is studied with concrete mix design following the previous works [7]. Steps were begun by pre-treatment of OPS using hot water at 50-°C based on [9]. In the second step, two samples of the identical beam with 15 × 25 × 300 cm^3 of dimension were cast at the same time as the cylindrical samples for characterisation tests. The size of the beam represents the typical dimension of the beam in one to two storey houses in Indonesia. In the third step, the cylindrical samples were tested to find the concrete compressive, flexural and tensile strength. Then, finally, beams were tested under four-point loading using force control hydraulic actuator. Study on the crack evolution and opening in OPS concrete beam is emphasised on the pure bending area. The crack opening on concrete is related to its tensile strength as this material is weak in tension. The results from the split test and flexural test are compared to see whether or not the OPS concrete has the same trend compared to normal concrete with natural coarse aggregates.

2. Material Characterization

OPS concrete has 1792-kg/m^3 of concrete density. Therefore, it fulfils the requirement of lightweight concrete as stated in the National Standard [10]. In this part, characteristic tests were performed to obtain the properties of concrete and steel bar. OPS concrete mix design is presented in Table 1 based on the previous research [7] with 7/20 water-cement ratio.

Table 1. The proportion of OPS Concrete [7].

Natural Fine Aggregates	0-5 mm (860 kg/m^3)
OPS coarse Aggregates	4.75-12.5 mm (273 kg/m^3)
Cement	PCC Portland Composite Cement (500 kg/m^3)

2.1. OPS Concrete Characteristics

Table 2. Mechanical properties of concrete materials.

Concrete compressive strength [11]*	f_c'	=	22.33 ± 0.42	MPa
Concrete flexural strength [12]*	f_c'	=	2.15 ± 0.20	MPa
Concrete tensile strength [13]*	f_t'	=	12.58 ± 1.89	MPa

* Average ± standard deviation of experimental results from five samples at 28 days after casting.

Concrete for material characteristic tests and beams were cast at the same time. After demolding the samples, cylindrical samples were cured in a pond with room temperature water. Meanwhile for RC beam, after casting, it was let harden inside a room with usual treatment to prevent early shrinkage. The test was done to know concrete's compressive, flexural and tensile strength based on standard [11-13]. Each test was performed on five cylindrical samples (15 cm of diameter and 30 cm in height).

2.2. Steel bar reinforcement

The nominal bar diameter is 13-mm for the ribbed bar (longitudinal reinforcement), and it corresponds to the cross-section of 132.73-mm^2. Tensile tests were performed according to SNI [14]. The yield strength of the bar is obtained at f_y = 419.28 ± 17.74-MPa, and its ultimate strength reached f_u = 568.60 ± 16.29-MPa (the average ± standard deviation of the tensile test results from four samples).

3. Experimental Campaign

3.1. Detail of experiment

Detail and configuration of OPS RC beam dimensions and steel bar reinforcement are presented in figure 1 and 2. Four-point loading tests were conducted on the beams after 44 and 48 days consecutively. The dimension of the beam is 15 × 25 × 300-cm^3, which leads to 2.7-m of support in regard to support distance. The beams were designed as under-reinforced beams according to the standard [3].

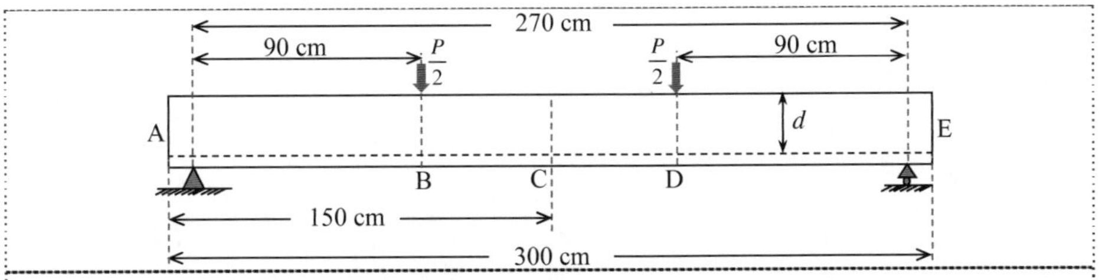

Figure 1. Experimental set-up of OPS RC beam specimen (in cm).

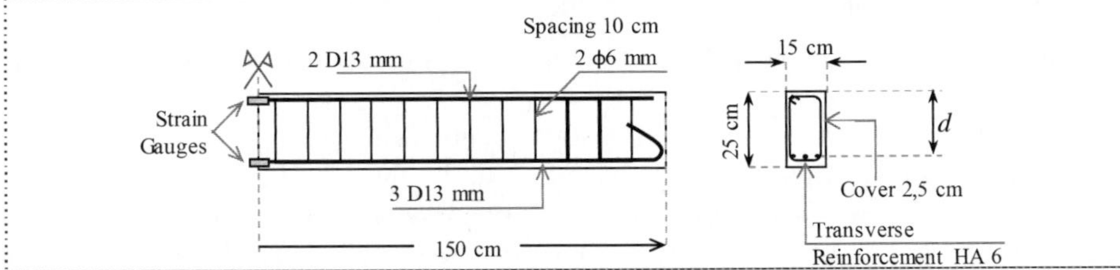

Figure 2. Detail of section (in mm).

By using a hydraulic actuator, loading charge was distributed into two points of loading within 0.9 metres leading to four-point flexure test with 200 kg of loading step application. For measurement, LVDT extensometer was employed to measure the displacement in the vertical direction at point C (see figure 1). Moreover, two strain gauges were attached to the tension bar, as shown in figure 2.

4. Experimental Results

4.1. Global Behavior: Force-Vertical Displacement

Global behaviour results of the OPS RC beam present the force application and deflection at the mid-span of the beam. Figure 3 below shows force-deflection curves for beam 1 (purple) and beam 2 (pink). The blue line in the curve presents the limit of allowable beam's deflection at the service limit condition (*l*/240) according to the standard [3]. At approximately 4.5 ton of loading application, OPS RC beam attains its service limit condition.

4.2. Crack Evolution and Crack Opening

By nature, concrete is weak in tension, so the first cracks occur when tensile strength is passed. In the material characterisation test, tensile strength test was performed by split test following standard [13]. The cylindrical specimen is charged from its lateral side. Meanwhile, flexural strength test was performed by a four-point flexural test following standard [14]. During this test, the bottom fibre of beam is under tension and failure of the beam happens due to the first crack, which starts from its bottom edge. Both tests normally give more or less the same results [2]. In the material characterisation part, a significant difference of flexural strength and split test occurred; the split test had six times the value of the flexural test. Some standards mention that tensile strength is approximately one tenth of its compressive strength [8]. In other words, a split test to obtain tensile strength may not give the first crack stress in OPS concrete. According to these results, OPS concrete does not have the same trend as normal concrete with natural coarse aggregates. Further investigations for split and flexural test on OPS concrete are needed.

For crack observation, our interest is to measure the crack opening and to observe the cracking pattern on the surface of the beam for each loading step application. Figure 4 presents the crack opening evolution curve for both samples. The presented crack width is the largest crack opening observed on the beam for each loading step, which is not necessarily measured from the same crack. Then, on the left part of figure 5, crack width measured by concrete crack measuring microscope is presented.

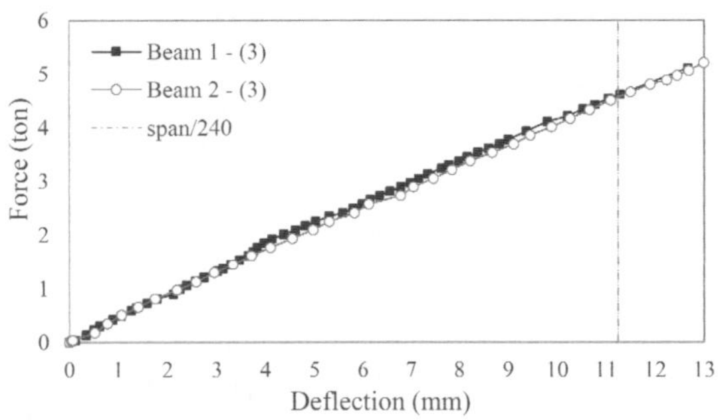

Figure 3. Force-Displacement Curve at Mid-span.

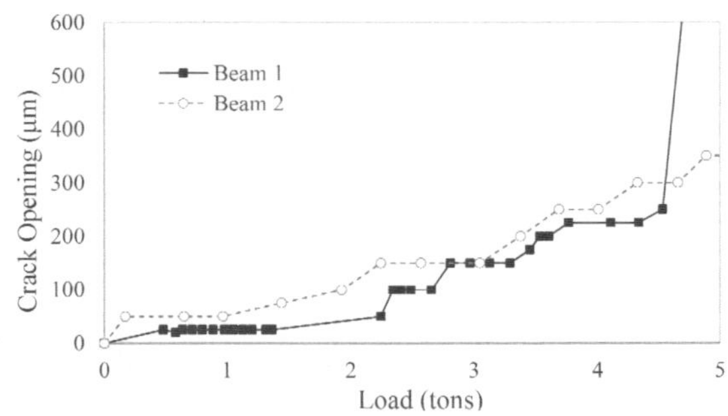

Figure 4. Crack Opening Evolution.

In figure 4 above, 300-µm of crack opening occurs when the loading application attains approximately 4.33 tons. According to [4], Eurocode 2 limits the crack opening at 300-µm for some utilisation of buildings. It means that, from both crack opening (figure 4) and maximum deflection at mid-span (figure 3), service limit states occur at approximately 4.33 tons of load (the smallest one is chosen). The crack opening for sample 1 and 2 at the maximum load is 2500-µm and 1350-µm, respectively (see left part of figure 5). Beam 2 has a larger crack opening since the beginning was ultimately due to larger shrinkage in the initial condition. Nevertheless, in the end, the largest cracks in beam 1 were more widely opened than the ones of beam 2 (figure 4 and 5).

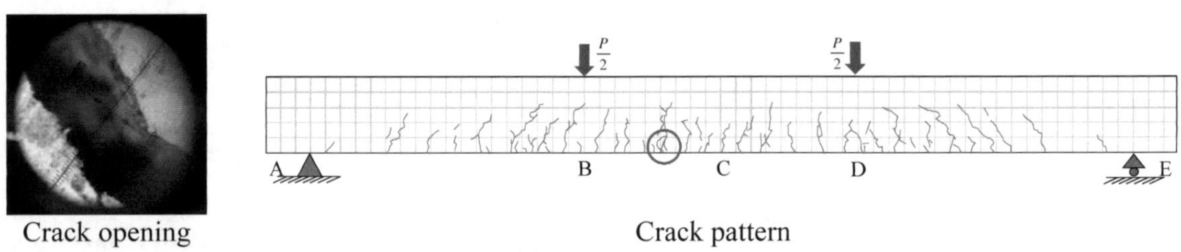

(a) Beam 1

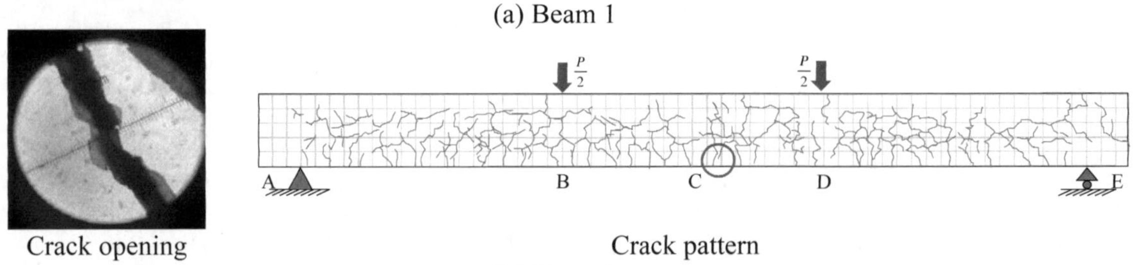

(b) Beam 2

Figure 5. Crack opening and crack pattern for both beams at the end of the test.

Cracking pattern of the two beams at the end of the tests is presented in figure 5. Based on visual observation, beam 2 has cracks that are scattered along the beam. During the experiment, after the first loading step, horizontal, vertical and leaning cracks spread on the front and back vertical surfaces. This condition may signify these cracks were induced by early shrinkage. Further loading application showed flexural cracks initiated from the bottom edge of the beam and occurred more or less at the same time for both beams. The horizontal cracks that occurred at the beginning did not increase their opening. Nevertheless, the force-displacement mechanical response evolution for the two beams presented in figure 3 shows a same tendency.

5. Conclusion

Tensile tests using split and flexural strength experiments have significantly different results. It means that OPS concrete does not have the same trend as normal concrete with natural coarse aggregates. It is recommended to be mechanically investigated as future research work. Experimental results on OPS concrete is better tested using the flexural test to find the stress of the first cracks.

Crack opening from OPS RC beam was observed at each loading step. The two beams have the same tendencies in force-displacement at mid-span although the crack pattern on the beam is very different, as beam 2 is influenced by the shrinkage based on experimental observation. Moreover, at 4.33 ton of loading application, crack width 300-μm occurred. Compared to the theoretical calculation for normal concrete with natural coarse aggregates based on standard [3], 4.33 ton is 74% of the ultimate load. Shrinkage of OPS concrete should be studied further in future research.

Acknowledgement

This work is supported by Hibah PITTA 2018 funded by DRPM Universitas Indonesia No.2428/UN2.R3.1/HKP.05.00/2018.

References

[1] Handika N 2017 *Multi-cracking of Reinforced Concrete Structures: Image Correlation Analysis and Modelling.* Doctoral Thesis INSA Toulouse (2017).

[2] Rahal S 2015 *Influence of Anisotropy Induced by cracks on the behaviour of Geomaterial Poromechanics.* Doctoral Thesis (in French) INSA Toulouse (2015).

[3] SNI 2847:2013 Requirements for Structural Concrete for Buildings. National Organisation for Standardization: 2013.

[4] Eurocode 2, 2005. *Eurocode 2, EN 1992-1-1. October 2005: Design of concrete structures - Part 1-1: General rules and rules for buildings.* Eur. Comm. Stand. CEN.

[5] Directorate General of Plantation. 2016. *Statistic of Indonesia Plantation* Three Crop Estate Statistics of Indonesia 2015-2017 Kelapa Sawit Palm Oil.

[6] Febriansyah H, Setiawan A A, Suryopratomo K and Setiawan A 2013 *Gama Stove: Biomass Stove for Palm Kernel Shells in Indonesia.* Conference and Exhibition Indonesia Renewable Energy & Energy Conservation [Indonesia EBTKE CONEX 2013].

[7] Tjahjono E, Fani A M, Dodi D D, Purnamasari E P, Silaban F A, Arijoeni E 2017 *Mater. Sci. Forum* **902**.

[8] Yew M K et al. 2014 *Effects of Oil Palm Shell Coarse Aggregate Species on High Strength Lightweight Concrete.* Kuala Lumpur: University of Malaya.

[9] Lee S T 2018 *Study on The Effect of Pre-Treatment of Oil Palm Shell As Coarse Aggregate Using Hot Water 50°C and Common Water 28°C to Lightweightconcrete Strength.* Bachelor thesis Universitas Indonesia.

[10] SNI 03-3449-2002 2002. *Procedures for Making Lightweight Concrete Mixtures with Light Aggregates.* Department of Public Works.

[11] ASTM C39 / *C39M – 18. Standard Test Method for Compressive Strength of Cylindrical Concrete Specimens United States of America*: ASTM Internasional.

[12] SNI 4431: 2011. *Test method for normal concrete flexural strength with two loading points.* Jakarta: Indonesia National Organisation for Standardization.

[13] SNI 03-2491-2022. *Test method for tensile strength using split test of concrete.* Jakarta: Indonesia National Organisation for Standardization.

[14] SNI 07-2529-1991 *Tensile Strength Testing Method of Reinforcement Steel Bar.* National Organisation for Standardization: 1991.

[15] SNI 1727:2013 *Minimum Load Standard for Building Planning and Other Structures.* National Organisation for Standardization: 2013.

Comparison on Thermodynamic Performances between Basic and Regenerative Sub-Critical Organic Rankine Cycles Coupled with Low Grade Heat Source

Pengfei Yuan[1, 2], Chao Zhang[1, 2*]

[1]Tianjin Key Laboratory for Advanced Mechatronic System Design and Intelligent Control, 300384 Tianjin, China
[2]National Demonstration Center for Experimental Mechanical and Electrical Engineering Education (Tianjin University of Technology), Tianjin 300384, China

Abstract. The coupling effects of low grade heat source on thermodynamic performances of both the basic and regenerative sub-critical Organic Rankine cycle (ORC) systems are investigated and compared. Turbine inlet pressure and turbine inlet temperature are treated as independent variables. The system exergy efficiency is selected as the thermodynamic performance criterion. The waste hot water with temperature range of 373.15K to 423.15K is selected as the low grade heat source. Optimization studies are carried out by using eight common organic working fluids to recover heat energy, and relevant operating conditions are obtained respectively. The results indicate that both hot fluid inlet temperature and the allowable minimum hot fluid outlet temperature influence the optimal working fluid, relevant turbine inlet condition and system exergy efficiency. Optimal system exergy efficiency increases monotonously with increasing of hot fluid temperature for both ORC systems above. In comparison to the basic cycle configuration, optimal system exergy efficiency for the regenerative system using the same working fluid is not changed if relevant turbine inlet vapor is saturated, but is significantly improved if relevant turbine inlet vapor is overheating. Besides, the optimal working fluid is dependent on the heat source temperature and specific cycle configuration.

1. Introduction

In the past few decades, organic Rnakine cycle (ORC) technology has drawn an increasing attention to harvest energy from the low grade heat sources. Generally, the low grade heat sources refer to the heat sources with temperature less than 473K, including the industrial waste hot water, the exhaust gas from internal combustion engines and gas turbines, the geothermal brine, the solar hot water, etc. The working principle of the ORC system is almost identical with the conventional steam Rankine system, except for the usage of alternative working fluid. In accordance with the lower temperature of heat sources, organic fluid with much lower critical temperature than water is selected as the working fluid.

For a new ORC power plant design, the first issue is to define the cycle configuration. Many cycle configurations such as sub-critical, trans-critical or super-critical, basic or regenerative, single-pressure or dual-pressure cycles [1, 2] have been proposed and investigated. In practice, the basic and regenerative, sub-critical and single pressure ORC systems are universally adopted in many state-of-the-art applications due to its allowable working pressure and system compactness [3, 4]. Therefore, the present study is focus on the basic and regenerative sub-critical ORC systems. In comparison to the basic cycle configuration, the use of regenerator can lower the heat load into the evaporator and thus improve cycle thermal efficiency. Relevant studies [5-7] showed that the benefit of using regenerator can be neglected in terms of net power output or specific net power output.

Although the layout of the basic or regenerative sub-critical ORC system is relatively simple, it is not an easy work to obtain the optimal performance and relevant working condition. Many efforts have been made to obtain the optimal working conditions and maximum thermodynamic performance of the ORC system. Yamamoto et al [8] concluded that saturated vapor for organic fluid at turbine inlet gives higher turbine power output than overheating vapor. Based on this conclusion, subsequent optimization works [9, 10] are mostly carried out on the evaporating pressure or temperature along the saturated vapor line. By contrast, Wang et al [11] and Satanphol et al [12] increased the investigation of operating conditions including saturated and superheated states at turbine inlet, and treated the turbine inlet pressure and turbine inlet temperature as independent variables. However, the previous studies [8-12] neglected the coupling effect of low grade heat source, i.e. the heat power recovery capability from the low grade heat source was not considered. The trade-off relation between cycle thermal efficiency and heat source recovery effectiveness was firstly discussed in the work of Liu et al [13]. They concluded that the cycle thermal efficiency alone may be an inappropriate criterion for thermodynamic performance evaluation, and a new criterion named system thermal efficiency was proposed. Borsukiewicz-Gozdur A [14] introduced two concepts about open and sealed type of heat sources, which highlight the difference between heat source capacity and heat power supplied to the cycle, the power plant exergy efficiency was finally proposed as a new criterion for the open type heat source recovery. This new performance evaluation criterion in the view of whole system is adopted by subsequent researchers [15, 16].

The aim of present paper is to perform parametric optimization and performance comparison between the basic and regenerative sub-critical ORC systems by using several common working fluids, especially considering the coupling effect of the low grade heat source.

2. ORC system modelling

2.1. ORC system description

The schematic layout of the regenerative ORC system is shown in figure 1. This ORC system consists of five components: an evaporator, a turbine, a regenerator, a condenser and a feed pump.

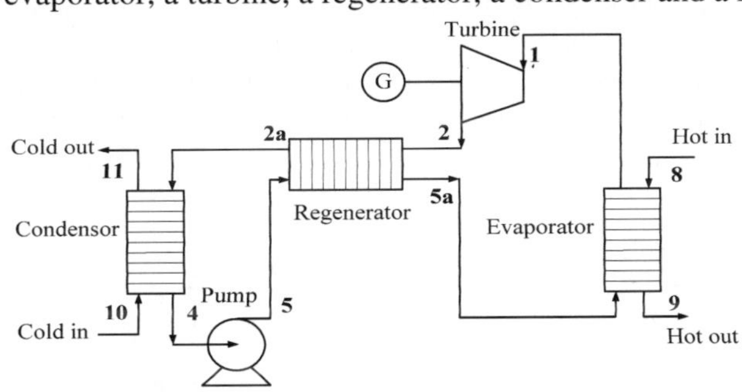

Figure 1. .The layout of regenerative ORC system

Figure 2 shows the thermodynamic processes in terms of temperature-entropy diagrams of the basic and regenerative ORC systems. The corresponding thermodynamic processes of working fluid in the basic cycle configuration are: process 1-2 (expansion process across the turbine), process 2-4 (heat removal by the cold fluid at constant low pressure), process 4-5 (pumping process), and process 5-1 (heat addition from the hot fluid at constant high pressure). Moreover, process 8-9 accommodated with process 5-1 in evaporator depicts the decreasing trend of hot fluid temperature, and process 10-11 accommodated with process 2-4 in condenser describes the increasing trend of cold fluid temperature. For the regenerative cycle configuration, the working processes of the working fluid are changed to process 5a-1 in the evaporator and 2a-4 in the condenser, respectively. In addition, evaporator pinch point temperature difference $\Delta T_{pp,e}$, condenser pinch point temperature difference $\Delta T_{pp,c}$, and regenerator pinch point temperature difference $\Delta T_{pp,r}$ are depicted in figure 2.

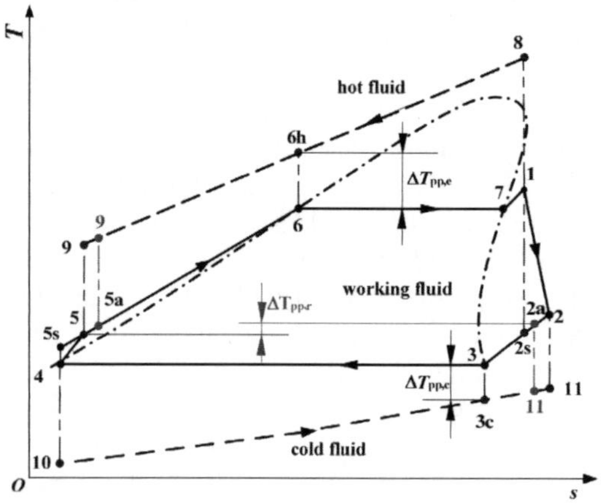

Figure 2. *T-s* diagrams for the basic and regenerative ORC systems. The blue lines indicate the regenerative cycle configuration.

2.2. Mathematical modelling

Based on the energy conservation under steady operating condition, the work output by turbine is given by

$$W_{t} = m_{wf}\left(h_{1} - h_{2s}\right)\eta_{t} \tag{1}$$

Where, W_t is the work output by turbine (W), m_{wf} is the mass flow rate of working fluid (kg/s), h_1,h_{2s} are the specific enthalpies at states 1 and 2s (J/kg), and η_t is the turbine isentropic efficiency.

The work consumption by the feed pump can be calculated as

$$W_{p} = m_{wf}\left(h_{5s} - h_{4}\right)/\eta_{p} \tag{2}$$

Where, W_p is the work consumption by the feed pump(W), h_{5s},h_4 are the specific enthalpies at states 5s and 4 respectively (J/kg), and η_p is the feed pump isentropic efficiency.

The low grade heat source recovery is usually constrained by the physical characteristics of hot fluid, hence the hot fluid outlet temperature is restricted by given minimum allowable value T_{9min}. Hence, system exergy efficiency, which is used to judge the overall utilization degree of the heat source, can be calculated as

$$\eta_{sym,ex} = \frac{W_{t} - W_{p}}{m_{hf}\left[\left(h_{8} - h_{9min}\right) - T_{0}\left(s_{8} - s_{9min}\right)\right]} \tag{3}$$

Where, $\eta_{sym.ex}$ is the system exergy efficiency, m_{hf} is the mass flow rate of hot fluid (kg/s), h_8,h_{9min} are the specific enthalpies at state 8 and with minimum allowable hot fluid outlet temperature (J/kg), s_8, s_{9min} are the specific entropies at state 8 and with minimum allowable hot fluid outlet temperature (J/kg·K), and T_0 is the reference temperature with value of 293.15K.

3. ORC system specifications

3.1. Specified conditions

The low grade heat source is waste hot water with the mass flow rate of hot water is 10kg/s. The minimum allowable hot water outlet temperature is set as 333.15K. The heat sink is cold water at 293.15K. The isentropic efficiencies of turbine and feed pump are set as 70% and 75% respectively. The pinch point temperature differences in evaporator, condenser and regenerator are 10K, 5K and 5K respectively. The condensing temperature is set as a typical value of 303.15K. All the above mentioned parameters are given with fixed values. Furthermore, the temperature of hot fluid varies from 373.15K to 423.15K to investigate the influence of heat source temperature.

The independent variables are turbine inlet pressure and turbine inlet temperature. Following the suggestion in [17], the valid working regions of independent variables (i.e. turbine inlet pressure and temperature) are given in the following:

$$p_{\mathrm{cond}} + 10\mathrm{kPa} \le p_1 \le p_{\mathrm{sta}}(\min[(T_{\mathrm{crit}} - 10\mathrm{K}), T_8 - \Delta T_{\mathrm{pp,e}}]) \tag{4}$$

$$T_{\mathrm{sta}}(p_1) \le T_1 \le T_8 - \Delta T_{\mathrm{pp,e}} \tag{5}$$

Besides, expansion process 1-2 across the turbine is constrained to locate at the superheated region, i.e. the dryness fraction of working fluid in process 1-2 is always higher than or equal to be 1.

3.2. Candidate working fluids

Based on the existing criteria for working fluid selection in [18-20], eight candidate working fluids R123, R245fa, R114, R236ea. R236fa, RC318, R227ea and R1234yf in present study are all dry or isentropic fluids. The main thermodynamic properties of the selected fluids are summarized in Table 1.T_{crit} and p_{crit} denote the critical temperature (K) and critical pressure (kPa) of the working fluid. All the thermodynamic properties of candidate working fluids are acquired from NIST REFPROP database. It is nothing that the condensing pressures are all higher than the atmosphere pressure to avoid the utilization of vacuum facilities for keeping vacuum conditions in condenser and turbine.

Table 1. Thermodynamic properties of candidate working fluids

Working Fluid	T_{crit}(K)	p_{crit}(kPa)	p_{cond}^{a}(kPa)
R123	456.83	3661.80	130.53
R245fa	427.16	3651.00	211.73
R114	418.83	3275.00	292.25
R236ea	412.44	3502.00	288.12
R236fa	398.07	3200.00	375.96
RC318	388.38	2777.50	425.15
R227ea	374.90	2925.00	610.79
R1234yf	367.85	3882.20	895.21

[a] Condensing pressure at T_3=303.15K.

4. Results and discussion

4.1. Optimal working condition

Figure 3 shows the optimal turbine inlet conditions with maximum system exergy efficiency using working fluids R114, R236fa and R1234yf with three heat source temperatures T_8=373.15K, 393.15K and 423.15K. It is noticed that all the optimal turbine inlet conditions always lie on the saturated lines for working fluids R114 and R236fa with either heat source temperature. But using working fluid R1234yf, the optimal turbine inlet conditions do not always locate on the saturated line. The optimal turbine inlet condition moves to the overheating region with the increasing of heat source inlet temperature. Besides, the optimal turbine inlet pressure and temperature increase with the increasing of heat source temperature for either specific working fluid. The regenerator has no effect on the optimal working conditions for working fluids R114 and R236fa, and influences the optimal working conditions for working fluid R1234yf.

4.2. Optimal system exergy efficiency

The change trends of optimal system exergy efficiency versus heat source temperature using working fluids R114, R236fa and R1234yf are shown in Figure 4. When optimal turbine inlet temperatures are kept at saturated states (see figure 3) using working fluids R114, R236fa, and R1234yf, optimal system exergy efficiencies monotonously increase with the increase of heat source temperature for either cycle configuration. The regenerative cycle using working fluid R1234yf also has increasing system exergy efficiency with the increasing of heat source temperature. But the basic cycle configuration using working fluid R1234yf shows a local maximum system exergy efficiency with heat source temperature T_8=403.15K. Compared with the system exergy efficiency for the basic cycle configuration, the system exergy efficiencies for the regenerative cycle configuration using working fluids R114 and R236fa show the same values. The regenerative cycle configuration exhibits more or less benefits when the optimal turbine inlet conditions are overheated (see figure 3) with heat source temperature T_8=403.15K - 423.15K.

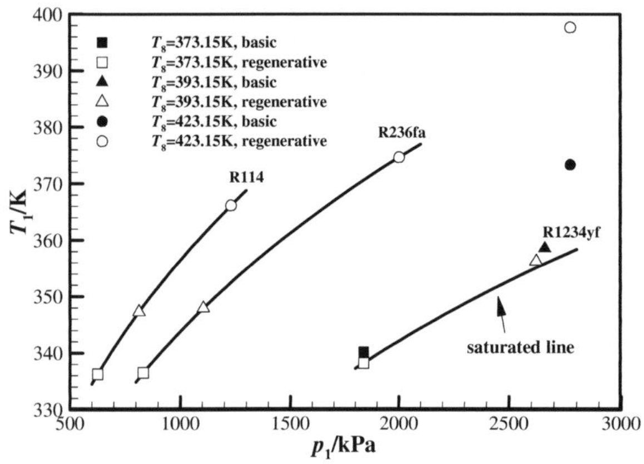

Figure 3. Optimal turbine inlet conditions for the basic and regenerative ORC systems. The solid lines indicate saturated lines for working fluids R114, R236fa and R1234yf.

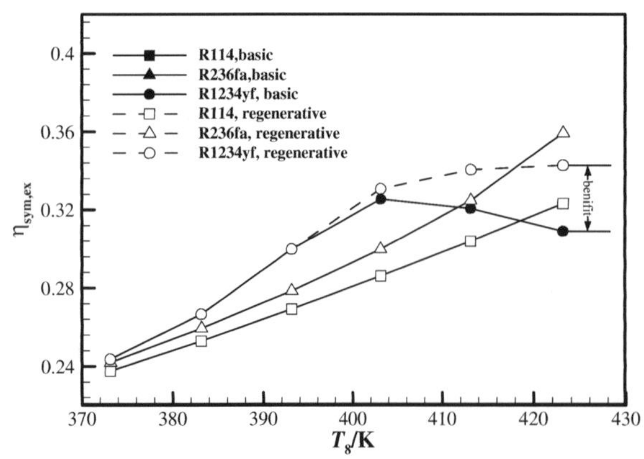

Figure 4. Optimal system exergy efficiency versus heat source temperature for the basic and regenerative ORC systems

The system exergy efficiencies for the basic and regenerative ORC system, and the benefits of regenerator using working fluids RC318, R227ea and R1234yf are listed in table 2. The maximum benefit of system exergy efficiency is 5.44% for working fluid RC318 with T_8=423.15K, and the minimum value is 0.53% for working fluid R1234yf with T_8=403.15K. The benefit is bigger with higher heat source temperature for a specific working fluid. With a certain heat source temperature, benefit of the regenerative ORC system decreases with the decreasing of critical point temperature of working fluid.

Table2. System exergy efficiency benefits of the regenerative ORC system.

Working fluid	T_8=403.15K			T_8=413.15K			T_8=423.15K		
	basic	regenerative	benefit	basic	regenerative	benefit	basic	regenerative	benefit
RC318							0.3509	0.4053	0.0544
R227ea	0.3401	0.3490	0.0089	0.3324	0.3701	0.0377	0.3173	0.3629	0.0456
R1234yf	0.3254	0.3307	0.0053	0.3207	0.3405	0.0198	0.3091	0.3428	0.0337

The working fluids with maximization of system exergy efficiency with different heat source temperatures for the basic and regenerative ORC systems are shown in figure 5. Note that the BSUB and RSUB are the abbreviations of the basic and regenerative sub-critical ORC systems, respectively. With the heat source temperature T8=373.15K-403.15K, best performances are obtained with R227ea as working fluid, and there are no differences of the system exergy efficiencies between the basic and regenerative ORC systems. System exergy efficiencies for the regenerative ORC system are larger than the basic ORC system with heat source temperature of T_8=403.15K-423.15K. With heat source temperature of T_8=413.15K, optimal system exergy efficiency is achieved with value of 35.41% for working fluid RC318 for the basic ORC system and 36.98% for working fluid R227ea for the regenerative ORC system. The maximum system exergy efficiency is 40.53% for working fluid RC318 for the regenertive ORC system, and 35.92% for working fluid R236fa for the basic ORC system with the hot fluid inlet temperature of T_8=423.15K.

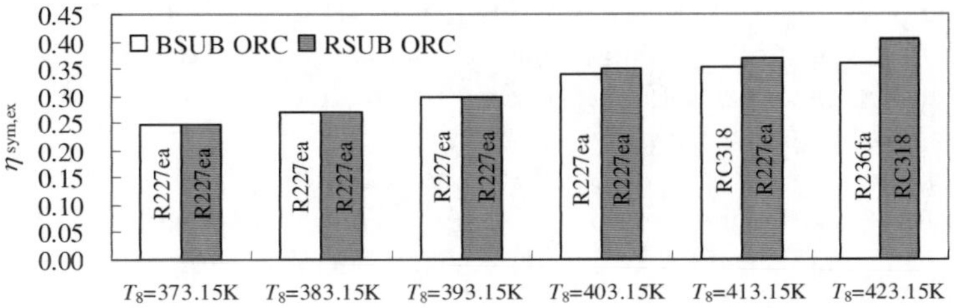

Figure 5. Best performance working fluids and relevant system exergy efficiencies.

5. Conclusion

Thermodynamic performance analysis and parameter optimization of the basic and regenerative ORC systems are performed considering the coupling effect of low grade heat source. The following conclusions can be drawn from the results of the presented work:

(1) Under the given conditions, whether saturated or overheated state for optimal turbine inlet condition is influenced by the heat source temperature, the critical temperature of working fluid and the limitation of minimum allowable heat source outlet temperature. If optimal turbine inlet vapor is saturated, optimal system exergy efficiency of the regenerative ORC system using the same working fluid is kept the same value as the one of the basic ORC system. While optimal turbine inlet vapor is located in the overheating region, optimal system exergy efficiency of the regenerative ORC system is higher.

(2) For a specific working fluid, the benefit is bigger with a higher heat source temperature. The benefit of the regenerative ORC system decreases with the decreasing of critical point temperature of working fluid when the heat source temperature is certain.

(3) Working fluid R227ea always shows best performances for both the basic and regenerative ORC systems with heat source temperature in range of 373.15K to 403.15K. The best performance working fluid is RC318 for the basic ORC system with hot fluid inlet temperature of 413.15K and for the regenerative ORC system with heat source temperature of 423.15K. Working fluids R227ea and R236fa give the highest system exergy efficiency for the regenerative ORC system with heat source temperature of 413.15K and for the basic ORC system with heat source temperature of 423.15K respectively.

Acknowledgments

The authors would like to acknowledge the financial support from Tianjin Science & Technology Pillar Program (Grant No.17YFZCSF01240) and the Natural Science Foundation of Tianjin (Grant No. 18JCQNJC07200).

References

[1] Tchanche B F, Lambrinos G, Frangoudakis A, and Papadakis G 2018 *Renew. Sustain. Energy Rev.* **15(8)** 3963-79.
[2] Vignesh P, Suresh S, and Grashin C J 2018 *Energy Convers. Manage.* **166** 474-88.
[3] Le V L, Kheiri A, Feidt M, and Pelloux-Prayer S 2014 *Energy* **78** 622-38.
[4] Astolfi M, Diega L N, Romano M C, Merlo U, Filippini S, and Macchi E 2017 *Energy Procedia* **129** 479-86.
[5] Dai Y P, Wang J F, and Gao L 2009 *Energy Convers. Manage.* **50(3)** 576-82.
[6] Maraver D, Royo J, Lemort V, and Quoilin S 2014 *Appl Energy* **117** 11-29.
[7] Javanshir A, Sarunac N, and Razzaghpanah Z 2017 *Appl. Therm. Eng.* **123** 852-64.
[8] Yamamoto T, Furuhata T, Arai N, and Mori K 2001 *Energy* **26** 239-51.
[9] Roy J P, Mishra M K, and Misra A 2010 *Energy* **35** 5049-62.
[10] He C, Liu C, Gao H, Xie H, Li Y R, Wu S Y, and Xu J L 2012 *Energy* ***38*** 136-43.
[11] Wang J F, Yan Z Q, Wang M, Ma S L, and Dai Y P 2013 *Energy* **49** 356-65.
[12] Satanphol K, Pridasawas W, and Suphanit B 2017 *Energy* **123** 326-39.
[13] Liu B T, Chien K H, and Wang C C 2004 *Energy* **29** 1207-17.
[14] Borsukiewicz-Gozdur A 2013 *Energy* **62** 73-81.
[15] Toffolo A, Lazzaretto A, Manente G, and Paci M A 2014 *Appl. Energy* **121** 219-32.
[16] Zhang C, Fu J L, Fu W C, and Liu J J 2017 *J. Eng. Thermophys.* **38(9)** 1837-43.
[17] Zhang C, Fu J L, Kang J, and Fu W C 2018 *J. Braz. Soc. Mech. Sci. Eng.* **40** 61.
[18] Mago P J, Chamra L M, and Somayaji C 2007 *IMechE Part A J. Power Energy* **221** 255-64.
[19] Vetter C, Wiemer H J, and Kuhn D 2013 *Appl. Therm. Eng.* **51** 871-9.
[20] Hærvig J, Sørensen K, and Condra T J 2016 *Energy* **96** 592-602.

Seismic Behaviour of Strap-Braced LWS Structures: Shake Table Testing and Numerical Modelling

Alessia Campiche[1], Sarmad Shakeel[1], Bianca Bucciero[1] ,Tatiana Pali[1], Luigi Fiorino[1], Raffaele Landolfo[1]

[1]Department of Structures for Engineering and Architecture, University of Naples "Federico II", Naples, Italy

Abstract. Nowadays Lightweight Steel (LWS) Constructions made with Cold-Formed Steel (CFS) profiles are currently used in seismic area, especially for residential buildings. The increasing use of these systems is accompanied by a large development of the research in this field. Indeed, the University of Naples "Federico II" has recently started an important collaboration with Lamieredil S.p.A. Company in order to investigate the seismic behaviour of strap-braced CFS systems through a series of shake-table tests, performed on two reduced-scale strap-braced structures. In addition, 3D numerical models were developed in OpenSees environment in order to simulate the dynamic/earthquake response of the whole structure, considering also the effect of the two different floor typologies. The present paper illustrates the main results of the shake-table testing and numerical modelling.

1. Introduction

In the last decade, many national and international research activities were undertaken in order to evaluate the seismic behavior of structures. In particular, the University of Naples "Federico II" dedicated specific researches to steel structures [1-13] and lightweight steel (LWS) constructions made with Cold-Formed Steel (CFS) profiles [14-20], focusing the attention on many issues. In this framework, an important cooperation between the University of Naples "Federico II" and Lamieredil S.p.A. company started in the last years. In particular, after monotonic and cyclic tests on full scale shear walls, carried out in past studies, the dynamic response of CFS strap-braced structures was analyzed by performing shake table tests on two reduced-scale (1:3) three-storeys CFS mock-ups. Two prototypes differ for the floor typology: (1) composite steel-concrete floors (CONCRETE) or (2) floors made of wood-based panels (OSB). In order to represent the experimental seismic response, a specific task has been devoted to develop 3D numerical models with OpenSees [19] software. Firstly, the numerical models of CFS strap-braced stud walls were calibrated on the base of the available experimental results, i.e. quasi-static cyclic tests on single walls [22], then numerical models of whole prototype were developed in order to simulate the non-linear dynamic response. The paper provides a description of shake table tests and numerical modelling of CFS strap-braced structures. Moreover, the experimental data and the numerical modelling results are compared and discussed in order to validate the developed models.

2. Shake-table testing

Two reduced scale mock-ups (figure 1) tested on the shake-table which representative of three-story and two bay systems, in which each storey was composed of three CFS strap-braced stud walls placed in one direction only. The two tested mock-ups differed with each other with respect to the floor typology, which was either composite steel-concrete floors (CONCRETE solution) or floors made of wood-based panels (OSB solution). Mock-ups tested on the shake table were scaled representations of the seismic force resisting zones of the case studies buildings, which were rectangular in plan with dimensions of 12.2 x 18.1 m and each storey 3 m high. The scaling procedure consists of fixing three independent scaling factors, which were length (S_L=1/3), stress (S_σ=1) and acceleration (S_a=1), while scaling factors of other parameters (dependent) are defined according to the similitude laws [23]. As a result of scaling process, the mock-ups had a rectangular plan, which covers an area of about 0.95 × 2.95 m, an inter-storey height of about 1.05 m with a total height of about 3.10 m and CFS strap-braced walls having dimension 800 x 900 mm (length x height). Response of the mock-ups was evaluated using the shake-table of Department of Structures for Engineering and Architecture, University of Naples "Federico II". Dynamic properties of mock-ups in shorter direction were evaluated through dynamic identification tests under white noise signals. In order to analyze experimental data obtained, experimental modal analysis technique was used, thanks to ARTeMIS Modal [24] software. Earthquake response was evaluated through shaking the mock-ups under natural ground motion record from 2016 Central Italy earthquake (Norcia station). The earthquake record was scaled at various intensities to replicate various ground motion hazard levels. Each test under an earthquake record was preceded by the random vibration test to track the progress of damage in structures through variation in natural period and damping ratios. Table 1 shows the scaling factors for earthquake records for both mock-ups.

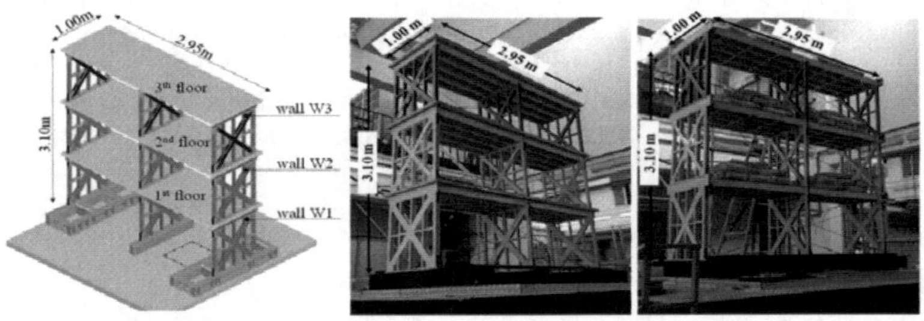

Figure 1. Reduced scale tested mock-ups.

Table 1. Scaling factors for earthquake records for both mock-ups.

Mock-up	Test label	Time history scaling factor
CONCRETE solution	TH1, TH2, TH3, TH4, TH5, TH6, TH7	9%, 12%, 38%, 49%, 100%, 120%, 150%
OSB solution	TH1, TH2, TH3, TH4, TH5, TH6, TH7	11%, 16%, 48%, 62%, 100%, 120%, 150%

3. Numerical modelling

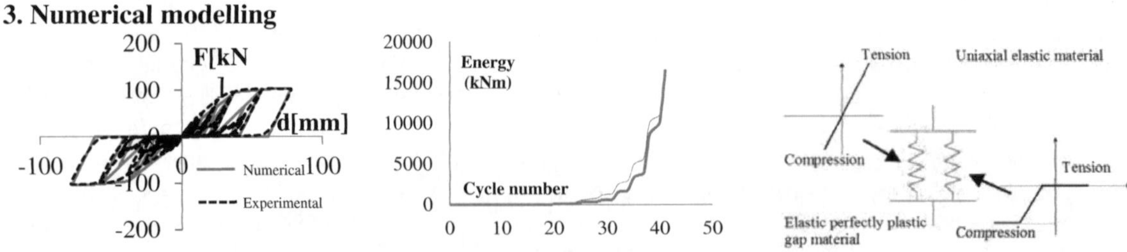

(a) Force-Displacement response of W3 wall (solid line – Experimental; dashed line- Numerical)

(b) Cumulative energy dissipation of W3 wall (solid line – Experimental; dashed line- Numerical)

(c) A two-node parallel spring element used for hold-downs

Figure 2. Modelling options and parameters used

OpenSees software [21] with a goal to better understand the contribution of several factors in the experimental seismic response of specimens and to evaluate the better options for modelling. The primary seismic force resisting system of the mock-ups, i.e. strap-braced stud walls was modelled using a pair of diagonal truss elements representing the straps of walls [25]. To simulate their hysteretic response, for the diagonal truss elements *Pinching4* material [26] was used. In particular, test data [22] was used to calibrate the backbone envelope and the parameters governing the cyclic behaviour in *Pinching4* material rule. *Pinching4* material can be defined through the set of 39 parameters, which 16 define the backbone curve, calibrated on the basis of available cyclic test data [22], and 23 control the cyclic behaviour. Model capacity to capture the wall behaviour was evaluated by comparing the experimental and numerical force-displacement responses (figure 2a) and the dissipated energy under same loading protocol (figure 2b). The definition of the position of numerical backbone points for respective walls were same in both specimen models (OSB and CONCRETE). In detail, the first point was positioned at 50% of the experimental yielding strength, the second point corresponds to the elastic limit, the third point represents the experimental peak strength and fourth point was used to capture the post peak behaviour, setting the displacement as maximum experimental displacement achieved and finding the corresponding force with an energy balance. Moreover, experimental results showed a vertical drop of strength after the peak. However, this steep drop of strength may lead to convergence problems, consequently post peak branch of the backbone envelope was aligned with little positive slope downwards using the fourth point. Four points of backbone envelope are listed in table 2. Additionally, among parameters controlling cyclic behaviour, the parameter governing ratio (*rDisp*) between the displacement of the reloading point and maximum positive displacement of preceding cycles was taken as 0.01, 0.05 and 0.1 for W1, W2 and W3 walls, respectively. Rest of the parameters required to define material rule for *Pinching4* material were taken as 0. Model capacity to capture the single wall behaviour is evaluated by comparing the experimental [22] and numerical responses. Figure 2a and 2b shows the comparison of force-displacement response and cumulative energy dissipation for W3 wall configuration. Connections present at the ends of each wall in specimens to transmit the axial forces from studs to floors or foundation were modelled with two parallel *Zerolength* elements (figure 2 c). The first *Zerolength* element was modelled with uniaxial elastic material representing the stiffness of hold down connection only, while the second element was assigned with *Elastic-perfectly plastic gap* material with a zero-gap representing the compression stiffness offered by foundation or floor. *Uniaxial elastic* material in case of the CONCRETE specimen had a stiffness of 20, 10 and 7.5 KN/mm for W1, W2 and W3 walls, respectively, whereas in case of OSB specimen it had a stiffness of 20, 20 and 15 KN/mm for W1, W2 and W3 walls, respectively. *Elastic perfectly-plastic gap* material had stiffness of 20 KN/mm for foundation, while for floors, stiffness was negligible in comparison with the foundation. The yield strength used for *Elastic perfectly-plastic gap* material was also extracted experimentally [22]. Floor was modelled by means of rigid beam-column elements in both CONCRETE and OSB solutions. Since the motion of shake table was uni-directional in x direction and the motion in the y direction was restrained, also in the model the motion in other degree of freedom was restrained. Seismic mass of the specimen, evaluated considering differences between CONCRETE and OSB solution, was distributed at the top end nodes of walls at each floor. Chord studs were modelled using truss elements having *Uniaxial elastic* material. Rayleigh damping was set to 5%, after a sensitivity analysis.

Table 2. *Truss* and *Zerolength* element properties.

MATERIAL- ELEMENT	WALL	CONCRETE	OSB
Pinching4 material - backbone envelope (stress in MPa, strain in mm/mm)	W1	(61379; 0.00053), (3068 95; 0.00877),	(283316, 0.01007), (245 516; 0.00368)
	W2	(46439,0; 000695), (232 194; 0.00911)	(185755; 0.003916), (22 2780,0.010483)
	W3	(31607,0.000643), (158 033, 0.00959)	(126426,0.003551), (15 8032,0.01079)
Elastic uniaxial material - zero length el ements stiffness (N/mm)	W1	20000	20000
	W2	10000	20000
	W3	7500	15000
Elastic-perfectly plastic gap material - z ero length elements stiffness (N/mm)	W1	20000	20000
	W2	1	1
	W3	1	1

4. Experimental vs numerical results

To check the validity of models, experimental and numerical dynamic properties obtained were compared in term of fundamental period and modal shapes, whereas the seismic behaviour was compared in term of peak roof drift ratio, defined as the ratio between the maximum top displacement and the height of the structure, and peak inter-storey drift ratio, defined as the ratio between the relative translational displacement difference between two consecutive floors and the storey height. The numerical models were subjected to modal analysis in order to find the fundamental periods of first (T1), second (T2) and third (T3) vibration modes and modal shapes, whereas experimental periods and modal shapes were defined via experimental modal analysis performed with ARTeMIS Modal software [24]. The experimental results showed that for the CONCRETE prototype the first period of vibration before Earthquake tests was 0.43 s, whereas for the OSB solution it was 0.50 s. The numerical fundamental periods are 0.40 s and 0.50 s for CONCRETE and OSB solution, respectively. The periods of vibration corresponding to the second and third mode were also evaluated and compared. Experimental and numerical values are summarized in table 3. The vibration modes obtained are compared and represented in figure 3. It is clear that numerical models give a good prediction of dynamic properties of prototypes for the first period of vibration (experimental-to-numerical ratio of 1.00 and 1.08 for OSB and CONCRETE solution, respectively), whereas the second and third periods of vibration are underestimated by numerical models (experimental-to-numerical ratio in the range 1.25 to 1.43).

Table 3. Comparison in term of fundamental periods of vibration.

	CONCRETE			**OSB**		
	EXP	**NUM**	**EXP/NUM**	**EXP**	**NUM**	**EXP/NUM**
T1 [s]	0.43	0.40	1.08	0.50	0.50	1,00
T2 [s]	0.15	0.12	1.25	0.18	0.13	1,38
T3 [s]	0.10	0.07	1,43	0.11	0.08	1,38

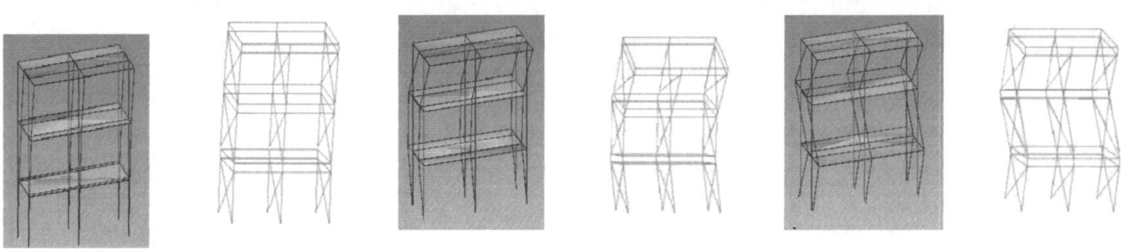

(a) First vibration mode **(b)** Second vibration mode **(c)** Third vibration mode

Figure 3. Comparison of experimental and numerical modal shapes (OpenSees).

Non-linear time history analysis was performed to evaluate the numerical dynamic behaviour of specimens and to make a comparison with the experimental response. As already introduced, in the experimental activity the earthquakes were applied in sequence to the specimens with increasing scale factors. In order to consider of the seismic sequence effects numerical models were subjected to the same strong multiple earthquakes really reproduced by the shake table. In this way, also the signal distortion (differences between signal really reproduced by the shake table and commanded signal) is considered by numerical models. Regarding the seismic response, it can be noticed that the maximum experimental inter-storey drifts were recorded for the input with maximum intensity (scaling factor of 150%) at the 3rd level: 3.49% for CONCRETE solution and 2.21% for OSB solution, whereas the maximum numerical drifts were 1.78% for CONCRETE recorded at 2nd level and 1.33% for OSB recorded at 3rd level. A comparison between experimental and numerical results in terms of peak inter-storey drift (PIDR) is given in figure 4. The experimental to numerical ratio is in the range 0.80 to 2.08 for CONCRETE specimen and 0.65 to 2.06 for OSB specimen. Models are able to capture

PIDR for low inputs, whereas they do not show good predictions for higher inputs. The reason behind the inaccurate response at high intensity earthquakes could be the accumulation of damage and rocking of stories of specimen, which could not be properly captured by numerical models and would be the focus of future studies. Moreover, a comparison between experimental and numerical results in terms of peak roof drift ratio (PRDR) is shown in figure 5.

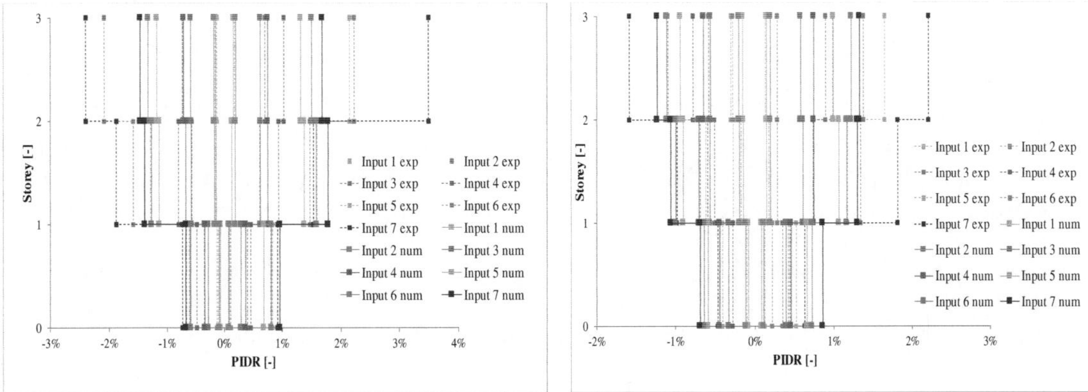

(a) CONCRETE (b) OSB

Figure 4. Comparison between experimental and numerical peak inter-storey drift ratio (PIDR).

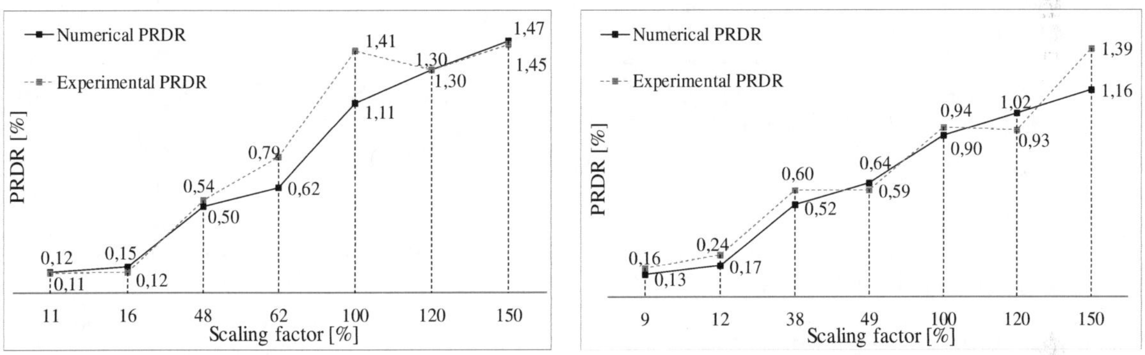

(a) CONCRETE (b) OSB

Figure 5. Comparison between experimental and numerical peak roof drift ratio (PRDR).

5. Conclusions

The paper illustrates the shake-table tests carried out at University of Naples "Federico II" on two three-storeys prototypes in reduced scale (1:3), as a part of cooperation with Lamieredil S.p.A. company. Two mock-ups, which differ only for the type of floors (composite steel-concrete and wood-based panels) were designed and opportunely scaled to be compatible with the available shaking table. The two mock-ups were subjected to white noise signal for dynamic identification and earthquakes for evaluation of seismic behaviour. Moreover, 3D numerical models for both structures were developed in OpenSees environment to represent the experimental response. A comparison between experimental and numerical results was done in term of fundamental periods, mode shapes, peak inter-storey drift and peak roof drift achieved during the time history analysis. The models were able enough to capture the dynamic properties of mock-ups, such as fundamental period and vibration modes of the buildings. Experimental and numerical peak inter-storey drifts (PIDR) and roof drifts (PRDR) were compared and the experimental to numerical ratios were in the range 0.80 to 2.08 (PIDR) and 0.80 to 1.27 (PRDR) for CONCRETE specimen and 0.65 to 2.06 (PIDR) and 0.91 to 1.41 (PRDR) for OSB specimen.

Acknowledgment
The study presented in this paper is part of wider research program coordinated by prof. Raffaele Landolfo with technical and financial support of Lamieredil S.p.A. Lamieredil S.p.A. is gratefully acknowledged by Authors for the consent to publish scientific results presented in this paper and in agreement with their common research scope.

References
[1] Formisano A, Matteis G. De, Mazzolani FM 2010, *Comput. Struct.* **88** 1348–60. doi:10.1016/j.compstruc.2008.09.010.
[2] Formisano A, Mazzolani FM, Matteis G. De 2007, *Int. J. Struct. Stab. Dyn.* **07** 273–94. doi:10.1142/S0219455407002289.
[3] Formisano A, Matteis G. De, Panico S, Calderoni B, Mazzolani F 2006 Full-scale test on existing RC frame reinforced with slender shear steel plates, Proc. 5th Int. Conf. Behav. Steel Struct. Seism. Areas - Stessa 2006.
[4] Tartaglia R, D'Aniello M, Rassati GA, Swanson JA, Landolfo R, 2018 *Eng. Struct.* **159** 155–71. doi:10.1016/j.engstruct.2017.12.053.
[5] Tartaglia R, D'Aniello M, Zimbru M, Landolfo R, 2018 *Steel Compos. Struct.* **27** 727–45. doi:10.12989/scs.2018.27.6.727.
[6] Tartaglia R,. D'Aniello M, Landolfo R, 2018 *J. Constr. Steel Res.* **148** 669–90. doi:10.1016/j.jcsr.2018.06.025.
[7] Tartaglia R, D'Aniello M, 2017 *Open Civ. Eng. J.* **11** 369–83. doi:10.2174/1874149501711010369.
[8] Tartaglia R, D'Aniello M, De Martino A, Di Lorenzo G 2018 *Ing Sismica.* **3** 104-20.
[9] Tartaglia R, D'Aniello M, Rassati GA, Swanson JA, Landolfo R 2018 *Key Eng. Mater.* **763** 818-25. doi:10.4028/www.scientific.net/KEM.763.818.
[10] Tenchini A, D'Aniello M, Rebelo C, Silva L Da, Lima L 2016 *Eng. Struct.* **124** 167–85.
[11] Cassiano D, D'Aniello M, Rebelo C, Landolfo R, da SL, 2016 *Steel Compos. Struct. An Int. J.* **21** 479–500.
[12] Dell'Aglio G, Montuori R, Nastri E, Piluso V, 2017 *Ingegneria Sismica,* **34(4)** 82-102.
[13] Nastri E, Vergato M, Latour M, 2017 *Earthq. Struct.* **12(1)** 13-21.
[14] Fiorino L, Shakeel S, Macillo V, Landolfo R 2017 *J. Constr. Steel Res.* **138** 324–39. doi:10.1016/j.jcsr.2017.07.014.
[15] Fiorino L, Shakeel S, Macillo V, Landolfo R, 2017 Seismic response of CFS shear walls sheathed with nailed gypsum panels: Numerical modelling, Thin-Walled Struct.
[16] Fiorino L, Macillo V, Landolfo R, 2017 *Eng. Struct.* **151** 633–47. doi:10.1016/j.engstruct.2017.08.056.
[17] Macillo V, Fiorino L, Landolfo R, 2017 *Thin-Walled Struct.* **120** 161–71. doi:10.1016/j.tws.2017.08.022.
[18] Terracciano MT, Bucciero B, Pali T, Macillo V, Fiorino L, Landolfo R, 2018 *Key Eng. Mater.* **763** 432–9. doi:10.4028/www.scientific.net/KEM.763.432.
[19] Campiche A, Shakeel S, Macillo V, Terracciano MT, Bucciero B, Pali T, Fiorino L, Landolfo R 2018 *Ing. Sismica.* **2** 106–23.
[20] Fiorino L, Pali T, Bucciero B, Macillo V, Terracciano MT, Landolfo R, 2017 *Thin-Walled Struct.* **116** 234–49. doi:10.1016/j.tws.2017.03.031.
[21] Mazzoni S, McKenna F, Scott MH, Fenves GL, 2009 OpenSees.
[22] Fiorino L, Terracciano MT, Landolfo R, 2016 *J. Constr. Steel Res.* **127** 92–107. doi:10.1016/j.jcsr.2016.07.027.
[23] Harris HG, Sabnis GM, 1999 Structural modeling and experimental techniques, CRC Press, Boca Raton, FL.
[24] Structural Vibration Solutions A/S, N.S. Park, 2018 ARTeMIS Modal 5.2.1.1.
[25] Macillo V, Shakeel S, Fiorino L, Landolfo R, 2018 *Adv. Steel Constr.* **14**.
[26] Lowes L N, Mitra N, Altoontash A 2004 *PEER Rep.* 2003/10. (2004).

Flow Mechanism of Cooling Effectiveness Improvement for the Cylindrical Film Cooling Hole with Contoured Craters

Linchao Bai[1, 2], Chao Zhang[1, 2]

[1]Tianjin Key Laboratory for Advanced Mechatronic System Design and Intelligent Control, 300384 Tianjin, China
[2]National Demonstration Center for Experimental Mechanical and Electrical Engineering Education (Tianjin University of Technology), Tianjin 300384, China

Abstract. In this paper, a typical cylindrical hole with contoured craters is investigated in terms of the downstream flow field by numerical method at blowing ratios of 0.5, 1 and 1.5. The flow field and the cooling effectiveness are analyzed using three-dimensional Reynolds-averaged Navier-Stokes equations with the shear stress transport turbulence model. In comparison to the cylindrical hole, the cratered hole shows a wider coolant coverage in lateral direction and a lower penetration into the mainstream flow due to the larger hole exit area. Moreover, the interaction between the coolant and the curved protrusion induces a new anti-kidney-shaped vortex pair, which can obviously improve the cooling coverage and thus the cooling effectiveness. The area-averaged cooling effectiveness for the cratered hole is always higher than that for the cylindrical hole at all three blowing ratios.

1. Introduction

Film cooling and thermal barrier coating (TBC) are widely used simultaneously to reduce the thermal load of gas turbine vanes/blades in order to enable higher turbine inlet temperature. Hence, new efficient but simple configurations based on the cylindrical hole may occur if the TBCs are sprayed with a special designed manner. Two kinds of typical novel configurations are the shallow trench slot [1] and the crater [2]. The geometrical difference is caused by a continuous slot-like mask around a row of holes for the trench slot but discrete masks around every discrete hole for the crater. In spite of higher cooling effectiveness for the trenched hole, the cratered hole may be a better choice especially in consideration of the thermal stress [3]. Another disadvantage of trenched hole is that the regions between cooling holes may be warmed by the invading hot gases [4]. Thus present study is mainly focused on the cratered hole.

Relevant studies [2, 5] showed that the adiabatic cooling effectiveness can improve by adopting a proper crater shape. The circular cratered hole performs better than the concentric elliptical cratered hole and the upstream offset ellipse cratered hole. Zhang et al [6] numerically evaluated five different cratered holes at a lower blowing ratio of 0.5, but found the concentric elliptical cratered hole performs best. Kim et al [7] carried out a design optimization of cratered cylindrical hole at blowing ratio of 0.5, and found an elliptical shape with wider lateral coverage showing the maximum area-averaged adiabatic cooling effectiveness. Khalatov et al [8] proposed two types of new cratered holes with hemispherical and triangular shape. Through their numerical analysis of a flat plate film cooling, the triangular shape showed better coolant lateral spreading, and thus improved adiabatic cooling effectiveness at blowing ratios higher than 1.0. Kalghatgi and Acharya [9] proposed a cratered hole with contoured shape and evaluated numerically at blowing ratios of 1.0 to 2.5. The key feature of this novel design was a blunt V-shaped protrusion toward the downstream end of cylindrical hole. An et al

[10] proposed a so-called near surface streamwise diffusion hole and investigated the adiabatic cooling effectiveness experimentally. The most improvement of this design was increasing the exit area with both lateral and streamwise expansions as the traditional shaped hole, which could lower the coolant momentum flux.

Recently, the double wall cooling concept [11] with thinner blade thickness has was proposed, which results in film cooling holes with short hole length-to-diameter ratio. Lutum and Johnson [12] analyzed the influence of hole length-to-diameter ratio of cylindrical hole ranged from 1.75 to 18 on adiabatic cooling effectiveness. They pointed out that due to the undeveloped flow in the hole and larger effective injection angle with respect to the wall, the adiabatic cooling effectiveness generally decreases with decreasing hole length-to-diameter ratio in the range of 1.75 to 5. Thus, the cratered hole with hole length-to-diameter ratio of 6 proposed in [10] is needed to be reassessed in condition of short hole length-to-diameter ratio.

Based on the novel cratered configuration proposed in [10], the present paper aims to investigate the flow mechanism of cooling effectiveness compared with the cylindrical hole. The adiabatic cooling performance at blowing ratios ranged from 0.5 to 1.5 will be also compared.

2. Physical model and numerical setup

2.1. Physical model

The physical model consists of a mainstream passage, a cratered cylindrical hole and a coolant plenum, which is shown in figure 1. The cylindrical hole has a diameter D=12.7mm, and the hole length-to-diameter is 3. The geometry of each component is also depicted in figure 1. The hole pitch has a typical value of 3D. The coordinates X, Y and Z denote the streamwise, lateral and vertical directions respectively.

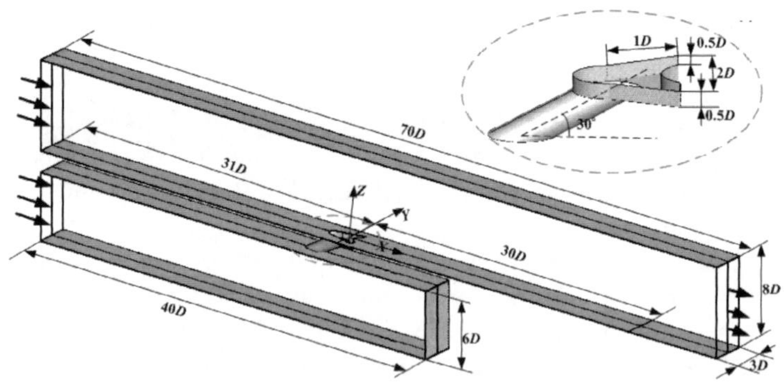

Figure 1..The physical model.

2.2. Numerical method and boundary conditions

Three dimensional RANS analysis of flow fluid and heat transfer are performed by software ANSYS CFX. This software solves the Reynolds-averaged Navier-Stokes equations (RANS) and uses the k-ω based shear-stress transport (SST) two-equation turbulence model to compute the averaged turbulent stresses. The k-ω SST turbulence model was validated for the film cooling simulations in the previous numerical works [13, 14]. Thus, in present simulations, the k-ω SST turbulence model is selected. The advection and turbulence terms are both solved with high resolution schemes. When the root mean square residuals of all variables are less than 10^{-6}, the simulations are regarded to achieve convergence.

The geometry of computational domain is created by using the software ICEM CFD. Due to the symmetric geometry by the plane Y/D=0, half of the physical model ($0 \leq Y/D \leq 1.5$) is adopted to save computational cost. The planes Y/D =0 and 1.5 are set as symmetric boundaries, and all other walls are treated as adiabatic no-slip conditions. The inlet boundary for the mainstream passage is a uniform velocity of 12.8m/s at a temperature of T_m=321K. The inlet temperature of coolant plenum is set as T_c=296K to obtain the coolant-to-mainstream density ratio of 1.08, which equals to the value in [5].

Also the inlet boundaries are set as uniform velocities of 0.2776m/s, 0.5552m/s and 0.8328m/s for the given blowing ratios M =0.5. 1.0, and 1.5 respectively. The outlet condition of mainstream passage is a constant static pressure of 0.1MPa.

The structured meshes are also generated by the software ICEM CFD. The grids are refined around all the walls including the hole and crater. The first grid point away from the wall is placed to satisfy the values of y^+ around 1. The detailed meshes around the hole are shown in figure 2. After the grid independence checking, the total number of meshes is 1.735 million.

Figure 2. Computational meshes around the hole.

2.3. Data reduction

Among the many results of air film cooling, the cooling effectiveness is used to evaluate the cooling effect. The adiabatic cooling effectiveness can be calculated as

$$\eta = (T_m - T_c)/(T_{\mathrm{aw}} - T_c) \tag{1}$$

Where, T_{aw} is the adiabatic temperature over the flat-plate wall (K).

As an important indicator to evaluate the cooling effect of the film, the higher the value of η, the smaller the temperature difference between the flat-plate wall and the inlet temperature of coolant plenum, and the better the cooling effectiveness.

Because the cooling effectiveness can't reflect the film cooling effect quantitatively, the area-averaged cooling effectiveness $\overline{\eta}$ and the intensity of the vorticity in X direction ω_X are introduced.

The area-averaged cooling effectiveness is calculated over the interested downstream region ($0 \leq Y/D \leq 1.5$, $0 \leq X/D \leq 30$), which is given by

$$\overline{\eta} = \frac{1}{1.5D \times 30D} \int_0^{30D} \int_0^{1.5D} \eta \, dY dX \tag{2}$$

The intensity of the vorticity in X direction ω_X is used to illustrate the strength of vortex pair, which can be calculated as

$$\omega_X = \partial U_Z / \partial Y - \partial U_Y / \partial Z \tag{3}$$

where U_Z, U_Y are the velocities in Z direction and Y direction, respectively.

In the following paragraphs, the vorticity distribution and the contour distribution cloud map of the streamline on the plane are generated to analyse the film cooling effect.

3. Results and discussion

3.1. Flow field

In order to better illustrate the flow mechanism of the cratered hole, a simple cylindrical hole is used for comparison. Figure 3 and 4 show the contours of velocity distribution in the Y and Z directions at the outlets of the hole and the crater with blowing ratio of 1 respectively. Note that the Y direction velocity determines the lateral expansion capability of the coolant, and the *Z* direction velocity represents the penetration ability into the mainstream flow. It is obviously seen in figure 3 that the coolant is contracted towards to the hole center by the depression of the mainstream flow for the cylindrical hole. But for the cratered hole, the upper half part of the outlet appears a positive high *Y* direction velocity region and the lower half part occurs a negative high *Y* direction velocity region. These two regions promote the coolant with a wider lateral expansion, which is benefit for the coolant coverage. There is a large difference for the *Z* direction velocity distributions, which can be clearly seen in figure 4. Due to the bending effect along the cylindrical hole, a high *Z* direction velocity region occurs at the windward side of the hole outlet, a highest *Z* direction velocity region with slim shape locates nearby the hole centerline. Moreover, the whole outlet exhibits positive *Z* direction velocity at every location. For the cratered hole outlet, the average *Z* direction velocity decreases due to a higher area. A higher *Z* direction velocity region locates at the leeward side due to the blocking effect by the curved protrusion. A negative *Z* direction velocity region appears just downstream the windward side by the mainstream flow ingestion, which is a disadvantage for cooling effectiveness improvement. The mainstream flow ingestion will heat the coolant, and thus reduce the cooling capability.

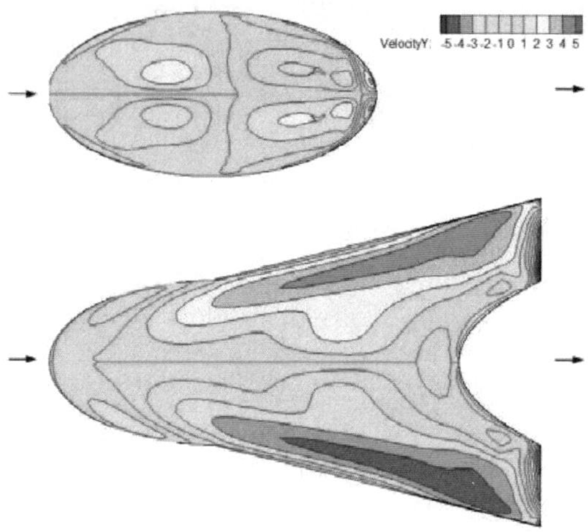

Figure 3. The contours of *Y* direction distribution at the hole outlets.

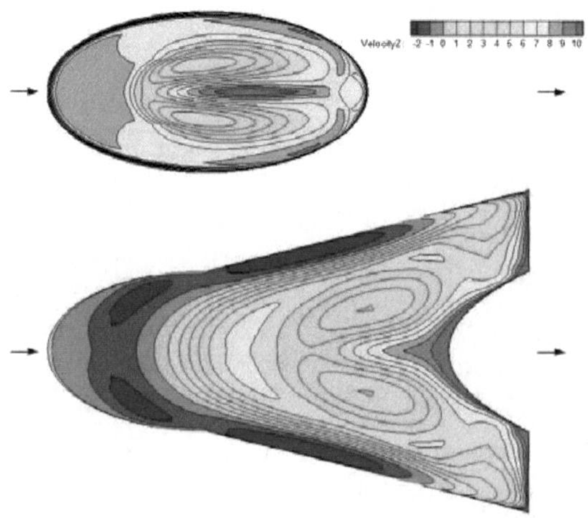

Figure 4. The contours of *Z* direction distribution at the hole outlets.

Figure 5 and 6 show the footprints of the vorticity ω_X and the streamlines over three planes X/D=0, 3.33 and 6.67 for the cylindrical hole and the cratered hole with blowing ratio of 1 respectively. It can be seen from figure 5 that a kidney-shaped vortex pair is generated by the mutual influence between the mainstream flow and the coolant injection. The coolant injection is lift away from the wall in the vertical direction, and thus cooling performance is poor. Besides, a much smaller and opposite vortex pair is located just below the kidney-shaped vortex pair at plane *X*/*D*=3.33 and 6.67, which means the coolant is reattached to the wall. For the cratered hole in figure 6, the contour of vorticity distribution is similar as the cylindrical hole, except for a small opposite vortex pair at the outside in the lateral direction of the diffusion edges. At plane *X*/*D*=3.33, an anti-kidney-shaped vortex pair is generated by the curved protrusion, and breaks the original kidney-shaped vortex pair into two segments. This anti-kidney-shaped vortex pair can drag the coolant to flow away from the hole centerline, which is benefit for a wider coolant coverage. But the intensity of the anti-kidney-shaped vortex pair is not high enough to overcome the entrainment effect by the mainstream flow, thus is disappeared at plane *X*/*D*=6.67. So the coolant coverage in the lateral direction is degraded for the far downstream region from the hole exit.

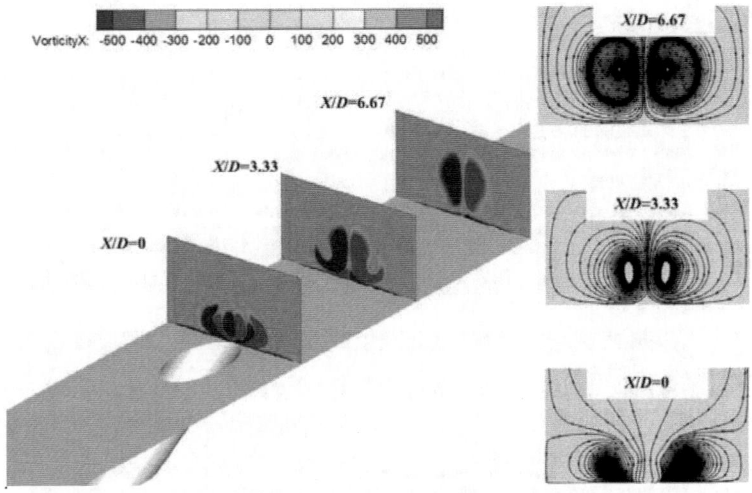

Figure 5. The contours of vorticity distribution and streamlines on planes *X*/*D*=0, 3.33 and 6.67 for the cylindrical hole.

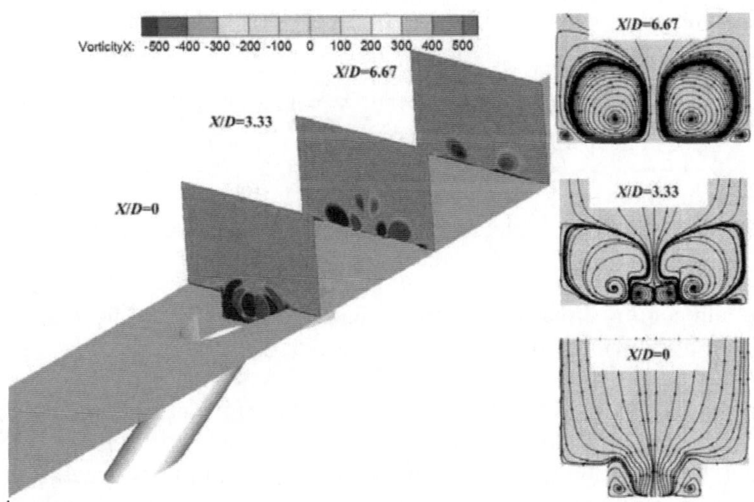

Figure 6. The contours of vorticity distribution and streamlines on planes *X*/*D*=0, 3.33 and 6.67 for the cratered hole.

3.2. Cooling effectiveness

The adiabatic cooling effectiveness distributions over the wall for the cratered hole and the cylindrical hole at three blowing ratios *M*= 0.5, 1 and 1.5 are shown in figure 7. It can see that the effective coolant coverage is much improved at any blowing ratio. A region of high cooling effectiveness is immediately observed just downstream the curved protrusion. At higher blowing ratios, the region high cooling effectiveness is still maintained, but the coolant flow towards away the centerline is much reduced. The reason is that the coolant penetration is higher into the mainstream flow, and less coolant is blocked by the curved protrusion, and thus the anti-kidney-shaped vortex pair with smaller intensity is generated.

In order to quantify the superiority of the cratered hole to the cylindrical hole, the area-averaged cooling effectiveness of both holes is given in table 1. Obviously, the cratered hole has positive influence on cooling effectiveness improvement at all three blowing ratios. The highest area-averaged cooling effectiveness with the value of 0.3025 is obtained at the lowest blowing ratio *M*=0.5. The area-averaged cooling effectiveness of cratered hole has a decreasing trend with the increasing of blowing ratio, which is consistent with the change trend of the cylindrical hole.

Figure 7. The cooling the effectiveness contours for the cratered hole and the cylindrical hole at three blowing ratios M= 0.5, 1 and 1.5.

Table 1. Area-averaged cooling effectiveness at different blowing ratios.

	M=0.5	*M*=1.0	*M*=1.5
Cylindrical hole	0.1837	0.0796	0.0357
Cratered hole	0.3205	0.2373	0.1177

4. Conclusion

The flow structure and cooling effectiveness are numerically investigated for both the cylindrical hole and the cratered hole at blowing ratio of 0.5, 1.0 and 1.5. The following conclusions can be drawn from the results of the presented work:

(1) In comparison to the cylindrical hole, the larger exit area for the cratered hole significantly alter the velocities distributions in *Y* direction and *Z* direction. This will help to improve the coolant coverage in the lateral direction and decrease the penetration height into the mainstream flow.

(2) The curved protrusion plays a key role on the generation of a new anti-kidney-shaped vortex pair. This will promote the increasing of coolant coverage in the lateral direction, and improve the cooling effectiveness significantly.

(3) The cratered hole always shows higher area-averaged cooling effectiveness than the cylindrical hole at blowing ratios of 0.5, 1 and 1.5. The area-averaged cooling effectiveness decreases with the increasing of blowing ratio.

Acknowledgments
The authors would like to acknowledge the financial support by the National Science Foundation of China (Grant No. 51506150) and the Natural Science Foundation of Tianjin (Grant No. 18JCQNJC07200).

References

[1] Bunker R S 2002 Film cooling effectiveness due to discrete holes with a transverse surface slot *Proc. ASME Turbo Expo 2002* GT2002-30178.
[2] Dorrington J R, Bogard D G, Bunker R S 2007 Film effectiveness performance for coolant holes imbedded in various shallow trench and crater depressions *Proc. ASME Turbo Expo 2007* GT2007-27992.
[3] Lee H L, Kim K M, Shin S, and Cho H H 2009 *J. Thermophys. Heat Transfer* **23(4)** 843-6.
[4] Davidson F T, Dees J E, and Bogard D G 2011 An experimental study of thermal barrier coatings and film cooling on an internally cooled simulated turbine vane, *Proc. ASME Turbo Expo 2011* GT2011-46604.
[5] Lu Y P, Dhungel A, Ekkad S V, and Bunker R S 2009 *J. Turbomachinery* **131(1)** 011005.
[6] Zhang Z W, Qian W J, and Hu J M 2016 *Key Eng. Mater.* **693** 491-7.
[7] Kim J H, and Kim K Y 2016 *J. Therm. Sci. Technol.* **11(2)** JTST0025.
[8] Khalatov A A, Panchenko N A, Severin S D 2017 *J. Phys.: Conf. Seri.* **891** 012145.
[9] Kalghatgi P, and Acharya S 2015 *J. Turbomachinery* **137(10)** 101006.
[10] An B T, Liu J J, Zhang X D, Zhou S J, and Zhang C 2016 *Int. J. Heat Mass Transf.* **103** 1-13.
[11] Ligrani P, Ren Z, Liberatore F, Patel R, Srinivasan R, and Ho Y H 2017 *J. Eng. Gas Turb. Power* **140(5)** 051901.
[12] Lutum E, and Johnson B V 1999 *J. Turbomachinery* **121(2)** 209-16.
[13] Zhang C, Liu J J, Wang Z, and An B T 2013 The effects of Biot number on the conjugate film cooling effectiveness under different blowing ratios *Proc. ASME Turbo Expo 2013* GT2013-94041.
[14] Wang Z, Zhang C, Chen Y, Du W J, and Xin G M 2017 *J. Propul. Technol.* **38(9)** 2038-45.

Presentation of the Exact Technique for Calculation of the Torsional Constant for the T-Section

N Jurkowska[1]

[1]Institute of Structural Mechanics, Tadeusz Kosciuszko Cracow University of Technology, Poland

Abstract. Both bending and torsional stiffness are the important factors affecting the spatial work of the reinforced concrete engineering structures. While bending stiffness is studied carefully, the torsional stiffness is not sufficiently investigated. This paper studies one of its components, the torsional constant. The accepted in the practice formulae of the strength of materials are appropriate only for thin-walled sections and give a significant error in other cases. That's why the exact technique has been tested on the example of the T-section, which is one of the most common cross-sections applied in engineering (for the reinforced concrete overlaps, bridge superstructures, etc.). The results of the calculation by different methods have been compared. The essential error has been proved.

1. Introduction

The change in values of both bending and torsional stiffness influences significantly the stress redistribution of the spatial structures, as has been proved both theoretically and experimentally [1]. However, the torsional stiffness of reinforced concrete elements is not taken into account in the Building Standards, and its components are not studied sufficiently. Also, in the Building Codes shear modulus of concrete is accepted depending on the Young's modulus, which is not true after passing the zone of elastic deformation.

The torsional stiffness depends on the Kirchhoff modulus and the torsional constant. First, the secant shear modulus of concrete was investigated [1]. The next important task is to determine the torsion constant for the T-section. So, the purpose of this paper is to test the exact technique for computation of the torsional constant for the concrete section, in order to apply it in the subsequent calculations of the torsional stiffness of the reinforced concrete elements when studying stress redistribution in the cross-ribbed systems.

The example of the T-section is considered, as one of the most common cross section, because the overlap is often divided into T-elements for calculation of the overlaps and bridges with consideration of spatial work. The stiffness of the T-elements with normal cracks can be determined by [2]. The torsional constant of the whole (uncracked) element is generally determined as the sum of values of stiffness for the shelf and ribs, which in the case of not thin-walled sections gives a significant error, and the torsional deformations of such a section cannot be considered as the equivalent to the deformations of the thin-walled section. The reason is that the Saint-Venant's principle is exact only for the thin-walled sections. Due to the absence of the exact solution for the case of the not thin-walled elements (T-sections, I-beams, etc.), the torsional stiffness should be found by simulation using the volumetric finite elements and further approximation (e.g., according to [2]).

Therefore, this paper studies the technique for calculation the stiffness of the T-section by the exact formulae of the theory of elasticity.

2. Desctiption of the applied technique

The formulae of the proposed technique have been derived from the basic equations of the theory of elasticity [3]. The stiffness of the T-section has been determined by the exact method, which considers the T-section element as a prismatic rod of the polygonal section, the sides of which form right angles (figure 1).

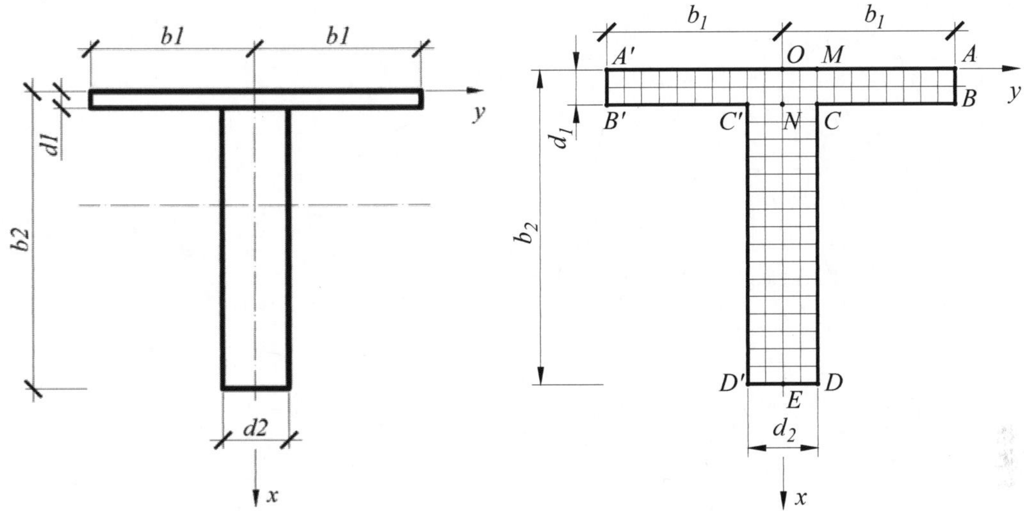

Figure 1. T-section: splitting into elements according to the accepted technique.

Torsional stiffness for T-beam is calculated by the formula derived from the elasticity theory formulae [3]:

$$C = G\left\{\begin{aligned}&\frac{d_2^3 b_2}{3}+\frac{2}{3}d_1^3\left(b_1-\frac{d_2}{2}\right)-\frac{d_1^3 d_2}{\pi^2}\sum_{k=1,3,\dots}^{\infty}\frac{z_{2k}}{k^3}\left[th\frac{k\pi d_2}{2d_1}+th\frac{k\pi}{2d_1}\left(b_1-\frac{d_2}{2}\right)\right]+\\&+\frac{d_2^3 d_1}{\pi^2}\sum_{k=1,3,\dots}^{\infty}\frac{z_k}{k^3}sh\frac{k\pi b_2}{2d_2}sch\frac{k\pi d_1}{2d_2}sch\frac{k\pi(b_2-d_1)}{2d_2}+\\&+\frac{64d_1^4}{\pi^5}\sum_{k=1,3,\dots}^{\infty}\frac{1}{k^5}\left[th\frac{k\pi d_2}{2d_1}-cth\frac{k\pi b_1}{d_1}+\operatorname{csc}h\frac{k\pi b_1}{d_1}sch\frac{k\pi d_2}{2d_1}\right]-\\&-\frac{32d_1^4}{\pi^5}\sum_{k=1,3,\dots}^{\infty}\frac{1}{k^5}\left[th\frac{k\pi b_2}{2d_2}+th\frac{k\pi d_1}{2d_1}\right]\end{aligned}\right\} \tag{1}$$

The expression (1) can be represented as

$$C = C_0\eta(b_1, b_2; d_1, d_2) \tag{2}$$

where C_0 is the sum of values of stiffness, which form the T-section, η is the influence coefficient for its stiffness.

Notation conventions in formulae (1-2):

b_1, b_2, d_1, d_2 are the geometric dimensions (see figure 1),

z_m are unknown coefficients which are determined from the infinite regular system of linear equations:

$$z_m = \sum_{n=1}^{\infty} C_{mn} z_n + \gamma_m, \qquad (m = 1, 2, \ldots) \tag{3}$$

where the components can be defined through the rows:

$$C_{2k-1,2p-1} = C_{2k,2p} = 0$$

$$\begin{cases} C_{2k-1,2p} = \dfrac{2(2k-1)d_1}{\pi d_2} \dfrac{sh\alpha_k d_1 sh\alpha_k (b_2 - d_1)\csc h\alpha_k b_2}{p^2 + (2k-1)^2 (d_1/d_2)^2} \\ C_{2k,2p-1} = \dfrac{-4kd_2}{\pi d_1} \dfrac{ch\dfrac{k\pi d_2}{2d_1} sh\dfrac{k\pi}{d_1}\left(b_1 - \dfrac{d_2}{2}\right)\csc h\dfrac{k\pi b_1}{d_1}}{(2p-1)^2 + k^2 (d_2/d_1)^2} \end{cases}$$

$$\begin{cases} \gamma_{2k-1} = \dfrac{4(2k-1)}{d_1 d_2} \dfrac{sh\alpha_k (b_2 - d_1)}{sh\alpha_k b_2} \left[-\dfrac{8d_2^2}{(2k-1)^3 \pi^3} + \dfrac{32 d_1^3}{\pi^4 d_2} sh\alpha_k d_1 \times \sum_{p=1}^{\infty} \dfrac{1 - sh\dfrac{(2p-1)\pi d_2}{2d_1}\csc h\dfrac{(2p-1)\pi b_1}{d_1}}{(2p-1)^2\left[(2p-1)^2 + (2k-1)^2 (d_1/d_2)^2\right]} \right] \\ \gamma_{2k} = \dfrac{8k}{d_1 d_2} \dfrac{sh\dfrac{k\pi}{d_1}\left(b_1 - \dfrac{d_2}{2}\right)}{sh\dfrac{k\pi b_1}{d_1}} \left[\dfrac{4d_1^2}{k^3 \pi^3} \dfrac{1 + (-1)^{k+1}}{ch\dfrac{k\pi b_1}{d_1}} + \dfrac{16 d_2^3}{\pi^4 d_1} ch\dfrac{k\pi b_1}{d_1} th\alpha_k \times \sum_{p=1}^{\infty} \dfrac{1 - sh\alpha_p d_1 \csc h\alpha_p b_2}{(2p-1)^2\left[(2p-1)^2 + k^2 (d_2/d_1)^2\right]} \right] \end{cases} \tag{4}$$

Substituting the found values of the unknown Z_k^+ and Z_k^- in the calculation formula, respectively, the upper and lower limits for the stiffness value C^+ and C^- for the T-section with an arbitrary ratio of the geometric dimensions of the shelves and the wall are obtained. The estimated value of the stiffness is $CC = GI = \dfrac{C^+ + C^-}{2}$, the error is $\delta(\beta) = \dfrac{C^+ + C^-}{C}$. The system (3) is regular. The set of free members $\{\gamma_m\}$ of the system is bounded from above and tends to zero at $m \to \infty$ as $O\left(\dfrac{1}{m}\right)$ which indicates its convergence and allows to determine the unknowns z_m with any accuracy.

3. Comparison of the results with another methods

The comparison for the calculation of torsional stiffness for the T-section by this technique with values obtained by the known formulae of the strength of materials as the sum of the values for rectangles, into which the section can be divided (considering the Saint-Venant's torsional coefficients) and the data of a similar calculation of the thin-walled section by the exact and approximate technique is given below.

The element of the T-section which models the elements of the bridge superstructure has been accepted. The geometric characteristics of the T-section are shown in table 1 and figure 2. Stiffness *C=GI* was calculated by the explained above exact technique and is equal to $G_c I_t = 1125 \times 227329 = 255744712$ kN/cm^2.

Table 1. Geometry of the T-section.

b_1, cm	b_2, cm	d_1, cm	d_2, cm
50	90	5	20

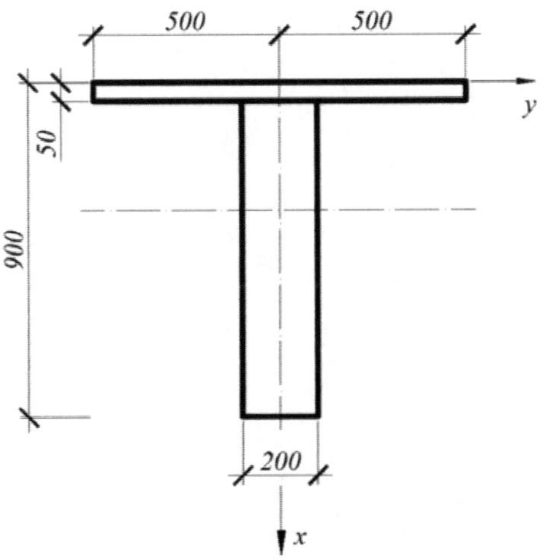

Figure 2. The cross section geometry.

The value of the torsional constant obtained by the formulae of the strength of materials is $I_t = 197111\,\text{cm}^4; G_c I_t = 221749407\,\text{kNcm}^2$ for the concrete of the shear modulus $G_c = 1125\,\text{kN/cm}^2$. The error is $\delta = 14.63\,\%$. When comparing the calculation with the more accurate technique of S.P. Timoshenko $I_t = 197088\,cm^4; G_c I_t = 221725002\,kNcm^2$, and the error is $\delta = 14.63\,\%$ (see table 2).

Table 2. Comparison of the values of the torsional constant calculated by different techniques.

The calculational technique	I_t, cm^4	GI_t, kNcm^2	δ,%
Strength of materials	197111	221749407	14.62
Method of Stephen Timoshenko	197088	221725002	14.63
FME (ANSYS)	217838	245067750	5.64
The exact technique [2]	210367	259721628	

The applied exact technique has been checked by the calculation of the thin-walled profile. Then, the shear stresses are distributed almost uniformly along the thickness of the closed thin-walled section eliminating in the midline of the profile. The torsion constant is determined by assuming that each section works as part of a rectangular cross-section, and its value can be obtained by replacing the integration along the middle line of the profile by approximate summation:

$$J_t = \frac{1}{3}\int_0^L \delta^3(s)\,ds = \frac{1}{3}\sum_{i=1}^{3} \delta_i^3 L_i \tag{5}$$

where L_i, δ_i are the length and the average thickness of the i -the section (figure 3).

In fact, the torsional stiffness is slightly higher than that determined by the formula (5) [4], which is explained by a number of factors for the studied section (the correction coefficient is 1.1-1.2).

If the thickness of the rib of the researched T-section will be decreased by a factor of four, then the thin-walled section will be obtained (see table 3 and figure 3).

Table 3. Geometry of the thin-walled T-section.

b_1, cm	b_2, cm	d_1, cm	d_2, cm
50	90	5	5

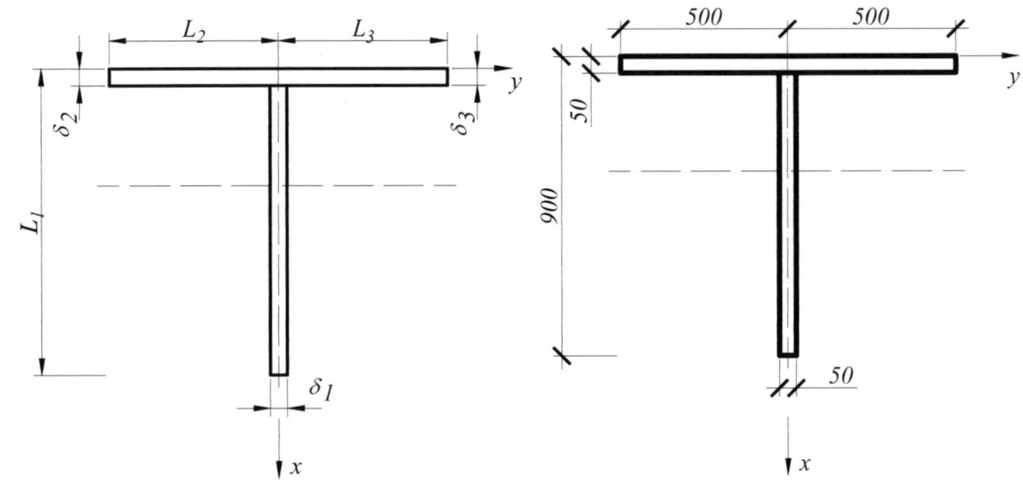

Figure 3. Geometry of the thin-walled T-section.

Value of the torsional constant is $I_t = 7572.56\ cm^4$. The error is $\delta = 1.848\ \%$ and $\delta = 1.675\ \%$ when compared to the generally accepted calculation methods (see table 4), which confirms the correctness of the exact technique for calculating the torsional stiffness of the T-section [3] and its effectiveness for different ratio of sizes of the T-section.

Table 4. Comparison of the values of the torsional constant calculated by different techniques.

The calculational technique	I_t, cm^4	GI_t, kNcm^2	δ,%
Strength of materials	7445.83	8662500	1.848
Method for thin-walled sections	7445.73	8671500	1.675
FME (ANSYS)	7963.00	8958375	5.156
The exact technique [2]	7572.56	8573625	

4. Conclusions and prospects of research

The torsional constant of the reinforced concrete T-sections should not be calculated by the traditional approximate technique, because for the nonsymmetrical elements with thick ribs and shelves the Saint-Venant's principle gives a significant error. Being the component of the torsional stiffness, an important factor affecting the spatial work of the engineering structures, the torsional constant should be calculated more precisely. The further studies will show the influence of the refined values of the torsional constant and the secant shear modulus of concrete as the components of the torsional stiffness on the strain redistribution in the elements of the spatial systems.

References

[1] Azizov T and Vildanova N 2013 Influence of the shear modulus of concrete on the structures' spatial work *3d International Conference on Science and Technology held by SCIEURO 21 22 October 2013* (London: Berforts Information Press Ltd) pp 97–106

[2] Azizov T and Stadnik V 2009 Torsional stiffness of reinforced concrete T-beams with normal cracks *Visn. Odes. derzhavnoi Akad. Budvn. ta arhtektury [Bulletin Odessa State Acad. Build. Archit.* 4–11

[3] Arutiunian N H and Abramian B L 1963 *Kruchenie uprugih tel [Torsion of elastic bodies]* (Moscow: Fizmatgiz)

[4] Walczak J 1977 *Wytrzymałość materiałów oraz podstawy teorii sprężystości i plastyczności [Strength of materials and foundations of the theory of elasticity and plasticity]* (Warszawa)

Optimization Design of Transmission Tower Based on Intelligent Selection

Junqi Chen[1,*], Shihao Cao[1] and Qingjun Xian[1]

[1]School of Civil Engineering and Architecture, Henan University of Technology, Zhengzhou, China

Abstract. In this work, the optimal design method for self-supporting transmission tower is put forward, whose procedure is that the tower body form is selected intelligently firstly and then the members types of tower are chosen. The case base of towers is established with application of object-oriented technology. Then, the knowledge of tower body is mined out with case based reasoning and data mining technology. At the same time, the structural regulations of tower in the standard are merged into the knowledge base. So, the intelligent selection system of tower structural scheme based on case reasoning and data mining is explored. A 500kV self-supporting tower is optimized according to above method and a real model test is conducted. Comparison between optimal tower and original tower indicates that weight of optimal tower is decreased by 3% and its ultimate bearing capacity is increased by 19%. Statistical analysis results of 20 optimal towers indicates that the weight is average reduced by 4.7%, the components of the optimal tower are stressed more uniformly and bearing capacity of the tower is enhanced 12% at least.

1. Introduction

The supports of EHV transmission lines are normally steel lattice towers. The cost of towers constitutes about 28 to 42 percent of the cost of transmission line and hence optimum tower design will bring in substantial savings [1]. Safety and economy are major issues in the design of civil engineering structures. The goal of structural optimization design is to achieve the best coordination of the units within the structure. At the same time, the performance of the material is most reasonable and meets the relevant requirements of the specification [2, 3].

At present, the transmission tower design procedure in China is usually type selection-analytical and calculation-full stress material selection-redesign. It is generally done in a few repetitive designs accounting to the designer's experience. In fact, it is still in the category of structural analysis, rather than the optimal design according to the active search optimal scheme.

2. Transmission line tower and design variables

One typical transmission tower is illustrated in figure 1. The tower is divided into three parts: the tower head, the tower body and the tower legs. In this work, the optimal design method for self-supporting transmission tower is put forward, whose procedure is that the tower body form is selected intelligently firstly and then components' types of tower are chosen.

Structural layout optimization is the intelligent selection of inter-section layout, including structural shape, section type, tower slope, upper and lower opening dimensions, and section height dimensions. There are many restrictions on the size of the tower head, including clearance, height limit of each cross arm, cross arm spacing limit, etc. The optimization effect of the tower head structure is not obvious for many reasons. Therefore, in this paper, the optimization design is mainly for the tower body.

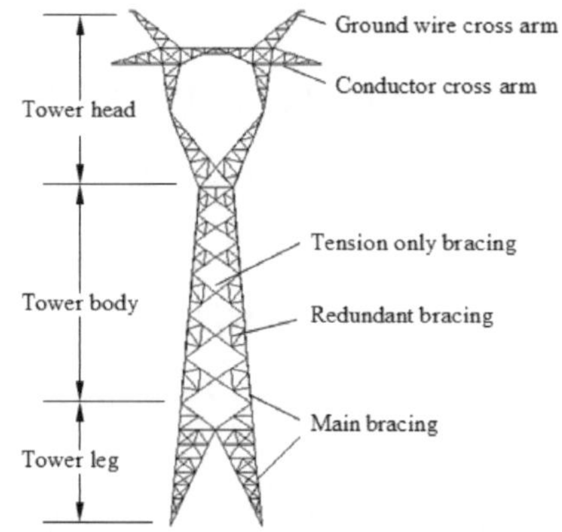

Figure 1. One typical transmission line tower.

The optimization of the tower body includes four aspects: the shape of the tower structure, the type of section layout, the type and size of the member section, and the material of the component. Optimization of the tower body includes optimization of tower structure layout (tower structure shape, section layout type) and component (section type, specification, material). The tower head and tower legs are only optimized for components.

3. Intelligent selection of transmission tower

The optimal methods of tower are mostly designed for the tower whose shape is finalized. Only the angle bar section or section arrangement type is optimized, but not for the whole structure optimization. In this paper, the tower body type is selected intelligently first. The structure of the tower body is adjusted through intelligent selection. The arrangement of the towers is more reasonable, and the stress condition of the tower is improved. Then the angle bars types are designed optimally including the member type and cross section.

Conceptual design, preliminary design and detailed design are the three stages of the engineering structure design process, and the most critical step is conceptual design. Conceptual design plays a decisive role in the function and quality of the entire project, which is directly related to the success of the project and the design. Structural selection is the most important part of the conceptual design stage, and it is also one of the important decision-making tasks in the early stage of building structure design. The intelligent selection of tower structure mainly includes three parts: knowledge acquisition, knowledge determination and scheme generation [4, 5].

3.1. Knowledge acquisition

Knowledge acquisition of tower structure is an analysis and design process used to model tacit knowledge to be incorporated into the tower structure intelligent selection system. It includes static and dynamic knowledge acquisition process.

In the initial stages of knowledge acquisition, access to the target level of knowledge is the main task. The target level of knowledge here refers to the program and its corresponding set of attributes. Programs set and attribute set are primarily established by domain experts, the literature knowledge, norms rules knowledge, design examples knowledge etc.

Dynamic knowledge acquisition process of tower structure is shown in figure 2. First, the initial knowledge base of the tower structure is generated by way of an interactive session. Then, a tower structure type program is generated by the tower selection program and submitted to the structure program evaluation module. At last, optimal solution is obtained by decision module of tower structure program if it runs successfully. The learning system model illustrated in figure 3 is used in knowledge acquisition which is defined by H. Simon [6, 7].

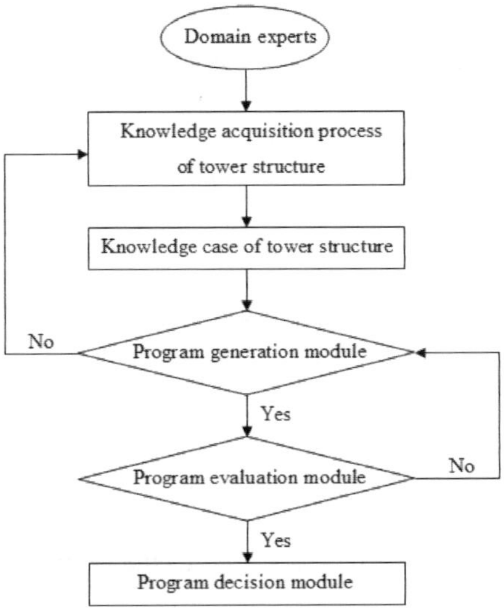

Figure 2. Dynamic knowledge acquisition process.

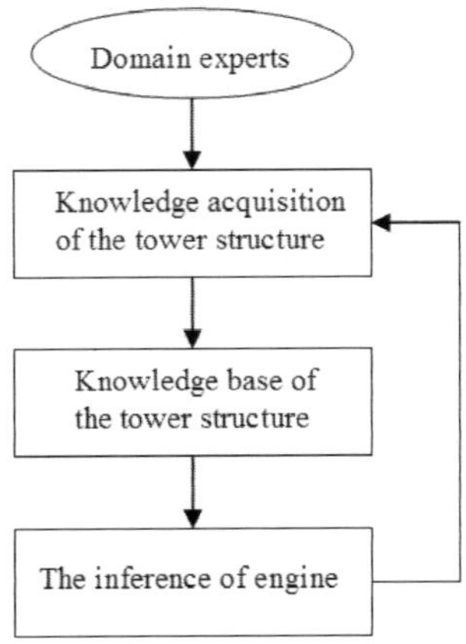

Figure 3. Knowledge acquisition model.

3.2. Knowledge identification

Knowledge identification here is to determine the source of knowledge of the various types of tower structures. There are various classifications to tower structure, such as materials, structural systems and other partition. Figure 4 lists three kinds of section arrangement types and figure 5 lists three kinds of diaphragm arrangement types [8].

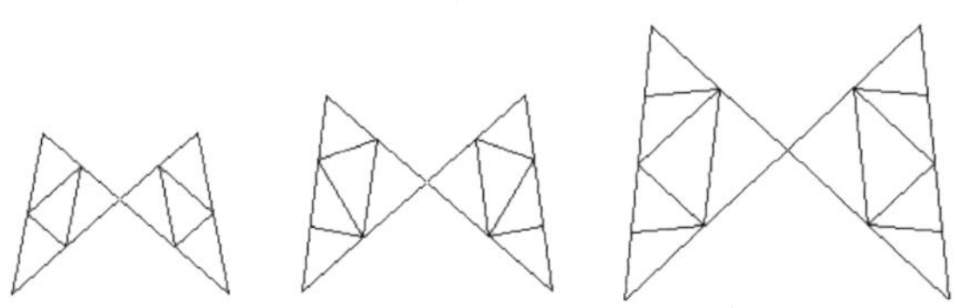

Figure 4. Three kinds of section arrangement.

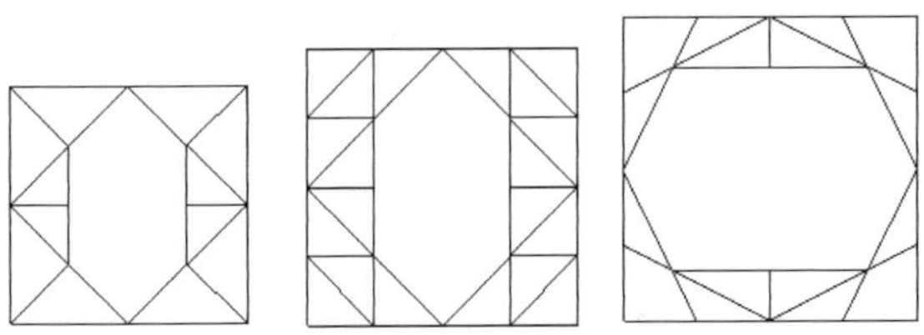

Figure 5. Three kinds of diaphragm arrangement.

3.3. Scheme generation

The scheme generation system of tower structure intelligent selection consists of three parts: conventional database module, structure type intelligent generation module, relations finding module.

The tower structural information can be increased, deleted, modified and inquired in conventional database module. The structure type intelligent generation module contains intelligent segmentation to tower, intelligent selection of internodes and diaphragm surfaces, tower legs modeling function. Association rule mining is used to relations finding module [9].

4. Optimization of tower body components

The design variables of transmission tower body components are discrete variables because the components are used angle bar whose specifications such as width, thickness, etc. are discrete [10]. The following function can be used to express the mathematical model of the discrete variable optimization design:

$$\begin{cases} \min f(X) \\ s.t. \;\; g_j(X) \le 0, \qquad j = 1,2,\cdots,m \\ \qquad x_i \in S_i; \;\; S_i \subseteq S, \;\; i = 1,2,\cdots,n \end{cases} \tag{1}$$

where $f(X)$ is the objective function and $X=(x_1, x_2, \ldots, x_n)^T$ is the vector of design variables. $g_j(X)$ is the constraint function.The constraints of transmission tower optimization mainly include strength constraints, compression stability constraints, and slenderness ratio constraints.

The stress constraint is

$$\sigma_j = \frac{N_j}{A_j} \leq f \tag{2}$$

where σ_j, N_j, A_j, f are stress, axial force, section area, design value of tensile(compression) strength of member j.

The stability constraint is

$$\sigma_{cr} = \frac{N_j}{\varphi_j A_j} \leq f \tag{3}$$

where σ_{cr} and φ_j are equivalent stability stress and axial compression buckling coefficient of member j.

The slenderness ratio constraint is

$$\lambda_j \leq \lambda \tag{4}$$

where λ_j is the slenderness ratio of member j. And λ is the allowable slenderness ratio of member j.

5. Optimal design and true type test of a 500kV self-supporting tower

ZB12 tower which is designed by Hunan Electric Power Design Institute is re-optimized design according to the above optimization method. Un-optimized and optimized structural scheme of tower is shown in figure 6.

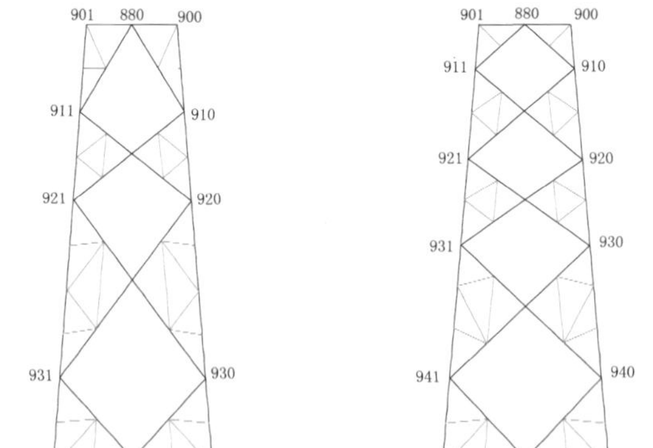

Figure 6. Un-optimized (left) and optimized (right) structural scheme of tower.

The weight of the former tower was optimized to 10.193 tons, and the weight of the optimized tower was reduced to 9.862 tons, which was reduced by 3.25% before and after optimization.

TTA which is programmed by Northeast Electric Power Design Institute is used to calculate the carrying capacity of tower. Then true type test of ZB12 tower is implemented in Beijing Electric Power Research Station. Figure 7 illustrate the true type test of ZB12 tower.

Among the measuring points of the tower body, only one section main member inner force increased by 5.33%, and the internal forces of the other main members and the oblique members were reduced, especially the internal force of the cross oblique members were significantly reduced. It shows that this method can better optimize the section arrangement layout and reduce the internal force, thus reducing the bar specifications and reducing the tower weight.

After optimization, the internal force of only 7 of the 37 tested members increased, and the internal forces of the other members decreased, which qualitatively showed that the internal forces of the rods were reduced.

After structural optimization, the tower failure test showed that the failure load of the tower was 155% of the normal working load. However, the failure load of the tower which was not optimized is 130% of the normal working load. The carrying capacity of tower is greatly enhanced under extreme conditions and its safety reliability is improved.

Figure 7. True type test of an optimal design 500kV self-supporting tower.

In order to further verify the optimization effect of the method, 20 towers were selected from the major power design institutes across the country for optimization design, which further confirmed that the method can reduce the tower weight by 3%-5%, and at the same time, the ultimate loads carrying capacity of the towers were at least increased by 12%.

6. Conclusions

Optimum design of transmission line towers has been discussed in this paper. It has been shown that the intelligent selection can be applied successfully to the tower structural design. The tower body form is selected intelligently firstly and then components types of tower are chosen. According to this method, the tower carrying capacity can be improved while its weight is reduced.

Acknowledgments

This work was financially supported by Key Research Project of Colleges and Universities in Henan Province(No.15A560019), Zhengzhou Scientific and Technology Project Plan(No.20140719) and Henan University of Technology Fund Project of High-level Talent(No.2014BS002).

References

[1] Souza R R, Miguel L F F, Lopez R H and Torii A J 2016 *Eng. Struct.* **111** 162-84.
[2] Tort C, Şahin S and Hasançebi O 2017 *Comput. Struct.* **179** 75-94.
[3] Couceiro I, París J, Martinez S, Colominas I, Navarrina F and Casteleiro M 2016 *Eng. Struct.* **117** 274-86.
[4] Camp C V and Farshchin M 2014 *Eng. Struct.* **62** 87-97.
[5] Ray T and Saini P 2001 *Eng. Optimiz.* **33** 735-48.
[6] Mirjalili S 2015 *Knowl-Based Syst.* **89** 228-49.
[7] Bailey S F and Smith I F 1994 *J. Comput. Civil Eng.* **8** 454-68.
[8] Soibelman L and Pena-Mora F 2000 *J. Struct. Eng.* **126** 733-42.
[9] Saa R, Garcia A, Gomez C, Carretero J and Garcia-Carballeira F 2012 *Expert Syst. Appl.* **39** 8784-92.
[10] Rajeev S and Krishnamoorthy C S 1992 *J. Struct. Eng.* **118** 1233-50.

IOP Conf. Series: Materials Science and Engineering **473** (2019) 012036 doi:10.1088/1757-899X/473/1/012036

Buckling Analysis of Transmission Tower Considering Ice Load

Junqi Chen[1,*], Qingjun Xian[1] and Peng Zhang[1]

[1]School of Civil Engineering and Architecture, Henan University of Technology, Zhengzhou, China

Abstract. Transmission lines are important lifeline projects. The stable operation of transmission lines is closely related to economic and social security and stability. Ice storm seriously affects the safety of transmission lines. The simulation analysis of ice damage of transmission tower structure has important theoretical significance and engineering value. In this paper, the finite element software ANSYS is used to analyze the buckling of the self-supporting tower based on eigenvalue buckling analysis and nonlinear buckling analysis. The effects of vertical span, uniform ice coating, uneven ice coating, and wind load were considered. The ice thickness of the tower buckling is reduced with the vertical span decreasing under uniform ice coating the transmission line. As the coefficient of unevenness increases, the thickness of the unstable ice coating of the tower drops sharply. The unstable wind speed decreases with the increase of ice thickness.

1. Introduction

Transmission lines are important lifeline projects. The stable operation of transmission lines is closely related to people's lives and economic security. In early 2005 and early 2008, China's transmission lines were seriously damaged by the ice disaster [1]. Two typical collapse towers in China induced by the ice storm in 2008 are illustrated in figure 1.

From February 7 to 17, 2005, a rare rain and snow disaster in history raided Hunan province in China. There have been accidents such as electric power line ice flashing, tower collapsed and wire broken. 24 towers of three 500kV lines and 18 towers of six 220kV lines collapsed due to overwhelming. At the beginning of 2008, the southern part of China suffered severe freezing disasters affected by the cold and warm air. 678 towers of 500 kV AC/DC transmission lines collapsed and 119 lines stopped running. 1432 towers of 220 kV transmission lines collapsed and 343 lines stopped running [2-4]. The ice disaster caused a major loss to the national economy and people's lives and property.

2. Buckling analysis

2.1. Eigenvalue buckling analysis

Eigenvalue buckling analysis, also known as linear buckling analysis, is used to analyze the buckling strength of an ideal structure. The advantage is that the critical load of the structure can be obtained without nonlinear analysis of the structure, and on this basis, it can provide a reference for nonlinear buckling analysis. The disadvantage is that the initial defects and geometric nonlinear deformation characteristics of the truss structure are not considered in the analysis process, and the calculation results are generally conservative [5].

a) Collapse tower in Leiyang, Hunan, china, 2008.

b) Collapse tower in Luan, Anhui, china, 2008.

Figure 1. Collapse towers in China induced by the ice storm in 2008.

In the stable equilibrium state, considering the influence of axial force or medium internal force on the bending deformation [6-8], the equilibrium equation of the structure is obtained according to the principle of resident potential energy, as shown in equation 1.

$$([K_E]+[K_G])\{U\}=\{P\} \tag{1}$$

where $[K_E]$ is the stiffness matrix of the structure, $[K_G]$ is the geometric stiffness matrix of the structure, $\{U\}$ is the node displacement vector, and $\{P\}$ is the node load vector. Equation 1 is also the equilibrium equation for geometric nonlinear analysis.

In order to obtain the indifferent equilibrium state [6-8], the second variation of the potential energy of the system should be zero, that is,

$$([K_E]+[K_G])\{\delta U\}=0 \tag{2}$$

The eigenvalue equation can be obtained [6-8], as shown in Equation 3.

$$([K_E]+\lambda_i[K_G])\{\varphi_i\}=0 \tag{3}$$

where λ_i is the *i*-th eigenvalue, $\{\varphi_i\}$ is the eigenvector corresponding to λ_i, i.e., the buckling mode.

2.2. *Nonlinear Buckling Analysis*

The nonlinear buckling analysis of the tower structure can be carried out based on the results of eigenvalue buckling analysis as its initial defects. In this paper, the non-linear equilibrium path is tracked by the equal arc method. In the iterative process of nonlinear buckling analysis, the load convergence criterion shown in equation 4 was chosen as the criterion for convergence [9, 10].

$$\sqrt{\{g\}^T\{g\}} \le \beta\sqrt{\{q\}^T\{q\}} \tag{4}$$

where g is the imbalance node force vector, q is the reference load vector, and β is the parameter, which is 10^{-5}.

3. Finite element model

The finite element model of the tower line system is mainly composed of three parts: power transmission lines (wires, ground lines), insulators and transmission towers. In this paper, the 5A-ZBC2 tower is used to build the finite element model of transmission tower line system. The geometric dimensions are shown in figure 2. The transmission line system in this study includes 3 towers, 2 ground wires and 3 groups of conductors, as shown in figure 3.

The structure of the transmission tower has strong geometric nonlinearity. The BEAM188 space beam element is used to simulate the poles of the transmission tower. The element is the characteristics of stress stiffening and large deformation. LINK 10 is used to simulate the transmission line. LINK10 is a 3D rod element which can only withstand axial tension or pressure. It can be used to simulate cables or gaps. The insulator is simulated by LINK8 element, which has the characteristics of plasticity, expansion, stress stiffening and large deformation.

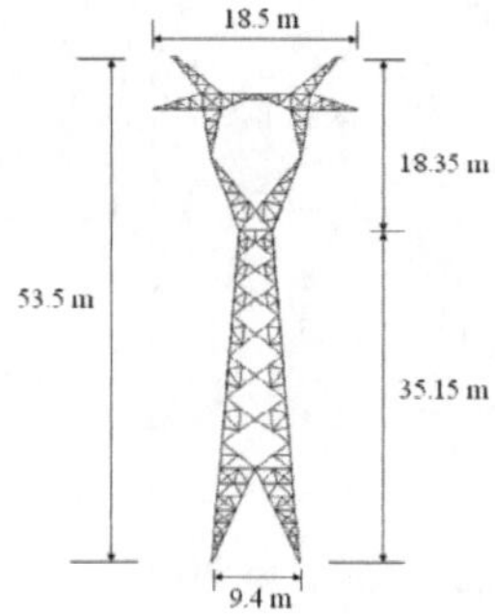

Figure 2. The geometry of the tower.

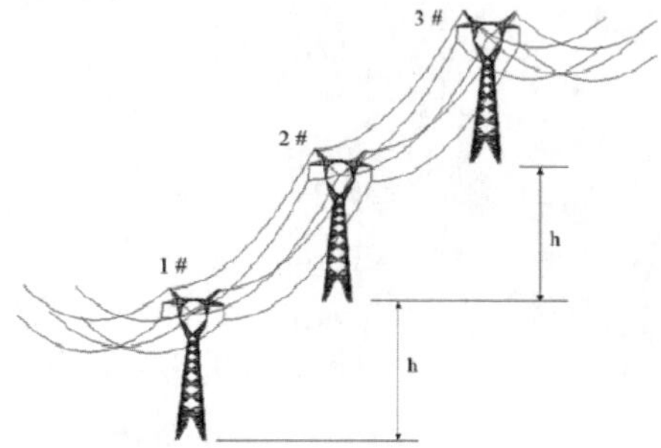

Figure 3. Target transmission line system.

4. Ice load and wind load of transmission line

4.1. Ice load of tower and insulator
The ice load (N/m^2) per unit surface area of the tower members and insulators is

$$q_t = 0.6b\alpha_1\gamma \tag{5}$$

where b is the basic ice thickness determined by engineering experience(mm), α_1 is the height increment coefficient of the ice coating thickness, and γ is the ice bulk density which is $9kN/m^3$[11].

4.2. Ice load of wire
Due to the action of the wind, the shape of the ice coating the wire is very irregular. It can be simplified to a circular shape when the ice load is calculated. The ice load (N/m) of the wire and ground wire is

$$q_l = \pi b\alpha_1\alpha_2(d + b\alpha_1\alpha_2)\gamma \tag{6}$$

where α_1 is the ice thickness correction factor related to the diameter of the guide (ground) line, d is the diameter of the wire or ground line(m) [11].

4.3. Wind load of transmission tower
The standard wind load value of the tower W_T (N) is as follows:

$$W_T = W_0 \cdot \mu_z \cdot \mu_s \cdot \beta_z \cdot A_s \tag{7}$$

where W_0 is the standard value of basic wind pressure (N/m^2), μ_z is the height transformation coefficient of wind pressure, μ_s is the shape coefficient, β_z is the wind load adjustment coefficient of tower, and A_s is the calculation value of withstand wind pressure area(m^2) [11].

4.4. Wind load of insulator
The standard wind load value of insulator string W_I (N) is determined by the following formula:

$$W_I = W_0 \cdot \mu_z \cdot A_I \tag{8}$$

where A_I is the calculation value of insulator withstand wind pressure area(m^2) [11].

4.5. Wind load of wire and ground line
The standard wind load value of wire and ground line W_L (N) is determined by the following formula:

$$W_L = \alpha \cdot W_0 \cdot \mu_z \cdot \mu_{sc} \cdot \beta_c \cdot d \cdot L_p \cdot \sin^2\theta \tag{9}$$

where α is wind pressure unevenness coefficient, β_c is wind load adjustment factor of 500kV line, μ_{sc} is shape coefficient of wire and ground line, d is the outer diameter of the wire or the calculated outer diameter of the wire(m), L_p is the horizontal span of transmission lines(m), and θ is the angle between the wind direction and the wire [11].

5. Analysis results

5.1. Buckling Analysis of Tower Under Uniform Ice Coating

The buckling mode of the tower under uniform ice coating on the transmission line is shown in figure 4. The relationship between the ice thickness of the tower buckling and the vertical span is shown in figure 5. The ice thickness of the tower buckling is reduced with the vertical span decreasing under uniform ice coating the transmission line.

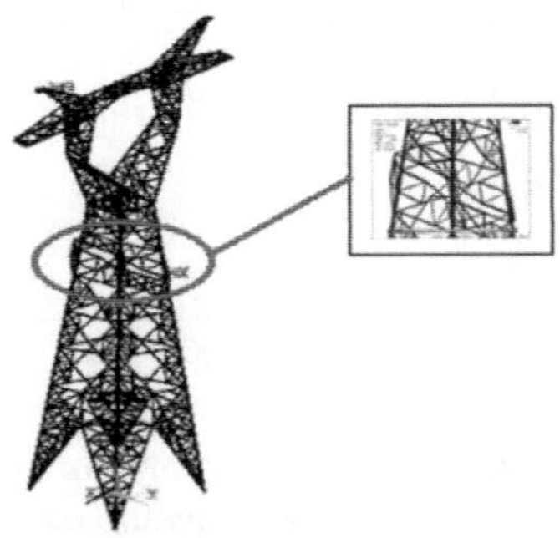

Figure 4. The buckling mode of tower under uniform ice coating.

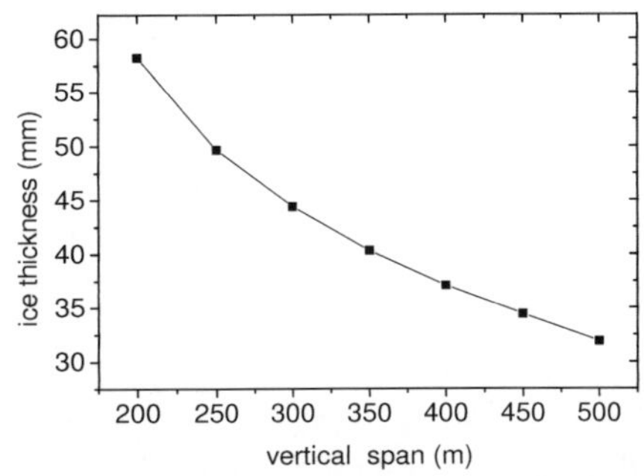

Figure 5. The relationship between the ice thickness of the tower buckling and the vertical span.

5.2. Buckling analysis of tower under uneven ice coating

In order to consider the influence of uneven ice coating on the stability of the tower in the same span, the unevenness coefficient ρ is used to indicate the uneven ice coating degree of the tower

$$\rho = \frac{\Delta l}{\sqrt{(L_C)^2 - (\Delta l)^2}} \tag{10}$$

where Δl is the horizontal offset along the line of suspension point of the wire on the suspended insulator string (m), and L_C is the length of suspension insulator string (m).

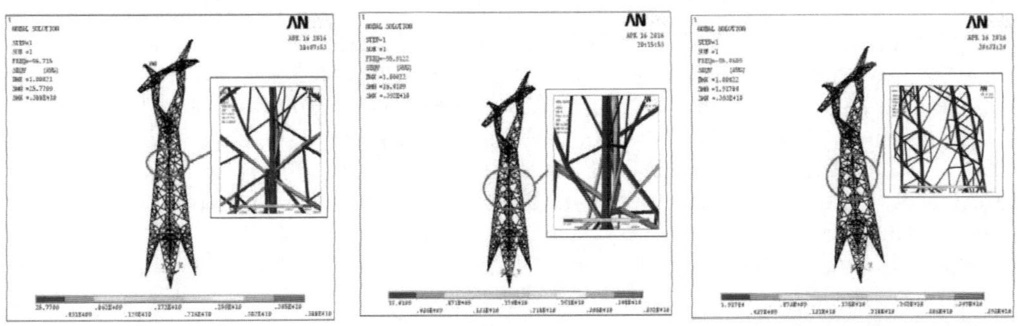

a) $\rho = 0.2$ b) $\rho = 0.6$ c) $\rho = 1.0$

Figure 6. The buckling mode of the tower under different unevenness coefficients.

Figure 6 shows the buckling mode of the tower under different unevenness coefficients. As the unevenness coefficient increases, the area of the buckling failure of the tower is expanding. The damage zone extends from one rod to multiple rods. The relationship between the ice thickness and unevenness coefficient is illustrated in figure 7. The unbalanced tension increases with the unevenness coefficient increases. So the thickness of the unstable ice coating of the tower drops sharply with the coefficient of unevenness increasing.

5.3. Buckling analysis of tower considering wind load

Wind load is also one of the causes of the collapse of the tower. The relationship between the ice thickness and unstable wind speed is illustrated in figure 8. The unstable wind speed decreases with the increase of ice thickness. When the ice thickness and wind speed of the transmission line exceed the curve of figure 8, the tower will be in danger of collapse.

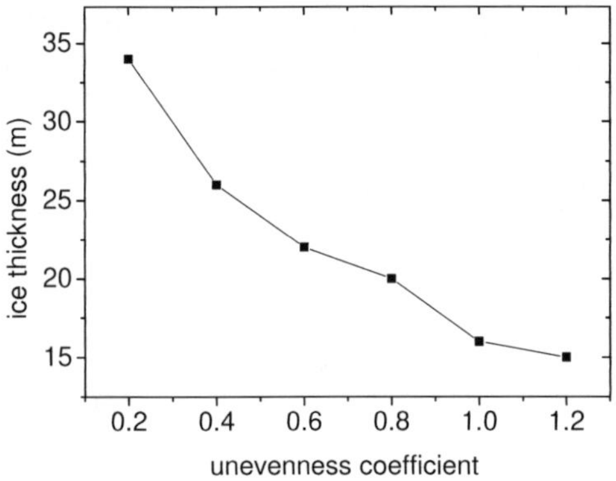

Figure 7. The relationship between the ice thickness and unevenness coefficient.

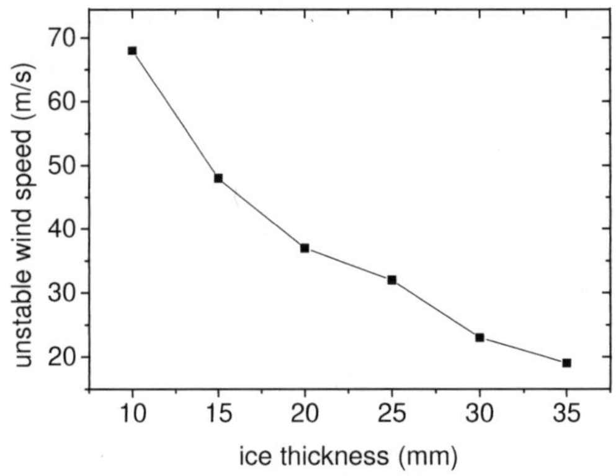

Figure 8. The relationship between the ice thickness and unstable wind speed.

6. Conclusions

The finite element simulation of the transmission tower buckling under ice coating is carried out in this paper. The results show that the ice thickness of the tower buckling is significantly reduced with the increase of uneven ice coating level, and the instability damage area is significantly expanded. As the wind speed increases, the ice thickness of the tower bulking decreases significantly.

Acknowledgments

This work was financially supported by Key Research Project of Colleges and Universities in Henan Province(No.15A560019), Zhengzhou Scientific and Technology Project Plan(No.20140719) and Henan University of Technology Fund Project of High-level Talent(No.2014BS002).

References

[1] Xie Q and Sun L 2012 *J Constr. Steel Res.* **74** 26-36.
[2] Shu Q, Yuan G, Huang Z, and Ye S 2016 *Eng. Struct.* **123** 166-80.
[3] Li J X, Li H N and Fu X 2018 *Int. J. Struct. Stab. Dy.* **18(03)** 485-96.
[4] Ji K, Rui X, Li L, Yang F, and McClure G 2016 *IEEE T. Power Deliver.* **31(5)** 2215-22.
[5] Shu Q, Huang Z, Yuan G, Ma W, Ye S, and Zhou J 2018 *Thin Wall. Struct.* **131** 619-30.
[6] Yang F, Yang J, Han J, and Fu D 2010 *Int. J. Struct. Stab. Dy.***10(03)** 461-81.
[7] Souza R R, Miguel L F F, Lopez R H and Torii A J 2016 *Eng. Struct.* **111** 162-84.
[8] Tort C, Şahin S and Hasançebi O 2017 *Comput. Struct.* **179** 75-94.
[9] Couceiro I, París J, Martinez S, Colominas I, Navarrina F and Casteleiro M 2016 *Eng. Struct.* **117** 274-86.
[10] Camp C V and Farshchin M 2014 *Eng. Struct.* **62** 87-97.
[11] Ye H S, Gong D V, Wei S Y, Yang C R and Li X L 2012 *Technical Code for Designing 110~500kV Overhead Transmission Line* (Beijing: China Electric Power Press)

IOP Conf. Series: Materials Science and Engineering **473** (2019) 012037 doi:10.1088/1757-899X/473/1/012037

Influence of Dissipative Joints on the Behaviour of Steel MRFs: FREEDAM vs Equal-Strength Bolted Joints

Roberto Tartaglia[1], Mariana Zimbru[1]

[1]University of Naples "Federico II" Department of Structures for Engineering and Architecture, via Forno Vecchio 32, 80134 Naples, Italy

Abstract. Modern practice and codes provide design rules of moment resisting frames (MRFs) based on the assumption that the dissipative zone is the end of the beams, which implies assuming full-strength joints. The recent EQUALJOINTS Research Project demonstrated the advantages of using equal-strength joints in MRFs in seismic areas. Further beneficial aspects can be attained when replacing the plastic deformation mechanism with a friction-based energy dissipation mechanism, as highlighted within FREEDAM Research Project, where friction connection have been developed and qualified. The current paper presents the comparison between the response of MRFs alternatively equipped with equal-strength joints and friction connections

1. Introduction

The modern seismic design of steel structures [1-18] focuses on the capacity-based principle, which fundamentally allows the plastic deformation of specially designed areas of the frame members, preserving the integrity of the remaining elements. Following the recommended guidelines, the safety requirements under design seismic events are met; however, in the earthquake aftermath, the repair costs can be large.

In the case of steel moment resisting frames (MRFs) it is expected to have plastic deformations in the beams while the joints are generally designed to be full strength, which imply costly solutions. The recent RFCS EQUALJOINTS project showed that equal-strength joints can be effective in seismic areas. The test results showed that, when the component design is properly addressed, and sufficient ductility is guaranteed, equal strength joints exhibit good behavior, being also cheaper than full strength joints [19-21].

Friction connections can be viable alternative to traditional steel joints. Friction connections dissipate the seismic input by means of relative displacement of surfaces held in contact by preloaded bolts [22-30]. The connection typology presented by [25, 26] is characterized by very stable cyclic behavior and ductility while localizing the damage in easily replaceable elements of the connection.

The fundamental difference between the traditional MRFs and the MRFs equipped with friction devices (FD-MRFs) is the dissipative element i.e. the beam or joint for the former, while the friction connection (or damper) for the latter. The initial steps are naturally identical with the design of traditional moment resisting frames, as the friction devices do not bring substantial alterations to the MRFs. However, the seismic design requirements differ consistently, especially due to the special application of the capacity design rules. Within the framework of the FREEDAM Research project, a modified design methodology of the MRFs equipped with friction devices has been proposed based on the EC8 recommendations [3].

The current work presents the main characteristics, design and response of two frame models: (i) frames equipped with the more traditional equal-strength bolted extended stiffened end-plate connection and (ii) frames equipped with friction devices.

2. Features of the considered joints

Equal-strength joints are designed to guarantee simultaneous yielding of both connection and beam (see figure 1a). The design resistance of this joint typology is assumed equal to the plastic resistance of the beam. The adopted connection is a bolted extended stiffened end-plate. The design of equal-strength joints can be done according to the Component method implemented in EN1993 Part 1-8. However, a complete design methodology is reported in [11].

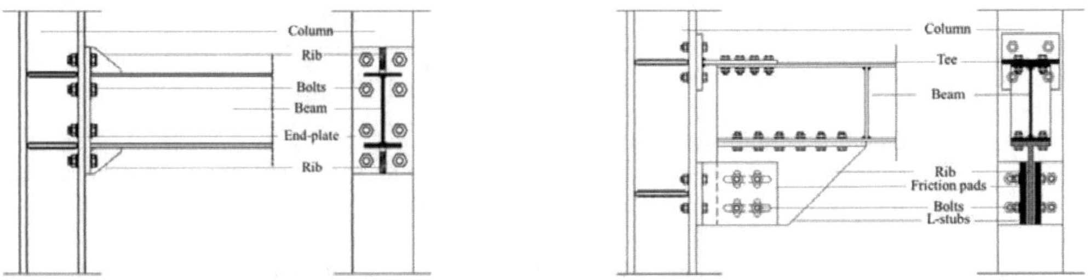

a) Equal-strength joint b) Friction joint

Figure 1. Joint geometrical configuration and main components.

The friction connections transfer the seismic energy input by means of relative sliding at the level of surfaces held together by means of preloaded bolts. Figure 1b presents the geometry and the components of the connection. The beam is connected to the column at the top by means of bolted T-stub and the bottom flange is connected to the rib. The rib is the mobile part hence, it is detailed with horizontal slotted holes while the friction pads and L-stubs have vertical slotted holes, in order to allow for the kinematics of the connection. The rib, friction pads and L-stubs are clamped together by means of high-strength bolts, and the whole assembly is connected to the column by the L-stubs' flanges.

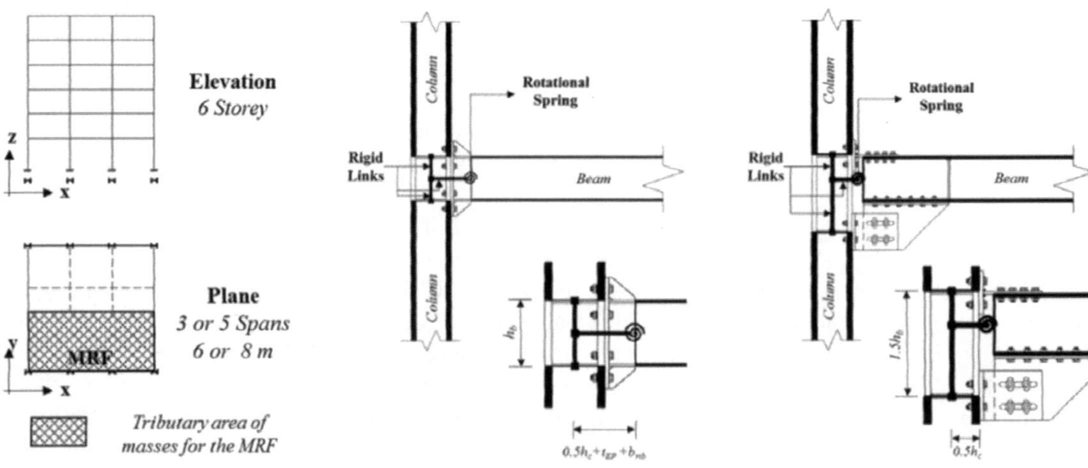

a) General geometry b) Equal-strength joint detail c) FREEDAM joint detail

Figure 2. Details of the case study frames.

3. Case study

The geometry of the frames selected for the case study is depicted in figure 2a. The structures were designed assuming S355 steel members. The permanent and live loads considered for design were chosen within a practical range (4.5 kN/m^2 and 2 kN/m^2 respectively). A High seismic Hazard (HH) level was considered for the design with PGA equal to 0.35g, assuming a Type 1 response spectrum and soil conditions Type C. The maximum behavior factor q = 6.5 was adopted for the design spectrum. The second order (P - Δ) effects as well as the accidental torsional effects are accounted for in the design. The interstorey drift for serviceability conditions was limited to 0.01hstorey. Global and local ductility requirements according to EN 1998-1 were checked. Both static and dynamic nonlinear analyses were performed in Seismostruct. Four structures were designed and analyzed, varying the number of MRF bays (3 and 5) and the span length (6 and 8m). The investigated structures are identified by a code: MRF-number of storeys-number of bays- and span length, therefore for instance the MRF-6-5-8 is a structure with 6 floors and 5 moment-resisting bays with a length of 8m.

Both Eurocode 8 compliant and the models of the FREEDAM frames were detailed with rotational springs at the column face (see figure 2b and c). These elements were used to model the bending capacity of the equal-strength joint and the friction device, respectively. The response of the traditional equal-strength joint was modelled based on the experimental tests performed within the EQUALJoints RFCS Research Project using a smooth model, while the bending resistance of the rotational spring for the FREEDAM joints was defined as given by equation (1).

$$M_{j,Rd|FD-B} = m \cdot M_{pl,b} \tag{1}$$

Where m is the utilization factor of the FREEDAM devices necessary for the obtained frame members and $M_{pl,b}$ is the bending plastic capacity of the connected beam. From the stiffness point of view, the joints were assumed as full rigid.

4. Results

The seismic design of the two types of structural configurations led to similar member sizes, obvious fromFigure 3a, which depicts the frame structural weight. The results show slightly lighter EC8-FD frames in some cases (due to the lower overstrength coefficient for the design of non-ductile elements) but the serviceability check (interstorey drift ratios) was generally governing the design, which resulted in similar solutions for the two design situations.

Figure 3b and c show that the frames equipped with equal-strength joints (EC8-EqStr) are severely overdesigned. In terms of redundancy (Figure 3d), both frame sets are in the same range, with better values for the EC8-FD, due to the sequential activation of the friction joints and the reduced degradation. Finally, as depicted inFigure 3e, it can be concluded that designing frames in which the resistance and the stiffness are uncoupled (like the frames equipped with friction joints) more reasonable results are obtained compared to traditional solutions even if optimized.

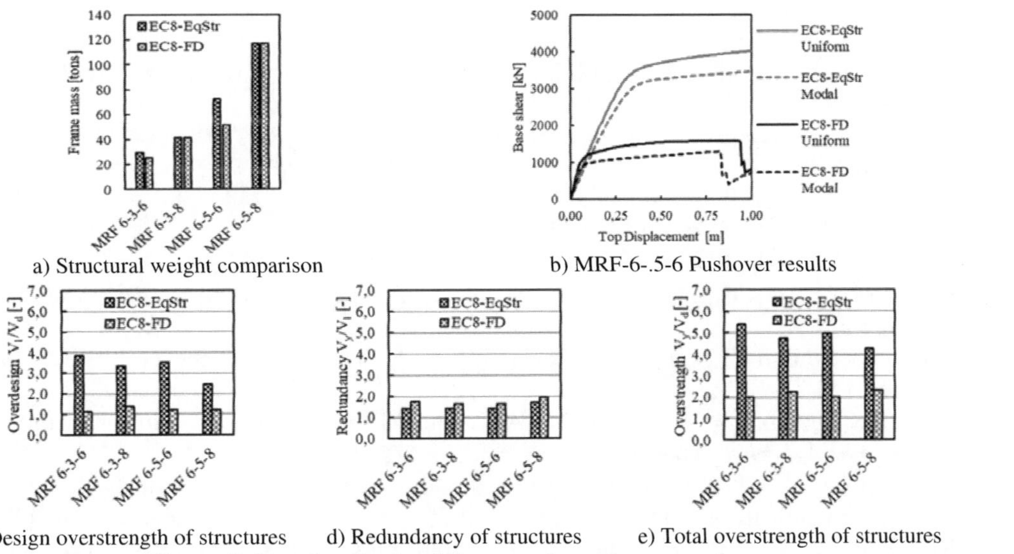

a) Structural weight comparison
b) MRF-6-.5-6 Pushover results
c) Design overstrength of structures
d) Redundancy of structures
e) Total overstrength of structures

Figure 3. Results obtained from static pushover analyses.

5. Conclusions

The design methodology currently implemented in EC8 leads to largely overdesigned moment resisting frames, even when equal-strength joints are used, with the main culprits being the severe deformability and global stability checks. Structural solutions that have disconnected strength and stiffness, like the frames equipped with friction joints, lead to optimized structures that have sufficient redundancy, without being severely overdesigned. Moreover, the benefits in terms of repair costs are expected to be significantly lower given the low-damage solution.

References

[1] Cassiano D, D'Aniello M, Rebelo C, Landolfo R, da SL, 2016 *Steel Compos. Struct. An Int. J.* **21** 479–500.

[2] Tenchini A, D'Aniello M, Rebelo C, Landolfo R, da Silva LS, Lima L, 2014 *J. Constr. Steel Res.* **101** 437-54.

[3] Montuori R, Nastri E, Piluso V, Streppone S, D'Aniello M, Zimbru M and Landolfo R, 2018 *The Open Constr. Build. Tech. J.* **12(1)** 140-53.

[4] Isaincu A, D'Aniello M, Stratan A, 2018 *The Open Constr. Build. Tech. J.* **12(1)** 124-31.

[5] Tartaglia R, D'Aniello M, De Martino A, Di Lorenzo G, 2018 *Ingegn. Sism.: Int. J. Earthq. Eng.* **35(3)** 104-20.

[6] D'Aniello M, Güneyisi EM, Landolfo R, Mermerdaş K 2014 *Thin-Walled Struct.* **77** 141-52.

[7] Güneyisi M, D'Aniello M, Landolfo E R, Mermerdaş K 2014 *Steel Comp. Struct.* **17(3)** 215-36.

[8] Tenchini A, D'Aniello M, Rebelo C, Silva L Da, Lima L 2016 *Eng. Struct.* **124** 167–85.

[9] Cassiano D, D'Aniello M, Rebelo C 2018 *An Int. J.* **26(5)** 621-34.

[10] Cassiano D, D'Aniello M, Rebelo C, 2017 *J. Constr. Steel Res.* **137** 77–92.

[11] D'Aniello M, Tartaglia R, Costanzo S, Landolfo R, 2017 *J. Constr. Steel Res.* **128** 512–27.

[12] Tartaglia R, D'Aniello M, Rassati G A, Swanson J A, Landolfo R, 2018 *Eng. Struct.* **159** 155–71.

[13] Tartaglia R, D'Aniello M, Zimbru M, Landolfo R, 2018 *Steel Compos. Struct.* **27** 727–45.

[14] Francavilla A B, Latour M, Piluso V, Rizzano G, 2018 *J. Constr. Steel Res.* **148**, 77–96.

[15] Fiorino L, Pali T, Landolfo R. 2018 *Thin-Walled Struct.* **130** 213-30.

[16] Fiorino L, Macillo V, Landolfo R 2017 *Eng. Struct.* **151** 633–47.

[17] Fiorino L, Shakeel S, Macillo V, Landolfo R 2017 *J. Constr. Steel Res.* **138** 324–39.

[18] Faggiano B, Formisano B A, Fiorino L, Castaldo C, Macillo V, Mazzolani F M 2017 *The Open Civil Eng. J.* **11** 464-74.

[19] D'Aniello M, Tartaglia R, Costanzo S, Campanella G, Landolfo R, De Martino A, 2018 *Key Eng. Mater.* **763** 406-13.

[20] Tartaglia R, D'Aniello M, Rassati G A, Swanson J A, Landolfo R, 2018 *Key Eng. Mater.* **763** 818-25.

[21] Tartaglia R, D'Aniello M, Landolfo R, 2018 *J. Constr. Steel Res.* **148** 669–90.

[22] Latour M, Piluso V, Rizzano G, 2014 *Constr. Build. Mater.* **65** 159-76.

[23] Piluso V, Montuori R, Troisi M, 2014 *Mech. Res. Commun.* **58**, 146-56.

[24] Latour M, Piluso V, Rizzano G, 2015 *Eng. Struct.* **85**, 219–33.

[25] Latour M, Piluso V, Rizzano G, 2018 *J. Constr. Steel Res.* **146**, 33-48.

[26] Latour M, D'Aniello M, Zimbru M, RizzanoG, Piluso V, Landolfo R, 2018 *Soil Dyn. Earthq. Eng.* **115** 66–81.

[27] Ferrante Cavallaro G, Latour M, Francavilla A B, Piluso V, Rizzano G, 2018 *J Constr. Steel Res.* **141** 145–55.

[28] Zimbru M, D'Aniello M, De Martino A, Latour M, Rizzano G and Piluso V, 2018 *The Open Constr. Build. Tech. J.* **12(1)** 154-69.

[29] D'Aniello M, Zimbru M, Landolfo R, Latour M, Rizzano G and Piluso V, Finite element analyses on free from damage seismic resisting beam-to-column joints in COMPDYN2017,, Rhodes Island, Greece, 12-15 June, 2017, Eccomas Proceedia COMPDYN 2017pp. 802-814.

[30] D'Antimo M, Zimbru M, D'Aniello M, Demonceau J F, Jaspart J P, Landolfo R, 2018 *Key Eng. Mater.* **763** 592-9.

IOP Conf. Series: Materials Science and Engineering **473** (2019) 012038 doi:10.1088/1757-899X/473/1/012038

Preliminary Finite Element Analyses on the Experimental Mock-Up Frames of FREEDAM Research Project

Mariana Zimbru[1], **Roberto Tartaglia**[1]

[1]University of Naples "Federico II" Department of Structures for Engineering and Architecture, via Forno Vecchio 32, 80134 Naples, Italy

Abstract. Recent research ventures focus on structural systems characterized either by low-damage or easily repairable solutions. In this context, moment resisting frames (MRFs) equipped with friction joints represent a viable option. Among the different types of friction beam-to-column connections, the arrangement with symmetric friction devices are very effective to provide stable and ductile friction hysteresis, thus reducing the damage in the structural members (beams and columns) to a minimum or even to zero. Numerical and experimental investigations on the beam-to-column assemblies within the RFCS FREEDAM project testify to the good seismic performance of the friction device when properly designed. However, within this research project full scale pseudo-dynamic tests will be performed for a better understanding of the influence of such joint on the overall structural behaviour of MRFs. In this paper the results of preliminary numerical simulations are presented to give the first insights on the response of the experimental mockup as well as to validate the accuracy of two software with largely different capabilities. Static pushover and nonlinear dynamic analyses were performed and the results for the benchmark structure (frame equipped with reduced beam section joint) and for the structure equipped with the friction devices are presented

1. Introduction

Steel moment resisting frames (MRFs) designed with full strength joints are structures that are widely used in construction practice [1-4]. Due to their high ductility the design allows for high expected damage in the aftermath of earthquakes [5-24] and this is of course linked to high repair costs. Replacing the traditional plastic hinge dissipation mechanism with a low-damage, friction-based energy dissipation solution [25-27], like the friction device developed for steel beam-to-column joints within the FREEDAM project [28-33], is believed to mitigate the repair costs due to the minimal damage in easily replaceable parts (parts belonging to the friction device).

The friction device proposed by the FREEDAM consortium was subjected to several experimental campaigns. In the preliminary phases, the research was focused on the friction component by means of extensive experimental campaigns on lap-shear specimens [25, 28]. A subsequent phase consisted in the tests on beam-to-column assemblies, considering two geometrical configurations, which proved the effectiveness of the design procedure and the very good seismic performance of the friction devices. The third experimental phase consists in a set of full-scale pseudo-dynamic tests to be carried out on a 2-storey 1-bay mockup equipped alternatively equipped with traditional (e.g. bolted end-plate with dog-bone or reduced beam section - RBS) and friction joints, as shown in figure 1. In order to design the test setup and to provide preliminary characterization of the response of the experimental mockup, finite element (FE) simulations have been carried out.

A secondary aim of this paper is to verify the effectiveness of two software, with largely different capabilities, in the assessment of the structural response. The first set of analyses were performed in

SAP2000, which uses simplified modelling approaches, offering nevertheless a fair level of accuracy of results. The second step was modelling the same structure with a more sophisticated software: ABAQUS 6.14. In this latter case, the model is more refined and can better catch the global and local response. However, the software can be used only by a skilled user, requires more input data and is computational demanding. The description of the mockup and the modelling assumptions are described hereinafter.

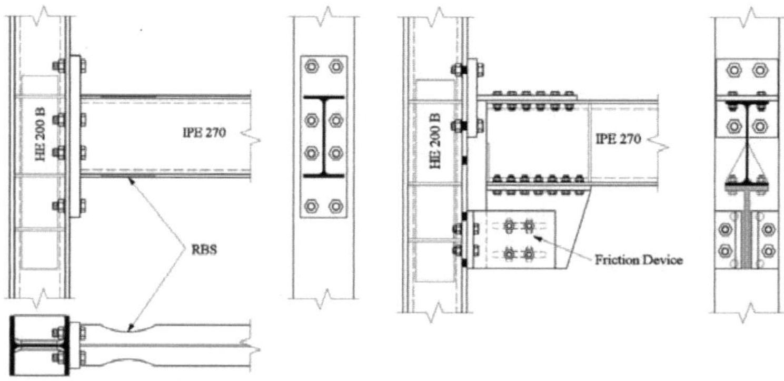

Figure 1. Joint details for the two frame typologies.

2. The Numerical Models: Simplified vs Advanced

The simplified modelling of the two frames (identified by their joint as RBS and FD, respectively) was developed in SAP2000. The members were modelled using linear elements and the experimental conditions, in terms of boundary conditions and loading, were duly set (see figure 2). Rigid offsets were used in the node area to model the higher stiffness. The plastic hinge properties were defined according to the expected behavior i.e. for the RBS the yield bending moment is calculated function of the section and material properties and for the FD the hogging and sagging flexural resistance were defined explicitly based on the expected values calculated analytically and validated experimentally.

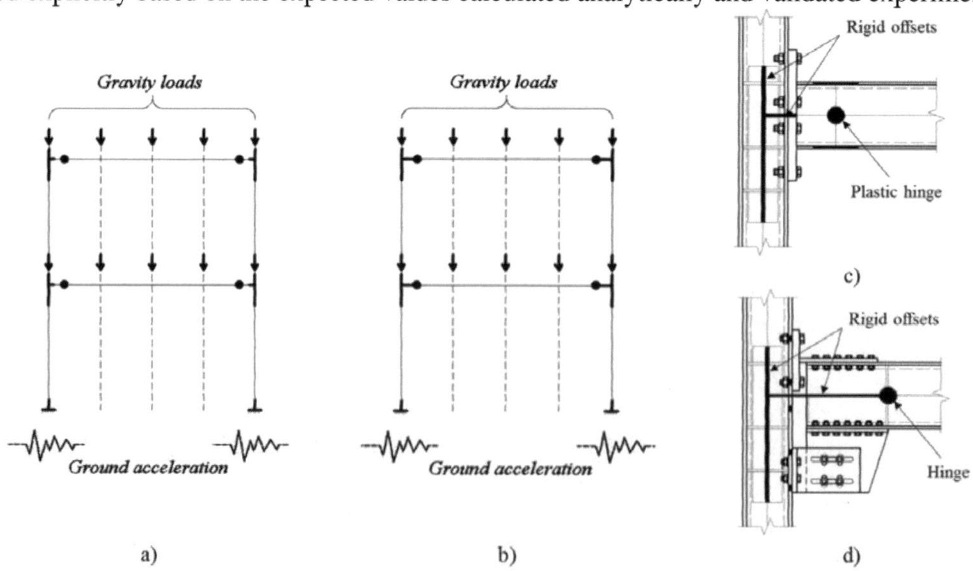

Figure 2. Simplified models in SAP a) RBS model; b) FD model; c) joint detail for RBS model; d) joint detail for FD model.

The advanced FE model was developed in Abaqus and it included wire elements only in the regions outside of the nodes, which were modelled in detail using 3D solid elements. Figure 3 depicts both the general frame parts and the details of the beam to column and column base joins. The material was modelled considering the actual post-yield material law obtained experimentally from coupon tests for all steel members. The boundary and loading conditions (gravity loads, bolt preloading, lateral restraints etc.) were assumed as similar as possible to the experimental conditions. C3D8R elements were used for the solid parts and B-31 beam elements are used for the wire elements. The bolts were modelled as shown in [15, 16].

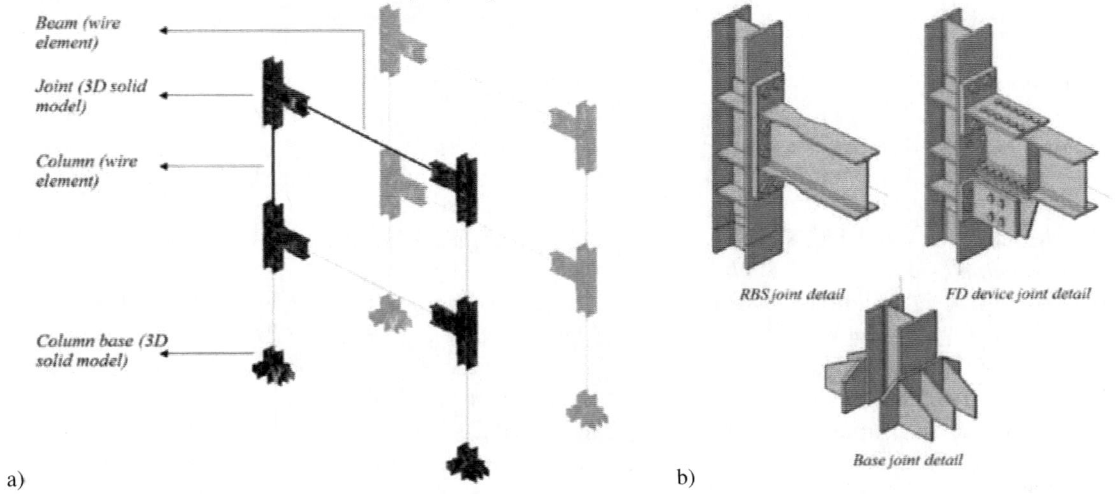

Figure 3. Advanced FE model.

3. Results

The results presented in figure 4 are relative to the two simplified and two advanced models. At a first look it's possible to observe that the monotonic response obtained using the two software is very similar for both frame configurations (figure 4 a and b). The SAP model shows that the maximum base shear for the planar frame is approximately 407kN for the RBS frame and 390 for frame FD. On the other hand, the time history results, evaluated considering an Italian earthquake (Castelnuovo –Assisi Mw: 6) (figure 4 d and e), show that the maximum base shear is underestimated by the simplified model and not achieved at the same time instant in the two models. Furthermore, the simplified model does not capture the residual drifts obvious in the advanced model. The comparison of monotonic and cyclic results of the RBS and FD advanced model (figure 4 c and f) show that the two models exhibit similar behavior, in line with what was the expected and desired response.

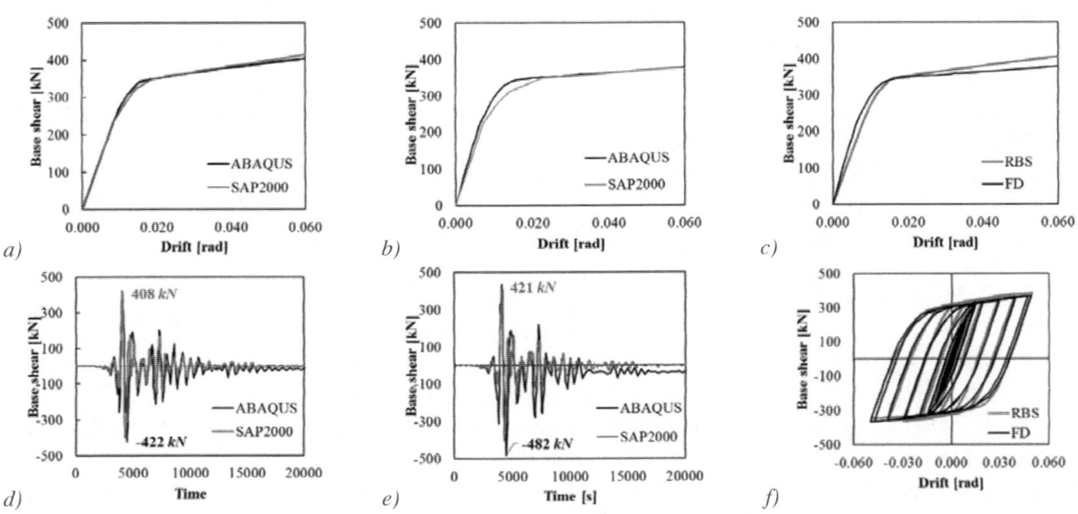

Figure 4. Simplified vs advanced models:
a) SPO for RBS frame b) SPO for FD frame, c) RBS vs FD frame.

4. Conclusions

The simplified model gives a good monotonic prediction of the overall response if compared to ABAQUS model, while the cyclic response underestimates the maximum base shear, overall response and residual drift evaluated with the refined FE model. However, the differences are acceptable and it can be concluded that simplified models can be used to perform demanding analyses and investigate the global response, while the advanced model is useful for the understanding of the local effects.

References

[1] Cassiano D, D'Aniello M, Rebelo C, Landolfo R, da SL, 2016 *Steel Compos. Struct. An Int. J.* **21** 479–500.
[2] Tenchini A, D'Aniello M, Rebelo C, Landolfo R, da Silva LS, Lima L, 2014 *J. Constr. Steel Res.* **101** 437-54.
[3] Montuori R, Nastri E, Piluso V, Streppone S, D'Aniello M, Zimbru M and Landolfo R, 2018 *The Open Constr. Build. Tech. J.* **12(1)** 140-53.
[4] Isaincu A, D'Aniello M, Stratan A, 2018 *The Open Constr. Build. Tech. J.* **12(1)** 124-31.
[5] D'Aniello M, Güneyisi EM, Landolfo R, Mermerdaş K 2014 *Thin-Walled Struct.* **77** 141-52.
[6] Güneyisi M, D'Aniello M, Landolfo ER, Mermerdaş K 2014 *Steel Comp. Struct.* **17(3)** 215-36.
[7] Tenchini A, D'Aniello M, Rebelo C, Silva L Da, Lima L 2016 *Eng. Struct.* **124** 167–85.
[8] D'Aniello M, Tartaglia R, Costanzo S, Landolfo R 2017 *J. Constr. Steel Res.* **128** 512–27.
[9] Tartaglia R, D'Aniello M, Rassati GA, Swanson JA, Landolfo R, 2018 *Eng. Struct.* **159** 155–71.
[10] Tartaglia R, D'Aniello M, Zimbru M, Landolfo R, 2018 *Steel Compos. Struct.* **27** 727–45.
[11] Francavilla AB, Latour M, Piluso V, Rizzano G, 2018 *J. Constr. Steel Res.* **148**, 77–96.
[12] D'Aniello M, Tartaglia R, Costanzo S, Campanella G, Landolfo R, De Martino A, 2018 *Key Eng. Mater.* **763** 406-13.
[13] Tartaglia R, D'Aniello M, Rassati GA, Swanson JA, Landolfo R, 2018 *Key Eng. Mater.* **763** 818-25.
[14] D'Aniello M, Cassiano D, Landolfo R, 2016 *J. Constr. Steel Res.* **124**, 77–90.
[15] D'Aniello M, Cassiano D, Landolfo R, 2017 *Steel Comp. Struct.* **24(6)**, 643-58.
[16] Fiorino L, Macillo V, Landolfo R 2017 *Eng. Struct.* **151** 633–47.
[17] Macillo V, Fiorino L, Landolfo R. 2017 *Thin-Walled Struct.* **120**, 161-71.
[18] Fiorino L, Terracciano MT, Landolfo R 2016 *J. Constr. Steel Res.* **127**, 92-107.
[19] Faggiano B, Formisano A, Castaldo C, Fiorino L, Macillo V, Mazzolani FM *Int. J. Earthq. Eng.* **33(3)** 42-51.
[20] Esposto M, Faggiano B, Mazzolani FM. 2006 Numerical vs. experimental results on a PTED beam-to-column connection for seismic resistant steel frames. In: F.M. Mazzolani, A. Wada. Behaviour of Steel Structures in Seismic Areas. p. 299-304, London:Taylor & Francis Group, Yokohama (Japan).
[21] Tartaglia R, D'Aniello M, De Martino A, Di Lorenzo G, 2018 *Ingegn. Sism.: Int. J. Earthq. Eng.* **35(3)** 104-120.
[22] Tartaglia R, D'Aniello M, Landolfo R, 2018 *J. Constr. Steel Res.* **148** 669–90.
[23] Macillo V, Campiche A, Shakeel S, Bucciero B, Pali T, Terracciano M T, Fiorino L, Landolfo R 2018 *Key Eng. Mater.* **763** 584–91. doi:10.4028/www.scientific.net/KEM.763.584.
[24] Campiche A, Shakeel S, Macillo V, Terracciano M T, *Ing. Sismica.* **2** 106–23.
[25] Latour M, Piluso V, Rizzano G, 2014 *Constr. Build. Mater.* **65** 159-76.
[26] Piluso V, Montuori R, Troisi M, 2014 *Mech. Res. Commun.* **58**, 146-56.
[27] Latour M, Piluso V, Rizzano G, 2015 *Eng. Struct.* **85**, 219–33.
[28] Latour M, Piluso V, Rizzano G, 2018 *J. Constr. Steel Res.* **146**, 33-48.
[29] Latour M, D'Aniello M, Zimbru M, RizzanoG, Piluso V, Landolfo R, 2018 *Soil Dyn. Earthq. Eng.* **115** 66–81.
[30] Ferrante Cavallaro G, Latour M, Francavilla AB, Piluso V, Rizzano G, 2018 *J Constr. Steel Res.* **141** 145–55.
[31] Zimbru M, D'Aniello M, De Martino A, Latour M, Rizzano G and Piluso V, 2018 *The Open Constr. Build. Tech. J.* **12(1)** 154-69.

IOP Conf. Series: Materials Science and Engineering **473** (2019) 012038 doi:10.1088/1757-899X/473/1/012038

[32] D'Aniello M, Zimbru M, Landolfo R, Latour M, Rizzano G and Piluso V, 2017 Finite element analyses on free from damage seismic resisting beam-to-column joints in COMPDYN2017,, Rhodes Island, Greece, Eccomas Proceedia COMPDYN 2017, 802-814.

[33] D'Antimo M, Zimbru M, D'Aniello M, Demonceau JF, Jaspart JP, Landolfo R, 2018 *Key Eng. Mater.* **763** 592-9.

The Effect of Spikes Occurrence on Surface Texture Parameter Assessments

P Podulka[1]

[1]Rzeszow University of Technology, The Faculty of Mechanical Engineering and Aeronautics, 35-959 Rzeszów, al. Powstańców Warszawy 12, Poland

Abstract. In this paper the influence of measurement errors (spikes) on surface topography parameter valuation was taken into consideration. More than 30 plateau-honed cylinder liner surfaces were taken into account. They were measured by white light interferometer Talysurf CCI Lite. The effect of two types of spikes (upper-spikes and/or lower-spikes) on areal form removal and/or surface texture parameter (from ISO 25178 standard) calculation was taken into account. It was assumed that surface topography parameters were falsely estimated when spikes appeared. Moreover, selection of reference plane was also disturbed when measurement errors (spikes) occurred.

1. Introduction

Surface topography measurement and/or analysis of cylinder liners is carefully considered; internal combustion engines are widely used in many applications. Assessment of this type of cylindrical elements is of a great importance especially with its improvement to air pollution, reduction of fuel consumption and emission. Since they consume fuel and contribute to air pollution, reducing of emission has an immense importance in environmental protection domain.

Surface topography is established in the final stage of a machining process. Its analysis is especially pertinent in assessment of functional properties (material contact, lubricant retention, wear resistance) [1]; the process of surface measurement should be subjected to strict control. Plateau-honed cylinder liner surface is an example of surface texture which consist of smooth plateau with valley. This type of surface is also characterized by highly acceptable sliding properties and lubricant maintains. Two-process surfaces have a decisive advantage over one-process surfaces [2].

Usually the surface topography parameters are calculated in relation to the reference plane [3] which is established with errors of form, waviness due to imperfections in manufacturing process. Surface topography areal form removal was often proposed by the cylinder fitting methods [4], polynomials [5], filtering methods [6-8], spline filters [9], wavelets [10] and other algorithms and/or procedures [11]. Many works were proposed for areal form removal [12, 13], some errors of two-process surface assessments with deep and wide valleys analysis were presented in [14-16].

The errors in surface topography analysis can be classified in measurement errors, the measured object errors [17], software and measuring method errors [18]. Moreover, there are some errors in measuring uncertainty, divided into: typical for measuring approaches, digitization process errors, errors obtained while data processing and/or other errors [19]. Sharp edges, inclusions, defects and other peculiarities of the surface can cause outliers and dropouts of data points in the topographical images measured with optical methods, more than with stylus technique [20].

There were many papers considering surface topography analysis of two-process surfaces [21]. However, the influence of measurement errors (spikes as an example) were not fully recognized when surface topography parameters were calculated. Moreover, the influence of individual peaks were not widely taken into account with filter appliances.

2. Materials and methods

In this paper plateau-honed cylinder liners with oil pockets created by the burnishing techniques were taken into account. The average values of diameter and depth of oil pockets were around 0.25 mm and 7 µm correspondingly. They were measured by white light interferometer Talysurf CCI Lite; height resolution 0.01 nm; the maximum size of analysed surfaces was 3.35 x 3.35 mm; the spacing was 3.27 µm. More than 20 surfaces were analysed but only few of them were showed in details. Textures from cylindrical elements were carefully analysed.

For extraction of surface topography features and/or parameters the following types of pre-processing techniques were applied: polynomials of 2nd (P_2) and/or 4th (P_4) degrees, digital filtering: Gaussian regression filter (F_{GR}) and robust Gaussian regression filter (F_{RGR}). For spikes removal the thresholding method (proposed by the author of this article in previous researches) was recommended.

The effect of spikes occurrence on the following parameters (from ISO 25178 standard) was taken into consideration: root mean square height *Sq*, skewness *Ssk*, kurtosis *Sku*, maximum surface peak height *Sp*, maximum valley depth *Sv*, maximum height *Sz*, arithmetic mean height *Sa*; *Sk* group parameters: reduced summit height *Spk*, reduced valley depth *Svk*, core roughness depth *Sk*, upper bearing area *Sr1* and lower bearing area *Sr2*.

3. Results and discussions

Two-types of spikes were taken into consideration; measurement errors (individual peaks) were classified into: upper-spikes (a, b) and lower-spikes (c, d) – examples were presented with profile exploration in figure 1. When F_{GR} was applied it was assumed that plateau-parts of profiles were distorted when valleys and/or upper-spikes were near-distributed (it was indicated by the arrows in figure 2). Application of F_{RGR} caused the minimization of reference line distortion in accordance to the non-robust filtering method.

In figure 3-a examples of isometric view of white-light-method measured surface containing lower-spikes was presented. From the analysis of isometric view of studied details it was assumed that only small differences were obtained when spikes were removed after P_2 or P_4 appliance. However, when lower-spikes were removed before pre-processing, the values of height parameters decreased: *Sp*, *Sz*, *Sk*, *Spk*. It was also found that the influence of spikes occurrence increased when the degree of polynomial also increased; the highest degree was applied the highest variance of height parameters (between surfaces after pre-processing and before/after spikes removal) was observed.

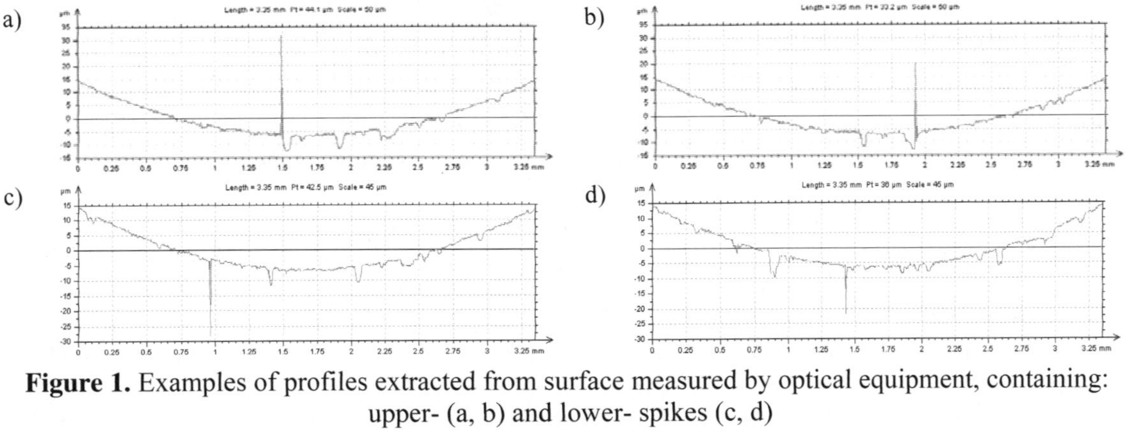

Figure 1. Examples of profiles extracted from surface measured by optical equipment, containing: upper- (a, b) and lower- spikes (c, d)

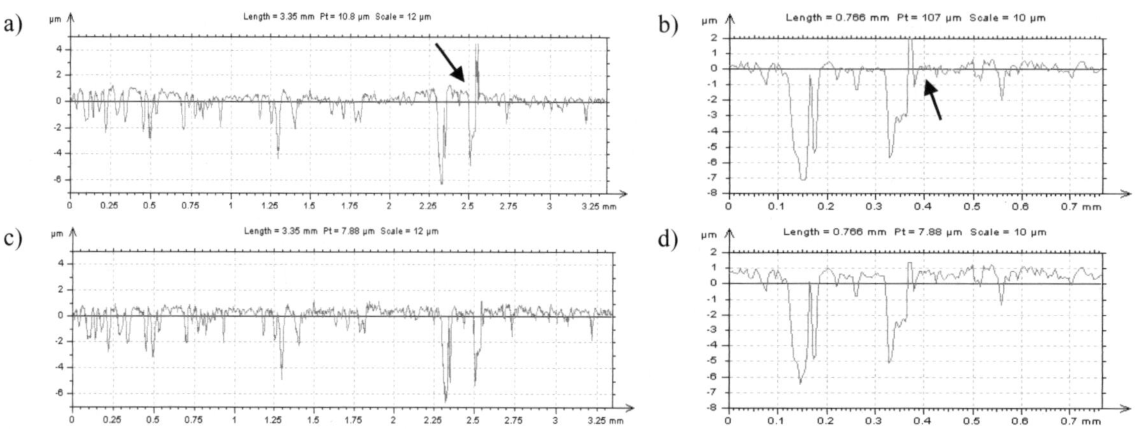

Figure 2. Profiles of surface after pre-processing by F_{GR} (a, b) and F_{RGR} (c, d) before (a, c) and after (b, d) upper-spikes removal; F_{COV} = 0.8 mm

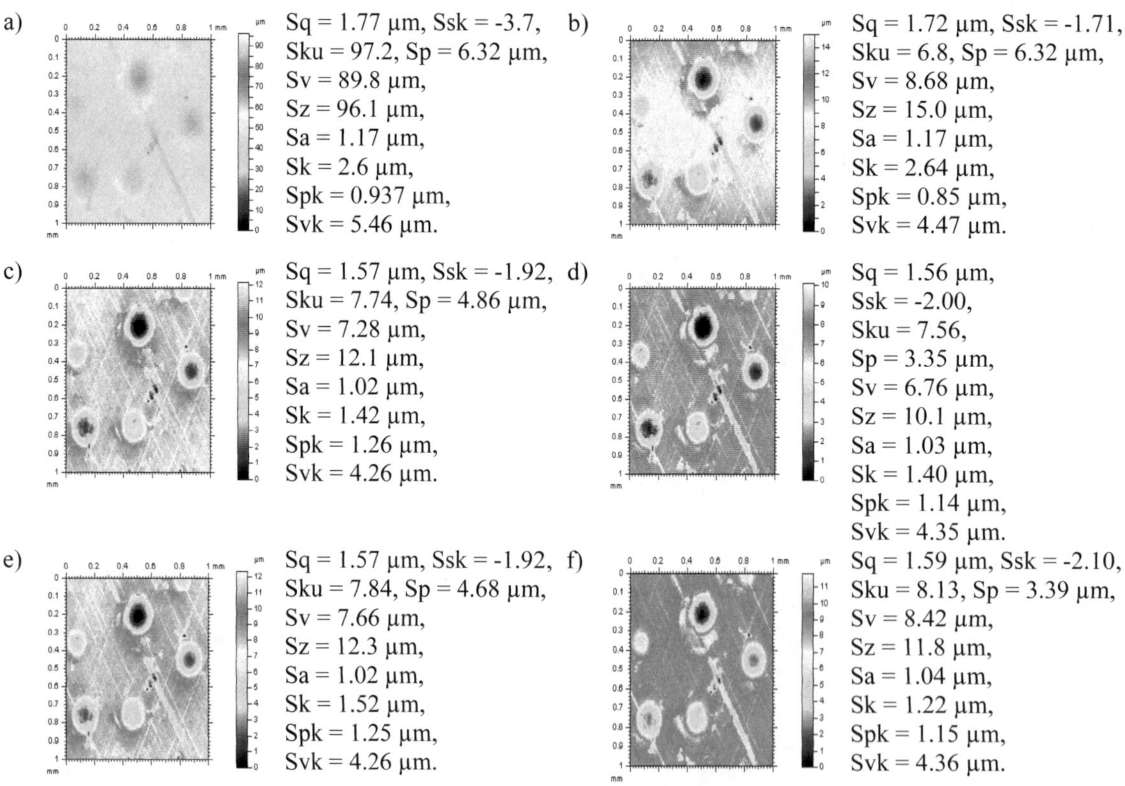

Figure 3. Isometric views of measured surface (a) and surface of lower-spikes removed (b) before (c, e) and after (d, f) pre-processing by: P_2 (c, d) and P_4 (e, f)

In figure 4 extracted free-of-dimples details with selected (likewise *Sk*) parameters were presented. The value of *Sq* parameter was minimized when pre-processing techniques were applied after spikes removal. Moreover, the value of *Sp* parameter decreased when spikes were removed before polynomial appliance. The value of *Sv* and *Svk* parameters did not changed or it was negligible. The maximum height of the surface was minimized with spikes removal before pre-processing techniques appliance; the minimum value of *Spk* parameter was received when P_4 was applied. When spikes were located near valleys/scratches, the reference line (plane) was non-accurate selected. Moreover, the spikes presence cause the increase of valley distortion when surface was pre-processed. For surface topography parameter calculation it is suggested to select the reference according to the *Sk* parameter value minimization.

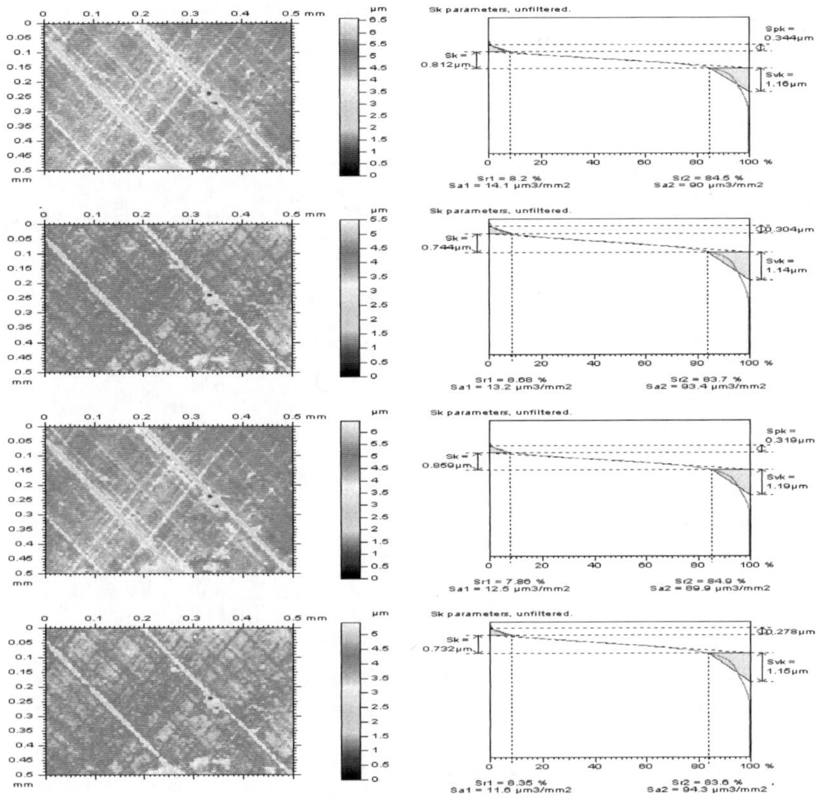

a) Sq = 0.504 µm, Ssk = -1.86, Sku = 9.06, Sp = 2.17 µm, Sv = 4.43 µm, Sz = 6.60 µm, Sa =0.338 µm.

b) Sq = 0.490 µm, Ssk = -2.07, Sku = 9.65, Sp = 1.07 µm, Sv = 4.44 µm, Sz = 5.51 µm, Sa = 0.326 µm.

c) Sq = 0.515 µm, Ssk = -1.85, Sku = 8.67, Sp = 2.04 µm, Sv = 4.42 µm, Sz = 6.46 µm, Sa = 0.350 µm.

d) Sq = 0.486 µm, Ssk = -2.13, Sku = 9,79, Sp = 1.0 µm, Sv = 4.42 µm, Sz = 5.42 µm, Sa = 0.323 µm.

Figure 4. Details (and their material ratio curves with selected parameters respectively) extracted from surface with upper-spikes removed before (a, c) and after (b, d) pre-processing by P_2 (a, b) or P_4 (c, d)

4. Conclusions

The following conclusions and/or remarks were noticed:

1. Surface topography measurement by white light interferometry is fast but more sensitive to extraneous effects contrary to stylus methods. One of the measurement errors are spikes (upper-spikes and/or lower-spikes).
2. When upper-spikes were detected, the values of parameters describing plateau-part of surface increased; usually *Sp*, *Sz* and *Spk* parameters were overestimated more than 100%. For lower-spikes occurrence errors in calculation of *Sv* and *Svk* parameters increased; describing of *Sk* family parameters is of a great importance, false estimation of this type of parameters can cause a classification of properly made parts as a lacks and its rejection.
3. When form was removed by application of F_{GR} it was assumed that plateau-parts of profiles were distorted when valleys and/or upper-spikes were nigh-situated. Application of F_{RGR} caused the minimization of reference line distortion in accordance to the non-robust filtering method.
4. For procedure with lower-spikes removal before pre-processing, the values of height parameters (*Sp*, *Sz*, *Sk*, *Spk*) decreased. It was also found that the influence of spikes occurrence increased when the degree of polynomial increased; the highest degree was applied the highest variance of height parameters (between surfaces after pre-processing and before/after spikes removal) was noticed.

5. The maximum height of the surface (*Sz*) was minimized with spikes removal before pre-processing techniques appliance; the minimum value of *Spk* parameter was received when P_4 was applied. It is recommended to calculate the surface topography parameters of two-process surface with spikes removal before form elimination; areal form removal before spikes detection can cause the increase of errors in surface texture specification. Therefor for selection of reference plane the 4th degree of polynomial with spikes removal by thresholding method is suggested.

References

[1] Pawlus P 1994 *Wear* **176** **(2)** 247-54.
[2] Jeng Y R 1996 *Tribol. Transact.* **39(2)** 354-61.
[3] Raja J, Muralikrishnan B and Fu S 2002 *Precis. Eng.* **26(2)** 222-35.
[4] Forbes A B 1989 Least squares best fit geometricelements. *NLP report DITC* **40(89)**, Teddington, UK.
[5] Podulka P, Dobrzański P, Pawlus P and Lenart A 2014 *Metrol. Measur. Syst.* **21(2)** 247–56.
[6] Brinkman S, Bodschwinna H 2000 Advanced Gaussian Filters. [in] *Advanced Techniques for Assessment Surface Topography*, Elsevier.
[7] Podulka P 2016 *Adv. Sci. Tech. Res. J.* **10(30)** 164–75.
[8] Janecki D 2011 *Precis. Eng.* **35(4)** 602–6.
[9] Muralikrishnan B and Raja J 2009 *Computational surface and roundness metrology*, Springer.
[10] De Chiffre L, Lonardo P, Trumpold H, Lucca D A, Goch G, Brown C A, Raja J and Hansen H N 2000 *Ann. Manuf. Technol.* **49(2)** 635–642, 644–652.
[11] Mainsah E, Greenwood J A and Chetwynd D G 2001 *Metrology and Properties of Engineering Surfaces*, Springer.
[12] Yuan Y-B, Qiang X-F, Song J-F and Vorburger TV 2000 *Prec. Eng.* **24(1)** 62–9.
[13] Stout K J, Sullivan P J, Dong W P, Mainsah E, Luo N, Mathia T and Zahouani H 1993 *Publication EUR 15178 EN Commission of the European Communities.*
[14] Podulka P 2018 Problem of selection of reference plane with deep and wide valleys analysis. *Journal of Physics: Conference Series* **1065(7)** 072017.
[15] Godi A, Grønbæk J and De Chiffre L 2017 *CIRP J. Manuf. Sci. Tech.* **16** 64-71.
[16] Podulka P 2018 Edge-area form removal of two-process surfaces with application of valley excluding method. Computational Metods in Engineering Science 2018 (CMES 2018), *E3S Web of Conferences.* (to be published)
[17] Stout K J (ed) 1994 Three-dimensional surface topography – measurement, interpretation and application. *Penton Press*, London.
[18] Thomas T R 1999 Rough Surfaces. Second Edition. *Imperial College Press*, London.
[19] Magdziak M 2017 *Aircraft Eng. Aerosp. Tech.* **89(6)** 953-9.
[20] Podulka P, Pawlus P, Dobrzański P and Lenart A 2014 Spikes removal in surface measurement *Journal of Physics: Conference Series* **483(1)** 012025.
[21] Podulka P 2018 Proposal of edge-area form removal of cylindrical surfaces containing wide dimples by application of various robust processing techniques. *Journal of Physics: Conference Series* **1065(7)** 072018.

The Material for Physical Simulation of Metal-Forming Processes in Super-Plastic State

O S Anishchenko[1], V V Kukhar[1], A H Prysyazhnyi[1], V V Agarkov[2], E S Klimov[3] and S M Chernenko[3]

[1] Department of Metal-Forming, Pryazovskyi State Technical University, 7 Universytetska Street, Mariupol, Ukraine, 87555
[2] State Enterprise "Kharkiv Regional Research and Production Center for Standardization, Metrology and Certification", Office 302, 36 Myronosytska Street, Kharkiv, Ukraine, 61002
[3] Department of Automobiles and Tractors, Kremenchuk Mykhailo Ostrohradskyi National University, 20 Pershotravneva Street, Kremenchuk, Ukraine, 39600

Abstract. In the article it was suggested to simulate the process of super-plastic deformation on optically transparent materials, created on diene and vinyl-aromatic carbohydrates, polar softener and achromatic stabilizer. It was shown that the index of rate hardening of such materials could change from 0.2 to 1.0 within the rate intervals of super-plastic deformation. The authors are convinced that prior to this article the alloy *Sn-38%Pb* was the best material for simulation of the processes of super-plastic deformation. Advantages and disadvantages of tin-lead alloy for simulation of super-plastic deformation were mentioned in the article. The article contains examples of chemical composition of new materials: the foundation (component content – 100 weight parts) – butadienesterene, isoprenesterene, butadiene-α-metylsterene; stabilizer (component content – 0.5 weight parts) – 2,6-ditretbutyl-4-metylphenol; polar softener (component content – 5-40 weight parts) – dibutylphtalate, dibutylsebacate. It was shown that introduction of polar softener reduces the effort of material flow. An unsufficient alternation of softeners content allowed modifying either behaviour of an alloy with different super-plastic structure or various temperature conditions of deforming.

1. Introduction

The phenomenon of super plasticity of metals (hereinafter SPM) was first observed by G.D. Bengough back in 1912, when he managed to reach elongation of a locally heated bronze sample by 160 % in a series of experiments, aimed at its slow elongation. Up to the present time such results are still taken into account and quoted in various works [1–3]. Elongation and deformation with local heating of the blank part are applied for manufacturing shaped and elongated parts of different materials [4–13]. Two decades later SPM ceased to be an object of exotic investigations and became an efficient foundation for creating new materials and technologies of their plastic forming, ensuring unique properties of manufactured parts and high engineering and economic efficiency of production processes [1, 14, 15].

Developing new technologies of super-plastic deformation presumes their simulation by application of the following ways: 1) by manufacturing industrial pilot batches [16–18]; 2) by computer simulation [19–21]; 3) by simulation of the processes with application of special materials [22–24]. The first variant produces the most precise results; however it requires substantial labour consumption that sometimes can't be justified. The second variant does not imply material expenses (except for purchasing expensive licensed computer software and training of personnel), however it largely depends upon the peculiarities of software, where a set of some or other materials models, methods of evaluation and formulae are used [25–27]. The third variant is a simplified variant of the first variant and presumes substantial reduction of labour consumption, due to selection of special materials for simulation, that are either relatively cheap, or they can simplify the process of simulation and raise its efficiency.

At present the second variant prevails, though computer simulation methods of deformation processes are often criticized [28]. That is why the search of new modelling materials, as an alternative to the first two variants of SPM processes simulation seems to be quite vital. These materials possess the following peculiarities: according to the firm opinion of many researchers [1, 29, 30] eutectic alloy *Sn-38%Pb*, this is the best material for simulating of super-plastic deformation processes. Selection of an alloy depends upon the following factors:

– the simplicity of forming an ultra-fine grained super-plastic structure in the alloy (an intensive deformation after advisably quick recrystallization of the melt);

– high sensitivity of the flow effort to deformation rate, usually determined by the value of the index of rate hardening $m = d(ln\sigma) / d(ln\xi)$ (where σ – is flow effort, ξ – is deformation rate);

– low flow rates and high ultimate degrees of alloy's deformation;

– the range of optimal temperatures of super-plasticity of *Sn-38%Pb* eutectics includes the room temperature, thus eliminating the problems with samples heating.

However, practical application of *Sn-38%Pb* alloy is connected with a series of difficulties that are not discussed very often. In [31] the fact that after intense deformation of a blank part, manufactured of *Sn-38%Pb* alloy it should be kept in a freezer, in order to avoid grain growth, occurring at the room temperature, was mentioned quite superficially (detailed explanation was missing there). Alloy deformation in optically transparent object (like, for instance, a transparent matrix at super-plastic forming) allows observing the dynamics of transformation of deformed state of the blank). Still it is true only for the surface of the blank, if it is covered with a coordinate grid. Deformation of the inner layers of the alloy can be investigated only after the process of deformation has been completed and subsequent mechanical separation of the sample.

A perspective of application of optically transparent polymeric materials that undergo loading (deforming) with subsequent investigation of the deformed state has but lately been found. Such materials were used in experimental process, described in [32, 33] and in the facilities elements in [34–36]. However, for selecting a group of materials, suitable for simulation of SPM processes it is necessary to investigate their composition and mechanical properties; it has not been performed up to now. The objective of the paper is to determine an opportunity and a degree of efficiency of application of optically transparent materials, similar to synthetic rubber for simulation of super-plastic deformation.

2. The Methods of Investigation

For reaching the desired goal a material is required, having beside the advantages of *Sn-38%Pb* eutectics possesses physical and chemical properties, stable in time and at room temperature, allowing to observe deformation of both surface and internal layers of the blank, i.e. it should be optically transparent.

Block-copolymer raisins, on the basis of diene a vinyl-aroma carbons satisfy nearly all these conditions. For simulation of pressing composite in diameter cross-section samples Ø 20x30 mm in dimensions were prepared of non-linearly viscous material of polybutadien – 25 % polystyrene composition. On the division plane a rectangular coordinate grid was marked with Indian ink, after that the blank was placed into an optically transparent stamping facilities and a deforming load was applied.

The character of the material flow and distortion of the coordinate grid during the process of deformation was registered by a video camera during the entire process of deforming. Unlike simulation on *Sn-38%Pb* alloy, in our case, beside the data, received by means of the coordinate grid, it was possible to build kinematic pictures of the pressing process along the flow lines, i.e. to increase the volume and the quality of the obtained data.

Simulation of other processes of plastic forming of other materials showed that it was advisable to add some quantity of polar softener and achromatic stabilizer into the block-copolymer raisin. For that purpose, a number of compositions of non-linearly viscous polymeric materials were developed for simulation of super-plastic deformation. Block-copolymer raisins, with achromatic stabilizers were rolled at 70–75 °C for 2 minutes until an elastic band was formed, then a polar softener (dibutylphtalate or dibutylcebacate). Then the mixture was mixed in rolls by its partial cutting off from the rolls. The samples for testing were pressed in hydraulic press at the die temperature 150 °C and pressure 7.5 MPa, after that they were cooled under pressure up to 30 °C.

Physical and mechanical tests of the samples were conducted in accordance with GOST 269, 270 for rubber materials, index *m* was determined by the method of strains relaxation [1, 30], on the samples Ø 20x30 mm in dimensions.

3. The Results of the Investigation and Their Discussion

The results of the tests are summarized in table 1.

As can be seen in table 1 sensitivity of flow effort to deformation rate of the materials proposed for simulation can vary within wide limits, mostly, because of some small additions of softeners Introduction and alternation of the softeners content causes changes of strength condition of the compound, when it undergoes elongation (simplification of simulation conditions), as compared to the materials without softeners, also changed are deformation characteristics of compounds, however relative elongation value until tear remains constantly high.

Besides, changes in *m* parameter can be reached by selecting one or another block-copolymer. However, the softener content should not exceed 40 % of the entire mass share of the block-copolymer, because: first, practically, there are no super-plastic metals and alloys with index $m = 1$, and second, the material softens and poorly preserves its shape.

The compounds, registered in table 1 were used for manufacturing blanks and further simulation of the process of metal setting in the state of super-plasticity. A composite blank part was cast into matrix, in parts, then a rectangular coordinate grid was marked with Indian ink along the cutoff plane. At simulation of the process of setting the prepared blanks together with the tool were placed into transparent rectangular bath, filled with water, for elimination of optical distortions, occurring as a result of changing of curvature of the surfaces of the blanks.

Table 1. Physical and mechanical properties of polymeric componds.

Index	Composition variant													
	1	2	3	4	5	6	7	8	9	10	11	12	13	14
Specific strenth at elongation, MPa	6.5	2.9	1.12	0.45	0.31	0.14	4.17	1.82	0.84	0.38	0.20	0.1	0.19	0.42
Relative elongation until tear, %	615	502	463	268	214	145	635	560	490	312	215	167	392	309
Relative residual deformation after tear , %	18	24	52	37	26	18	22	29	56	47	32	21	41	27
Plasticity acc. to Carrer	0.29	0.51	0.59	0.64	0.72	0.90	0.34	0.54	0.63	0.70	0.77	0.92	0.68	0.60
Coeficient of rate sensitivity *m* at deformetion rates $10^{-3}…10^{-4}$ sec^{-1}	0.29	0.37	0.42	0.51	0.56	1.0	0.26	0.39	0.52	0.72	1.0	1.0	0.49	0.61

The conducted experiments made it also possible to build the $\sigma = f(\xi)$ diagrams in logarithmic coordinates and determine *m* index by the tangent of inclination angle of the graphs to ξ axis (see figure 1).

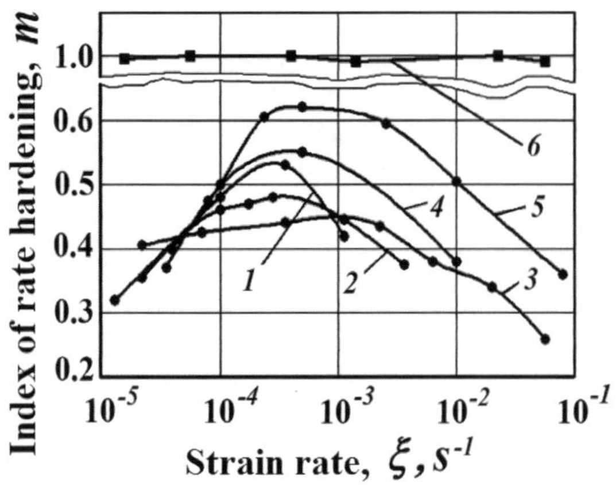

Figure 1. Dependence of rate hardening index from strain rate for optically transparent block-copolymers: (1) Butadienesterene block-copolymer; (2) Isoprenesterene block-copolymer; (3) Butadiene-α-metylsterene block-copolymer; (4) 2,6-di-tret-butyl-4-metylphenol; (5) Dibutylphtalate; (6) Dibutylsebacate.

The analysis of the results, summarized in figure1 proves that just like for super-plastic metals and alloys, deformation of samples, made of block-copolymers is characterized with high rate sensitivity of the flow effort, index of rate hardening *m* has its maximum value within the same range of deformation rates with metallic materials. Slight changes of the softener content allow to modify either alloy's behavior with one or another super-plastic structure, or some or other temperature conditions of deforming.

4. Conclusions

The simulation of the processes of super-plastic deformation with the help of optically transparent non-linear viscous block-copolymers, on the basis of diene and vinyl-aromatic carbohydrates of polar softener and achromatic stabilizer makes it possible to observe physically the process of deformation along the entire process of samples deforming. In prospects, by means of a coordinate grid, placed upon the diametric plane of such samples and video recording of the process it is possible to analyze quantitative and qualitative and qualitative changes along the entire volume at any time span.

References

[1] Giuliano G 2011 *Superplastic forming of advanced metallic materials* (Oxford: Woodhead Publishing Limited).
[2] Marinho EP, Ribeiro FC and Batalha GF 2012 *ABCM Symposium in Mechatronics* vol **5** Section II – Control Systems, ed S C A Alfaro, J M S T Motta and V J De Neg (Rio de Janeiro: ABCM – Brazilian Society of Mechanical Sciences and Engineering) 380–389.
[3] Divya HV, Laxmana NL, Niranjan HM, Vasundhara MG and Yogesha B 2013 *Int. J. Adv. Eng. Tech.* **6(4)** 1829–35.
[4] Furushima T and Manabe K 2007 *J. Mater. Proc. Tech.* **191(1)** 59–63.
[5] Furushima T and Manabe K 2010 *J. Chinese Soc. Mech. Eng. Trans. Chinese Inst. Eng. Seri. C* **31(2)** 99–105.
[6] Milenin A, Kustra P, Furushima T, Du P and Nemecek J 2018 *J. Mater. Proc. Tech.* **262** 65–74.
[7] Kukhar V, Artiukh V, Prysiazhnyi A and Pustovgar A 2018 *E3S Web of Conference*, (HRC 2017) **33** 02031.
[8] Kukhar VV 2015 *Metall. Min. Ind.* **6** 122–32.
[9] Kukhar V, Artiukh V, Serduik O and Balalayeva E 2016 *Proc. Eng.* **165** 1693–704.
[10] Kukhar V, Prysiazhnyi A, Balalayeva E and Anishchenko O 2017 *Modern Electrical and Energy System* MEES'2017 (Kremenchuk: Kremenchuk Mykhailo Ostrohradskyi National University, IEEE) 404–407.
[11] Kukhar V, Balalayeva E, Prysiaznyi A, Vasylevskyi O and Marchenko I 2018 *MATEC Web of Conferences* **178** 02003.
[12] Tandogan M and Eyercioglu O 2017 *Int. Adv. Res. J. Sci., Eng. Tech.* **4(12)** 62–73.

[13] Jones J J and Mears L 2010 *Proceedings of the 2010 International Manufacturing Science and Engineering Conference MSEC2010* (Erie, Pennsylvania: ASME), DRAFT-MSEC2010-34144.
[14] Langdon TG 2009 *J. Mater. Sci.* **44** 5998–6010.
[15] Langdon TG 2016 *Mater. Sci. Forum* **838–839** 3–12.
[16] Davis B and Hryn J 2007 *Innovative forming and fabrication technologies: new opportunities. Final Report*, www.ipd.anl.gov/anlpubs/2008/01/60753.pdf.
[17] Lihui L, Kangning L, Cai G, Yang X, Guo C and Bu G 2014 *Manuf. Rev.* **1(9)** 1–20.
[18] Abu-Farha FK and Khraisheh MK *2008 Int. J. Sustain. Manuf.* **1(1/2)** 18–40.
[19] Alabor E, Putman D, Reed RC 2015 *Acta Mater.* **95(15)** 428–42.
[20] Tisza M 2005 *Mater. Sci. Forum* **473–474** 135–40.
[21] Wood RD, Curtis R, Bonet J, Said R, Gil A, Garriga-Majo D and Odendahl S 2004 *Mater. Sci. Forum* **447–448** 131–8.
[22] Kuwaharada S, Nakanishi K, Mure Y and Matsumoto Y 2011 *Mater. Trans.* **52(8)** 1589–94.
[23] Abdullah AB, Muda MR and Samad Z 2008 *Int. Edu. Stud.* **1(1)** 101–9.
[24] Pertence AEM and Cetlin PR 1998 *J. Mater. Proc. Tech.* **84** 261–7.
[25] Tekkaya AE 2000 *Simulation of Metal Forming Processes* (In: Banabic D (eds) Formability of Metallic Materials, Engineering Materials, Springer, Berlin, Heidelberg).
[26] Kukhar V, Artiukh V, Butyrin A and Prysiazhnyi A 2018 *Adv. Intell. Syst. Comput.* **692** 201–11.
[27] Grushko AV, Kukhar VV and Slobodyanyuk Yu O 2017 *Solid State Phenom.* **265** 114–23.
[28] Ravindra Reddy PVR., Chandra Mohan Reddy G and Radhakrishna Prasad P 2012 *Int. J. Mod. Eng. Res. (IJMER)* **2(4)** 2326–30.
[29] Korshak VF, Shapovalov YA and Vaselenko NN 2015 *Metallofiz. Noveishie Tekhnol.* **37(12)** 1633–42.
[30] Rudskoi AI and Rudaev Ya I 2009 *Mech. Dyn. Superplast. Aluminum Alloy.* (Sankt-Petersburg, Nauka, 218 p.) (rus).
[31] Sheikh-Ali Askar D 2004 *A Study of Deformation Mechanisms of Creep and Superplasticity in Zinc* (PhD Thesis, McGili University, Montreal, Québec, Canada).
[32] Tu ZH, Shim VPW and Lim CT 2001 *Int. J. Sol. Struct.* **38(50–51)** 9267–79.
[33] Shams I and Yano H 2015 *Scientific Reports* **5** 16421.
[34] Balalayeva E, Artiukh V, Kukhar V, Tuzenko O, Glazko V, Prysiazhnyi A and Kankhva V 2018 *Adv. Intell. Syst. Comput.* **692** 220–35.
[35] Kukhar V, Balalayeva E and Nesterov O 2017 *MATEC Web of Conferences* **129** 01041.
[36] Artiukh V, Nikitchenko A, Ignatovich I and Prykina L 2017 *IOP Conf. Ser.: Earth Environ. Sci.* **90** 012191

IOP Conf. Series: Materials Science and Engineering **473** (2019) 012041 doi:10.1088/1757-899X/473/1/012041

Study on the Effect of Vibratory Stress Relief on the Quality of Gravity Die Casting-Theory and Justifications

Nazzal Salem[1]

[1]Zarqa University, Al Zarqa 13132 Jordan,

Abstract. Experimental studies of improvement of the quality of gravity die casting by a low-frequency vibration treatment were carried out. The objective of the study was to establish in principle the possibilities of applying the method of vibration effect in the gravity die casting process for improving the production efficiency. Different testing methods established the advantage of the proposed method comparing to the basic one. The use of the technology of interest reduces a number of casting defects and can be recommended for use in conditions of large production capacities.

1. Introduction

At the current stage of development of the machine engineering, one of the most important and relevant tasks is to develop the resource-saving technologies for reducing the amount of waste in the manufacture of parts made of iron-carbon alloys, and especially of non-ferrous metals. In casthouse production it is of critical importance to reduce casting defects in billets and to obtain parts with minimal allowances for machining.

Foundry is the blank production base for mechanical engineering. Despite the long process of improving the production of castings, the technology of the manufacture thereof has many disadvantages reducing the quality of casting [1]. The analysis of foundries reveals that the casting production suffers significant losses from casting defects.

The number and types of casting defects can be judged on the defective casting storage sites. Here it is necessary to establish a kind of defects, since the correct determination of cause of their occurrence and elimination method depends on it. This is a rather difficult process, because the defect appears due to:

- Variations of process parameters (temperature, humidity, strength, gas permeability, mix gas-generation value etc., totally - more than two hundred parameters)
- Deficiencies in the casting production method
- Seasonal changes in air temperature and humidity
- Inadequate expertise of process staff in foundry shops.

There are some castings with defects formed due to violation of process discipline among the bulk of the sound castings. A distinction is made between casting defects of noncompliance with geometry (misrun, shift, variation in wall thickness, warpage, etc.) or shallow discontinuity (hot cracks, cold cracks, gas porosity, shrinkage cavity, shrinkage porosity, sand inclusions, etc.), noncompliance of casting metal with required structure (chilling, liquation, etc.) and impurity (metallic/nonmetallic inclusions, etc.).

Defects are divided into two groups – non-repairable and repairable defects. Non-repairable defects (usually large ones) cannot be repaired or repair is uneconomical, so a casting with such defects is considered unfit for use (defective) and is sent for remelting. Repairable defects (usually small ones) are eliminated in order to make the casting fit for further treatment and use. As the experience of many foundry shops producing shaped castings has shown the most common defects in castings are cavities and cracks. There are only 7 types of blow holes with specific characteristics. Usually two, at best three kinds of gas shells are mentioned. But it is even worse, when the defect is not detected in the billet immediately, but by non-destructive testing method in the already finished part after having passed all the stages of machining and treatment. This indicates that there is a real need to develop a method that helps to assist the foundrymen to reduce various defects and eliminate the causes thereof. As such a method, the use of vibro-impulse exposure is proposed. The main task of experimental studies on mechanical vibration is to determine its rational parameters and overall effectiveness of appliance of this technology for improving the quality of products and confirming the theoretical and laboratory research data of vibration effect in full-scale conditions.

2. Justification of the Need for Vibration Effect

Due to the uneven metal solidification, the phase transformations (during crystallization and cooling), related to the increase or decrease in the volume of individual grains, generate significant residual stresses in the casting process. Reference layers with different densities can be formed in the casting: a thin outer layer consisting of equiaxed crystals, followed by a zone of large elongated columnar crystals, and a central zone of large non-oriented crystals. When the ingot solidifies in its upper part a shrinkage cavity is formed. The near-surface layers solidify faster and create stresses that cause the tensile strain, and the inner layers generate compression stresses that balance the tensile stresses. For this reason and for a number of other reasons, there is a complex pattern of the strain-stress state in the finished product, which negatively affects the final shape of the billet. When calculating the strength of casting, according to the stress diagrams of the destructive forces, the outer layers of the casting experience the maximum stresses and displacements, and the median layers – the minimum stresses and displacements.

One of the most common casting processes, in which defective castings often occur, is the gravity die casting, namely, the casting using a permanent mold – f long-life mold with a melt filling by gravitational forces, and with natural or forced cooling [2]. In many cases, the permanent molds have nonmetallic parts, usually sand cores and inserts, but most of the surface of the casting is formed by the metal elements of the permanent mold, which significantly affects its quality. The disadvantages of gravity die casting, leading to casting defects, include:

- Rapid melt quenching when filling a metal mold, which makes it difficult to produce thin-walled castings, especially with extended walls
- Rigidity of permanent mold which can lead to the casting internal stresses, and as a result to warpage and even cracks
- Limited possibilities for air and gas exhaust from the cavity of the metal form.

To minimize the probability of forming these defects and to reduce the stress gradients, it is proposed to apply the mechanical vibration during metal solidification [3]. The method of low-frequency mechanical vibration is designed to reduce mechanical residual stresses in metal products with high requirements to their quality characteristics. When using the methodology, the natural frequency of the product is determined, after which the structure is subjected to mechanical vibration at a resonance frequency for 15-20 minutes, resulting in a redistribution of mechanical stresses.

3. Method Parameters and Determination of the Engineering Tools

The technology of vibration effect is based on the use of low-frequency vibrations generated by a special device - electric vibrator and it is characterized by absolute ecological purity, low energy costs and high efficiency.

In low-frequency vibration systems, it is most rational to use an eccentric actuator [4, 5]. This type of actuator is able to create large amplitude vibrations at low frequencies. Low-frequency machines also have less dynamic load on the actuator, which allows working with some detuning from the natural frequency of the vibration system, providing greater stability. When using variable speed motors, such as direct current motors or three-phase commutator motors, an actuator with a wide range of frequency control can be obtained. Using rotary eccentrics, it is possible also to adjust the vibration amplitude.

The difference between machines with a forced eccentric vibroactuator with a rigid connecting rod and machines with a kinematic eccentric vibroactuator in that the vibrations of their working member have an amplitude that does not depend on the dynamic parameters and load of the machine, but is only a function of the kinematic parameters of the actuator.

To select the type of the vibration effect device, it is necessary to calculate the diagram of the machine with an eccentric actuator.

Displacement of the movable operating element of vibration machine shall be according to the law [6-8]:

$$x = A sin\omega t \tag{1}$$

where A – is a displacement amplitude ($A = r$); r – is a actuator eccentricity.

Substituting the values $\frac{d^2x}{dt^2}$, $\frac{dx}{dt}$ and x into the equation of motion of the vibration machine, we obtain the following:

$$mA\omega^2 sin\omega t + cA\omega cos\omega t + kA sin\omega t = P sin(\omega t + \varphi) \tag{2}$$

Hence the following amplitude value of the driving force:

$$\mathrm{P} = \mathrm{r}\sqrt{(p^2-\omega^2)^2 + 4n^2\omega^2} \tag{3}$$

where n – is a viscous resistance coefficient, $2n = c/M$; p – is a natural frequency of vibrations of the product with mass M; $p^2 = k/M$.

The phase difference between the displacement of the working member and the driving force:

$$\varphi = arctg\frac{2n\omega}{p^2-\omega^2} \tag{4}$$

The power supplied by the vibroactuator to overcome resistances in elastic constraints:

$$N = mn\omega^2 r^2 \tag{5}$$

Minimum driving force at $\omega = p$.

In the steady state, under the eccentric vibroactuator operation, the elastic element – vibration mount is deformed, and the deformation depends not only on the parameters of the eccentric, but also on the law of motion of the working member. Due to the fact that the vibration system displaces x with the speed of $\frac{dx}{dt}$, the elastic constraint of the actuator is deformed by $x_0 - x$ with the speed of $\frac{dx_0}{dt} - \frac{dx}{dt}$. Here, x_0 and $\frac{dx_0}{dt}$ are, respectively, the displacement and displacement velocity of the vibration mount.

The dimensionless vibration amplitude $\frac{A}{\chi r}$ depending on the adjustment coefficient $z = \frac{\omega}{p}$ for various damping coefficients of the elastic vibration mount $\nu = \frac{n}{p}$ at different stiffness ratios and resistance coefficients of vibroactuator and vibration mount $\chi = \frac{p_0^2}{p^2} = \frac{n_0}{n}$ is shown in figure 1.

Experimental studies on the maximum withstand load were carried out to select the type of vibration mounts. The vibration mounts were uniaxially loaded in terms of linear increasing strain. The P-50 pressing machine was used as the power unit, the stiffness of which was provided by elastic elements, extending the support plates of the loading device. As such elements, which compensate the response time of pressing machine, two hydraulic jacks fed from a high-pressure oil pumping station were used. In this case, with an increase in the load on the vibro-support to the ultimate strength, the increased stiffness of the testing machine redistribute the load between the hydraulic jacks and vibration mount, which allows to control the deformation process even beyond the limit of strength.

Mounted on the press displacement and force sensors were sending the signal to the computer through the analog-to-digital converter.

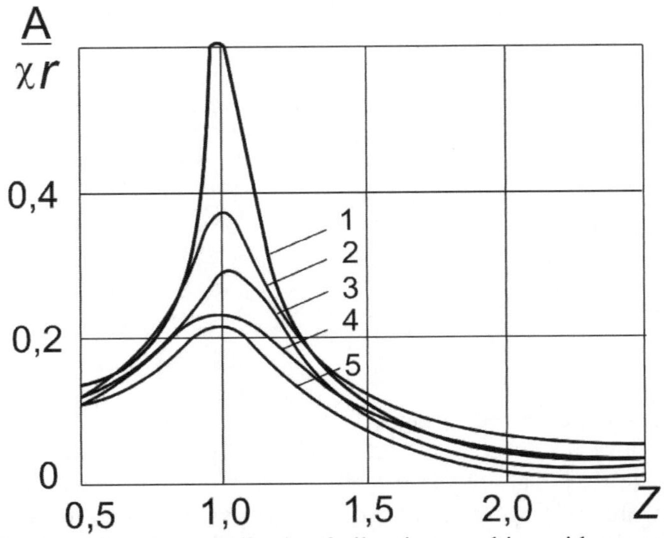

Figure 1. – Dimensionless vibration amplitude of vibration machine with eccentric vibroactuator — adjustment system coefficient at various damping coefficients curves.

Figure 2 shows the resulting stress-strain curve. Analyzing the curve, we see that the entire period of loading of the vibration mount can be conventionally divided into four stages:

1) – period of elastic strain;
2) – period of elasto-plastic strain;
3) – period of transition to the area of the ultimate strain;
4) – period of unloading.

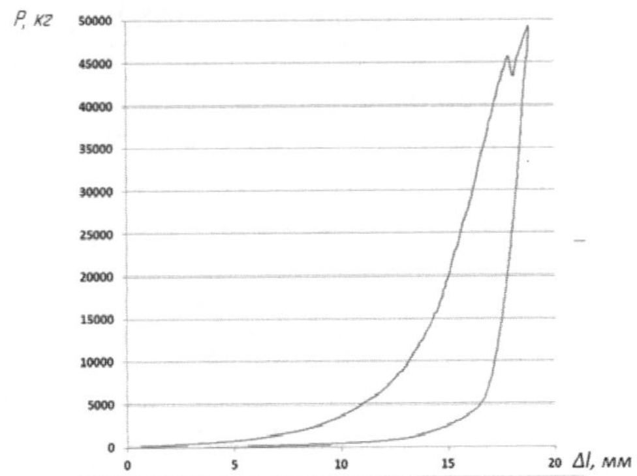

Figure 2. – Vibration mount loading diagram
(Strains [mm] [*along the axis OX*] — load [kg] [*along the axis OY*] curve).

At the first stage (up to 10 tons) the vibration mount is compressed wherein smaller values of displacements in the lateral direction correspond to certain value of the relative displacement in the longitudinal direction, as the particles approach the total friction surface is increased. At this stage, the material becomes consolidated, increasing the differential modulus of elasticity caused by compaction, and as a result - by an increase in the slope of the time-load curve.

The transition from the first to the second stage is shown on the stress curve by changing the sign of curvature - from convexity to concavity, which leads to a decrease in the differential modulus of elasticity and gradual decrease in the slope of the time-load curve, [9].

The transition from the second to the third stage is marked by a maximum of strains (about 46 tons). In this case, the vibration mount is destroyed in the critical areas. After that, there is still some increase in the load up to 49 tons.

After unloading there is a transition to the fourth stage – the stage of the vibration mount shape recovery according to below curve.

Knowing the mass of the product *M*, the parameters of the vibration exposure can be selected, which minimize the casting defects and reduce maximally the residual mechanical stresses.

Based on the results of calculations, the main parameters of the complex of vibration exposure are the following:

– maximum shaking force [kN]	1.8
– synchronous speed [rpm]	1,500
– operating speed range [Hz]	1-200
– rated power [W]	120
– voltage applied to vibrator [V]	220

4. Conclusions

Vibration method widely manages operating effect parameters due to changes in the values of vibration velocity and vibration acceleration, more qualitatively and less costly equalizes the residual compressive and tensile stresses that arise in the billet during gravity die casting. The use of the test technology reduces a number of casting defects and can be recommended for use in large production capacities. The method of low-frequency vibration effect allows processing not only small-sized structures but also structures with a mass of up to 100 tons. In the serial production, the mechanical vibration can be automatically operated on a vibration exciter by a timer. Generally, the cumulative savings at each stage can provide tangible economic benefits as a whole.

References

[1] Kalpakjian S, Schmid S 2006 *Manuf. Eng. Tech.* (5th ed.), Pearson.

[2] Todd Robert H, Allen Dell K, Alting Leo 1994 Manufacturing Processes Reference Guide, Industrial Press Inc.

[3] Pat. 2010-069514 JP, B22D 27/08. Casting method with vibration solidification, casting mold for the same method, and casting apparatus with vibration solidification: National Institute of Advanced Industrial Science & Technology Kotobuki Kinzoku Kogyo KK / Miwa Kenji, Omura Naoki, Yokoi Mitsuyoshi, Harada Masayuki and Furukawa Hideki. 02.04.2010.

[4] Junkichi I, Yoshiaki A, Yuklmasa O, 1975 On the Selfsynchronization of Mechanical Vibrators (Part 4, A Number of Unbalanced Rotors Arranged in Space). Transact, of the Japan Society of Mechanical Engrs., 1975, 41, N. 350.

[5] Reynolds Douglas D 2016 Engineering Principles of Mechanical Vibration (4th ed.). Bloomington, Indiana, USA: Trafford On Demand Publishing. p. 485.

[6] Thomson WT 1971 Vibration theory and applications. London, Allen and Unwin, 384.

[7] Anderson R 1967 Fundamentals of vibrations. New York, The Macmillan Company London, Collier-Macmillan Limited, 412.

[8] Schmidt G. 1975 Parametererregte Schwingungen. Berlin, Deutscher Verl. der Wiss. 313.

[9] DIN EN ISO 6892-1-2009. Metallic materials - Tensile testing - Part 1: Method of test at room temperature.

Aerosol Jet Printing of Platinum Microheaters for the Application in Gas Sensors

P V Arsenov[1], I S Vlasov[1], A A Efimov[1*], K N Minkov[2] and V V Ivanov[1]

[1]Department of Physical and Quantum Electronics, Moscow Institute of Physics and Technology, Dolgoprudny 141701, Russia
[2]Moscow Institute of Electronics and Mathematics, National Research University of Higher School of Economics, Moscow 123458, Russia

Abstract. This paper presents the results of characterization of microheaters fabricated on the surface of ceramic substrates by aerosol jet printing with commercially available platinum ink containing colloidal particles with the average size of about 100 nm. The microheater under study represents planar microstructure with variable cross-sectional area comprising narrow segment capable of being heated up to few hundred degrees when current is passed through. The narrow segment is fabricated in the form of single line with the width and height of 50 μm and 6.7 μm, respectively. To remove the organic substance from the deposited material (dry residue of the ink), the printed microheaters were fired at 150°C for 20 min. The dependence of the operating temperature on the power consumed by the microheater was obtained from the measurements of direct current resistance as a function of consumed power with the use of previously determined temperature coefficient of the material ($3.9 \cdot 10^{-3}$ K^{-1}).

1. Introduction

The aerosol jet printing method (AJP) is a universal technology for printing on substrates of various materials using a large selection of functional inks [1]. The use of AJP in the field of printed electronics becomes more and more popular, since the print resolution reaches about 10 microns [2]. This resolution is not possible while using alternative methods, for example, inkjet or screen printing. One of the AJP development areas is a creation of low cost sensors for determination of gas or liquid chemical composition [3]. The basis of this sensors making technology is the application of functional layers of a nanocomposite (gas sensitive material) on a special silicon or ceramics substrates [4]. As a result, the sensor has sufficient selectivity due to the use of nanocomposite material and it has inexpensive mass production capabilities. The main and most difficult task for creating such sensors is the printing of a microheater, which locally heats the gas sensitive layer. Such microheater must accord to a number of geometrical, mechanical, temperature and electrical requirements. For the above reason, in this paper we investigated the printing parameters and characteristics of a microheater for potential gas detectors applications. Also we managed to explore the contactless method of aerosol printing, which allows to form Pt microheaters on thin (<100 μm) ceramic substrates without loss of expensive material. Such aim is an impossible task while using traditional methods of contact manufacturing of microheaters, such as photolithography and screen printing [5].

2. Experiment

The gas sensor microheater was printed using a commercial AJ 15XE aerosol jet printer (Neotech AMT GmbH). The platinum metal nanoparticles ink with the average size of about 100 nm was

IOP Conf. Series: Materials Science and Engineering **473** (2019) 012042 doi:10.1088/1757-899X/473/1/012042

sprayed with a pneumatic atomizer. The flow of aerosol particles in the form of microdroplets contains metallic nanoparticles inside it. In our experiments it was focused on a ceramic substrate based on aluminum oxide (LTCC) with the size of 2.5×0.5×0.1 mm using an argon gas flow. The optimal printing parameters [6], at which the deposited material did not spread on the substrate and was conductive after sintering, were investigated and selected for platinum ink. The aerosol and focusing gas flows were 20 and 5 sccm, respectively. The precipitated ink was dried at 150 °C (20 min) and then sintered at 900 °C (30 min). After that the four-probe method was applied to make electrical measurements. For this purpose, Agilent U1253B multimeter and Keithley 2401 current source and voltage source were used. According to the obtained dependencies of resistance on the applied voltage and formula (1), the temperature of the microheater T was calculated, and the dependence of the temperature on the power consumption P was constructed:

$$T = T_0 + \frac{R - R_0}{\alpha R_0} \quad (1)$$

where T is the temperature of the microheater, T_0 is the room temperature, R is the resistance of the microheater at temperature T, R_0 is the resistance of the microheater at room temperature, α is the temperature coefficient of resistance ($3.9 \cdot 10^{-3}$ K^{-1}) [7]. The formula (1) was taken from [7].

Optical images of the printed structures and their 3D profiles were also obtained using Keyence VHX 1000 microscope and Leica DCM 3D optical profilometer, respectively.

3. Results

Figure 1 shows the microheater images in the form of a curved line on the surface of the LTTC substrate, measured using optical microscope and 3D profilometer.

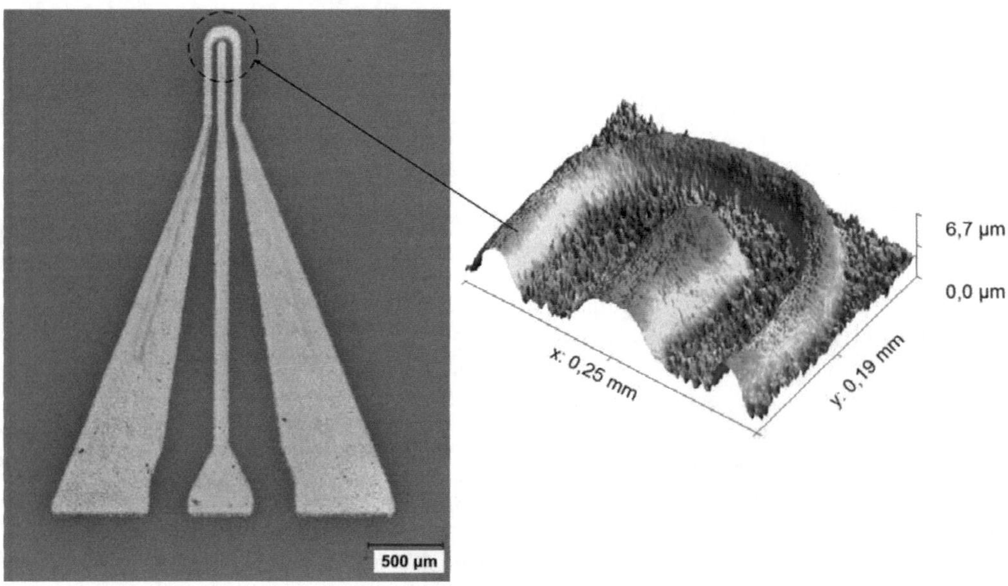

Figure 1. Optical image (left) and 3D profile (right) of a platinum microheater in the shape of a curved line on a LTTC ceramic substrate, measured with optical microscope and 3D profilometer, respectively.

From the image analysis presented in figure 1, information about the width, height, and profile of the deposited microheater structures was obtained, and it was carried out the monitoring of the microheaters for the presence of closures and spreading of the deposited material. Typical widths and heights were 50 and 6.7 µm, respectively.

Figure 2 shows the dependence of the calculated temperature of the microheater, according to formula (1), on the power consumption.

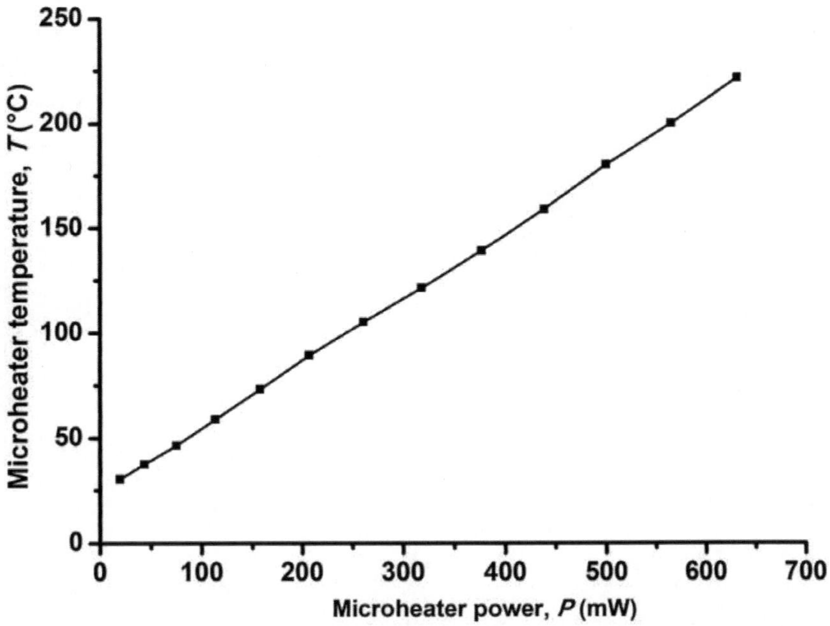

Figure 2. Dependence of the heating temperature on the power consumption of the platinum microheater printed by aerosol jet printing.

From figure 2 it can be seen that the dependence of temperature on power consumption is linear, which is characteristic of this type microheaters. From the analysis of the dependence presented in figure 2, it was established that the recommended power consumption with continuous heating of the microheater to 225 °C should be about 650 mW. In the case of its exceeded, there is an irreversible degradation of the microheater contacts is observed.

4. Conclusions

The paper presents the possibility of aerosol jet printing method usage for the manufacturing of platinum microheaters for potential applications in gas sensors. As a result, printing parameters were selected and microheater templates on ceramic substrates from LTCC were obtained. The dependence of the heating temperature on the power consumption of the microheater was obtained by using the four-probe method. It was established that the recommended power consumption of the microheater is about 650 mW at its heating temperature up to 225 °C, and a further power increase leads to overheating of the microheater.

Acknowledgements

This work was supported by the grant of the President of Russian Federation for young scientists MK-2302.2017.8 and carried out using the Shared Use Equipment Center for high-precision measuring in photonics (ckp.vniiofi.ru, VNIIOFI).

References

[1] Deiner L and L. Reitz T 2017 *Adv. Eng. Mater.* **19** 1600878.
[2] Mahajan A, Frisbie C D and Francis L F 2013 *ACS Appl. Mater. Interfaces* **5** 4856–64.
[3] Clifford B, Beynon D, Phillips C O and Deganello D 2017 Sens. Actuators B Chem. **255.**

[4] Vasiliev A, Varfolomeev A, Volkov I, Simonenko N, Arsenov P, Vlasov I, Ivanov V, Pislyakov A, Lagutin A and Jahatspanian I et al. 2018 *Sensors* **18(8)** 2600.
[5] Bucknall D G 2005 Nanolithography and patterning techniques in microelectronics *Taylor & Francis.*
[6] Arsenov P V, Efimov A A and Ivanov V V 2018 K*ey Eng. Mater.* **779** 159164.
[7] Zhang X, Xie H, Fujii M, Ago H, Takahashi K, Ikuta T, Abe H and Shimizu T 2005 *Appl. Phys. Lett.* **86** 171912.

IOP Conf. Series: Materials Science and Engineering **473** (2019) 012043 doi:10.1088/1757-899X/473/1/012043

Study on the Relevance of Lightweight Steel Structures and Thermal Hazard During Fires

Chung-Chyi Chou*, Chia-Chou Tsai, Chi-Chung Huang, Yu-Jhih Ou

Da Yeh University of Department of Environmental Engineering
Da-Yeh University of Department of Fire and Safety, 515, Taiwan (ROC)*

Abstract. Lightweight Steel Structures with high application targets, rapid construction and low cost. However, the influence of heat radiation on the internal temperature is very large, and in order to reduce the heat conduction caused by the sun, the user often attaches a flammable heat insulating material inside. In this study, the fire simulation software FDS was used to simulate the lightweight steel structure building fire, and the research on the thermal buoyancy, heat conduction, smoke distribution and building structure collapse limit of this type of building form was calculated. The research found that the steel thermal fluctuation and the fire flashover temperature of 600 °C were achieved after 520 seconds of fire. This study provides lightweight steel structure construction based on numerical calculation results. It is recommended to set the smoke exhaust window appropriately or use fireproof materials for interior decoration to reduce the heat accumulation inside the building to prolong the time of steel thermal fluctuation and make the lightweight steel structure It is not easy to flashover (backdraft) in the building, and the steel structure is not easy to collapse. It can improve the safety factor and make the life of fire rescue personnel more secure.

1. Introduction

In view of the recent serious accidents caused by fires in lightweight steel structures in the international community. This study explores the fire case of "Bakery Store"[1] in Changhua County on April 14, 2014, this disaster caused five people to die in a fire in a lightweight steel structure, Not only aroused the attention of the public to the fire rescue of lightweight steel structures, but also burned the disaster prevention problem of lightweight steel structures.

Lightweight steel structure buildings are seriously inadequate due to structural durability and fire performance, In addition, the accumulation of smoke heat is prone to flashover, backdraft and the risk of building steel structural strength damaged and collapse. Causes people and fire rescue personnel in the building to lose their lives.

The purpose of this study is to understand the occurrence time of flashover and steel thermal desuperheating temperature in the heat distribution of lightweight steel structures, and to analyse the countermeasures to avoid the occurrence of structural durability and fire performance. The collapse of the life of the fire rescue personnel.

2. Methods

This study collects relevant literature to explore fire risk factors for lightweight steel structures. With the "Fire Dynamics Simulator" (FDS) as the case reduction, the difference between the smoke from the roof opening and the smoke flow on the floor exhaust, the improvement plan for the fire collapse damage of the steel-type iron-type building is proposed. The Fire Dynamics Simulator (FDS) [2] is a computational fluid dynamics (CFD) model of fire-driven fluid flow. The software solves

IOP Conf. Series: Materials Science and Engineering **473** (2019) 012043 doi:10.1088/1757-899X/473/1/012043

numerically a form of the Navier-Stokes equations appropriate for low-speed, thermally-driven flow, with an emphasis on smoke and heat transport from fires.

2.1. Research process

According to the actual building to draw of the fire case (figure 1) and situational simulation (figure 2). The fire building is a double storefront open shop, brick wall, iron sheet to build a 2-storey building.

Figure 1. Rescue situation at the beginning of the fire.

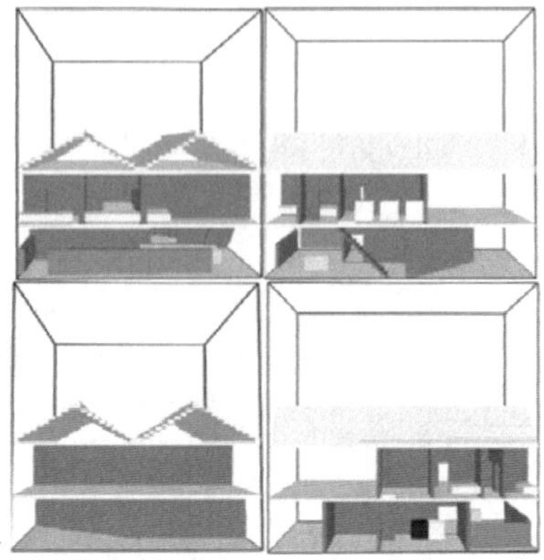

Figure 2. Rebuilding a fire building with FDS.

2.2. Research tool

This study used the fire simulation software (FDS). A software that uses the CFD (Computational Fluid Dynamic) method to build a fire case model and uses the mixed component combustion mode and equation to calculate the fire high temperature smoke and the heat flow field change in the building space. The direct numerical equation of DNS (Direct Numerical Simulation) can be used to describe the simulated fire dynamic behaviour driven by thermal buoyancy.

In terms of the thermal fire rate of electrical fires. Mei [4], mentioned that: FDS simulates the electric fire case caused by the short-circuit of the extension line in the betel nut booth in Kaohsiung City, in order to allow the simulated fire to generate its own heat, in order to conform to the example situation, the fire source The heat release rate per unit area is set to 1500 kW/m^2, and the fire is extinguished after 600 seconds. In addition, Feng [5], pointed out that in the study of fire hazard analysis, a high-voltage cable with a diameter of 12 mm of 600V is used, and the conductor is copper. The insulation layer and the outer skin were both PVC. The actual combustion test was carried out. It was observed that the maximum unit heat release rate was 422 kW/m^2 at 315 seconds, and the total heat release was 106 MJ/m^2. In addition, Monideep (2002) pointed out in the report of the International Expert Group on Cable-line Fire in Nuclear Power Plants: According to the experimental data, the distribution of the heat release rate of a cable with a length of 15m, a width of 0.6m and a thickness of 0.1m in a well-ventilated environment It is about 0.8 MW to 2 MW, and its growth characteristics follow the t-squared fire model. It takes about 10 minutes to grow to the highest pyrolysis rate. In Shen (2004) research report: set the fire source size of the cable fire to 1 MW and 2 MW, the thermal release rate parameter set HRRPUA = 222 kW / m^2, the fire growth time is set to 10 minutes, and then maintain the highest heat release rate of 10 minutes.

The combustion parameters of this study are based on the fire damage investigation and appraisal of Changhua County Fire Department. The fire case of this study is the mixer of the middle section of No. 589 on the 1st floor. The cause of the fire does not rule out the possibility of fire caused by electrical factors. No. 589 In the middle section of the first floor, a surface fire source is set up. In the fire cause investigation and appraisal book, the building is placed with a large number of display cabinets and commodity packaging materials. The flammable items such as wooden bed compartments and wardrobes on the 2nd floor have a large amount of fire load, and their growth. The feature also follows the t-squared fire model, which takes about 10 minutes to grow to the highest pyrolysis rate (Shen, 2004). For the initial temperature parameter setting, refer to the Central Meteorological Administration observation data query system. At 04:45 on April 14, 2014, the temperature of the member city was 22.4°C, the humidity was 84%, the wind speed was 0.6 m/sec, and there was no rain.

3. Results

3.1. Fire simulation lightweight steel structural roof smoke and heat flow condition

Based on the field measurement data of this research case and the research reference of Chen (2014) [3], the fire simulation was carried out by FDS. A 3D graphical analysis was carried out to compare the vertical and horizontal spread of smoke.

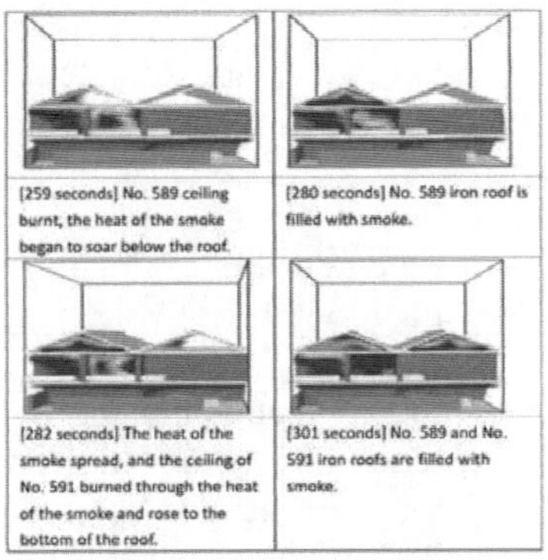

Figure 3. Fire simulation [no opening on the roof] Heavy smoke spread.

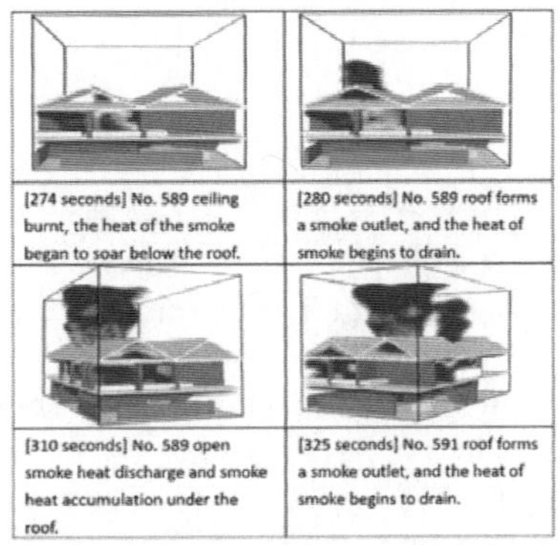

Figure 4. Fire simulation [opening on the roof] Heavy smoke spread.

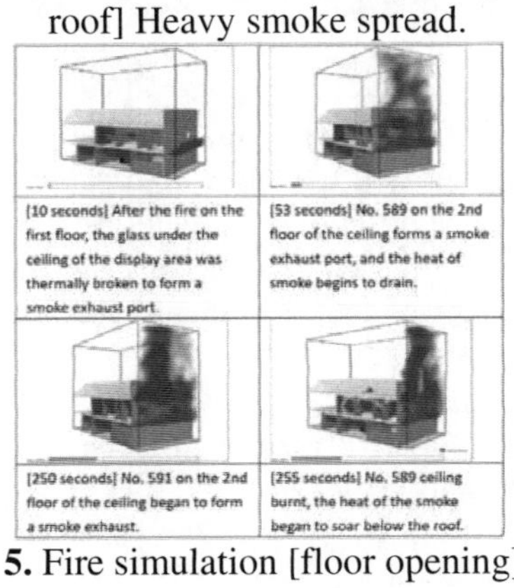

Figure 5. Fire simulation [floor opening] smoke exhaust situation.

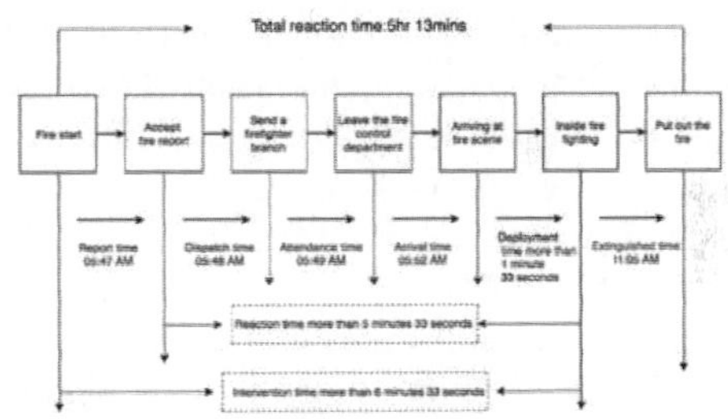

Figure 6. Flow chart of fire-fighting reaction intervention in the fire.

Lightweight steel structural polyol depending on the use, upholstery, bedding, and other items plurality bedroom fire load of greater, a large amount of smoke after the fire, due to poor adaptation metal buildings, good air circulation, quickly jumped smoke. In this study, the fire of the simulated object was located in the middle of the first floor of No. 589. The fire broke through the ceiling at 259 seconds, and the heat began to accumulate under the iron roof. The fire was filled under the iron roof in 280 seconds; the 282 seconds of fire smoke spread to 591 No. The ceiling began to burn through and the smoke was filled under the roof of the entire iron building in 301 seconds.

The results of this study show that the ceiling of No. 589 is burned to the heat of the 599, which is only 23 seconds. The time below the roof of the iron roof is full of smoke for 42 seconds.

3.2. Changes in sheet metal structure after fire-fighting intervention

From the beginning of the fire until it is extinguished, the entire continuous process is displayed in time-distance mode, as shown in figure 6.

3.3. Analysis of roof temperature of lightweight steel structure

According to Zhuang (2004), the behaviour of steel structures under normal temperature and high temperature is affected by their mechanical properties, such as the yield strength, the modulus of elasticity, which will decrease as the temperature increases. This study explores the temperature of the front, middle and rear sections of the sheet metal roof as a whole, and whether the temperature changes after the intervention of fire-fighting cause the collapse of the steel structure building and threaten the life safety of fire rescue personnel (figure 7).

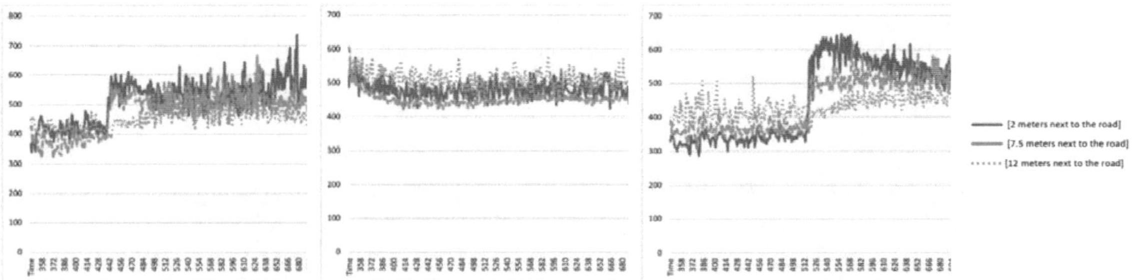

Figure 7. The temperature curve diagram of the sheet metal roof.

Temperature after fire-fighting intervention, (1) At the detection point 2 meters next to the road on the sheet metal roof [NO opening on the roof], the temperature reached 595°C at 446 seconds, the temperature reached 603°C at 451 seconds. The detection point 7.5 meters next to the road at 446 seconds, the temperature reached 559°C, the temperature reached 627°C in 569 second. The detection point 12 meters next to the road reached 564°C in 538 seconds and 609°C in 625 seconds.

3.4. Comparison of fire-fighting intervention time to flashover time

Based on the study of the dangers of flashover and backdraft to firefighters (Su, 2007) [6], from the moment of fire to the moment when the fire is extinguished, this is a question of time series. The following is a discussion of the experimental dimensions of each scholar and the heat flux and temperature at the time of flash fire:

According to NFPA265-2007, the following five items are provided as indicators of flashover. Conforms to the two to be possible to determine the flashover occurrence:

(1) Heat release rate is greater than 1MW
(2) The heat flux of the floor surface reaches 20kW/m²
(3) The temperature of the hot smoke layer above the space is above 600°C
(4) The fire has been burned to the emergency entrance
(5) The inflammable in the space reaches the lower limit of autoignition.

Time-temperature curve diagram of the sheet metal roof after fire-fighting intervention (figure 7). At the temperature of "No opening on the roof", the roof temperature exceeds the flashover condition of 600°C. According to Chen (2014) [3], etc. The research on the steel skeleton structure of steel structural members up to 538°C above the heat resistance, the steel reinforce is easy to be hot bending, the entire sheet metal house will be unstable due to the center of gravity, lost balance and collapse.

4. Conclusion

The main reason is that in the simulation of lightweight steel structure building fires, the distribution of smoke heat is the main point of analysis for the flashover temperature and the temperature of the steel thermal declination temperature. The safety time of personnel and buildings is shown in figure 8:

1. According to the fire investigation data, in the situation no opening on the roof of the original case, the temperature of the sheet metal roof "2 meters next to the road" reaches 603°C at 451 seconds, and the temperature of "7.5 meters next to the road " reaches 606°C at 568 seconds. "12 meters next to the road " reached 609°C at 625 seconds.

2. Simulate fire safety equipment on each floor to set the Smoke Vent. After the fire-fighting intervention, the temperature of the sheet metal roof is about 400°C. The structure of the lightweight steel is relatively stable and not easy to collapse, but if the firefighters arrive at the scene of the fire. After more than 510 seconds, the temperature of the iron sheet began to rise, and the strength of the steel structure began to be challenged by high temperatures, posing a risk of collapse.

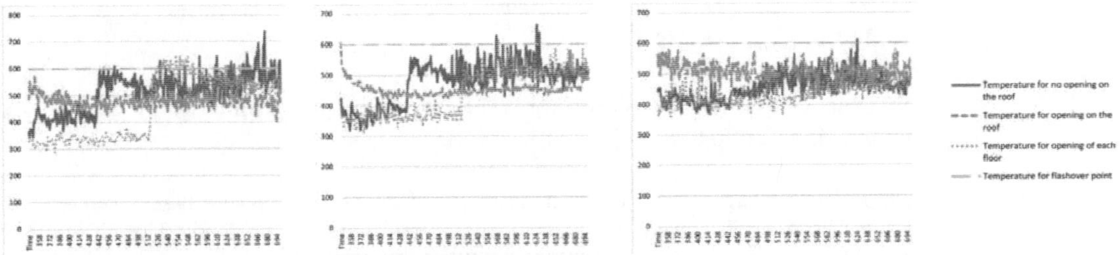

[2 meters next to the road] [7.5 meters next to the road] [12 meters next to the road]

Figure 8. Temperature curve diagram and flashover temperature of each module roof.

The situation temperature of "no opening on the roof" is higher than the other two scenarios. Repeatedly exceeding the flash point of 600°C and the steel softening temperature of 538°C, the impact on the lightweight steel structure is the most serious, easily lead to collapse of buildings, endangering the safety of fire rescue personnel. The decorate , the packaging materials of the products and the large space high-capacity machines placed in the studio space cause high fire load in the building and high use of fire to form a highly dangerous place. In order to improve the fire safety of buildings, firewalls, fireproof floor and fireproof windows and door should be used to divide the partitions, and the basic fire compartment should be used to reduce the burning speed of the fire.

References

[1] Fire Bureau of Changhua County, “Fire identification report”, 2014.
[2] McGrattan K et al. Fire Dynamics Simulator Technical Reference Guide Volume 1: Mathematical Model, *NIST Special Publication* 1018-1 Sixth Edition.
[3] Chen et al,. 2014 Study on Fire Safety and Fire Fighting Countermeasures for Lightweight Steel Structure Buildings (Iron House) - Data Collection and Analysis Report, *Architecture and Building Research Institute, Ministry of the Interior,* ROC (Taiwan).
[4] Mei et al. 2012 Discussion on the promotion of fire cause identification technology and the cooperation mode of the third impartial institution. *The Central Police University commissioned a research project report.*
[5] Feng 2007 The Atomic Energy Commission of the Executive Yuan commissioned a research report for the study, *Fire hazard analysis.*
[6] Su, et al., 2007 Research on the dangers of flash fire and deflagration to firefighters. Master's thesis of Department of Environmental and Safety Engineering, *National Yunlin University of Science and Technology.*
[7] Peacock R D et al., 1999 *Fire Safety J.* **32** (1999).
[8] Yang P Z, Tan X, Xin W, 2011 Experimental study and numerical simulation for a storehouse fire accident, *Building and Environment.*
[9] Quintiere J G.and McCaffrey B J 1980 The burning of wood and plastic cribs in an enclosure, *I.NBS (U.S.), NBSIR 80-2054,* 1980.
[10] Spearpoint M, Mowrer F W and McGrattan K 1999 Simulation of a compartment flashover fire using hand calculation, zone model and field model, October 4-8, 1999, *Chicago, IL, Society of Fire Protection Engineers,* Boston, MA, 3-14 pp, 1999.
[11] McGrattan K, McDermott R, Hostikka S, Floyd J 2014 Fire Dynamics Simulator(Version 5) Users Guide, NIST Special Publication 1019-5, *National Institude of Standards and Technology.*

CFD Study on the Windage Power Loss of High Speed Gear

A He[1], R Deng[1], Y Xiong[1]

[1]Department of Wind Power Development, Nanjing High Accurate Drive Equipment Manufacturing Group Co., Ltd., Nanjing, China

Abstract. The windage power loss is main factor of the power loss in a high speed gearbox. CFD simulation method has been developed for a detailed understanding of the physical mechanism of windage power loss. Comparisons are made with experimental data from the open literature. The CFD results are in good agreement with experiment. The study focuses on the influence of the shroud on the windage power loss for rotating speed ranging from 5000rpm to 7000rpm. It is critical to capture the energy efficiency prize by shrouding the gear. Smooth shroud and grooved shroud have been selected to reduce the windage power loss. It is observed that the better performance occurs with shroud, whereas it is found that the grooved shroud provides a considerable increase in energy efficiency at about 10% over the smooth shroud.

1. Introduction

Gearbox is the key part of high speed train, it mainly responsible for the transmission of motor power to the wheel. With the improvement of train speed, the power loss has become the key problem in the high speed gearbox. The power loss directly affects the equilibrium temperature of the gearbox, which is directly related to the working conditions of the lubricating oil, the sealing system and the bearing lubrication. As a result, it greatly affects the reliability and life of the gearbox [1].

The splash lubrication is generally adopted in high speed gearbox. It is commonly overheated by the windage power loss. The windage power loss is strongly related to the rotating speed. With the increasing of rotating speed, the windage power loss increases significantly and it becomes an indispensable part in the power losses of high speed gearbox. The equilibrium temperature of the gearbox is very high at high rotating speed. The equilibrium temperature can be reduced by decreasing of windage power loss.

The numerical simulation of windage power loss of high speed gear has been focused in this work. According to previous studies, some works were done to exam the efficiency losses of the gearing systems. Massini did numerical and experimental investigations to study windage power losses resulting by a single spur gear rotating in a free oil environment [2]. The numerical results were compared with experimental data in terms of resistant torque as well as PIV measurements, achieving a good agreement for all of the rotating speeds. Two different theoretical approaches were developed to predict windage losses in high speed gear by Diab [3]. It is found that both theoretical approaches give good results in comparison with the experimental evidences. A numerical approach was investigated and validated to develop physical understanding of the aerodynamics of windage power loss by Matthew [4]. Absolute and relative frame CFD simulation, overset gridding, multiphase flow analysis, and sub-layer resolved turbulence modeling were brought to bear in developing the numerical method. Sylvain applied a CFD code to three-dimensional simulations of windage power loss generated by spur gears rotating in air [5]. Emphasis was placed on the simplification which can be made in the numerical approach in order to gain cost and time needed for reaching a converged solution when using both fine mesh and important angular velocity. Although there were several previous studies on the windage power loss, the control methods of windage power loss had been seldom studied. The windage power loss causes general efficiency losses of gearing systems. It is thus important to control the windage power loss.

The purpose of the present study has been to explore the effects of smooth and grooved shrouds on windage power loss. Numerical models have been established to replicate the flow field surrounding a single spur gear. The gear geometry in present study has been built based on the existing experiment [6]. The effects of smooth and grooved shrouds on the windage power loss have been indicated in terms of the resistant torque.

2. Validation of numerical model

The existing experiment of high speed gear has been conducted by Marchese for measuring windage power loss [6]. A frequency converter has been used to adjust the rotating speed. Windage power loss has been found to increase with increasing the rotating speed. In this study, numerical model has been applied to replicate the existing experiment. Geometry of the high speed gear as well as operating conditions of the numerical model has been made based on the existing experiment. The gear is characterized by a module of 5mm, a pressure angle of 20degree, a pitch diameter of 150mm and a face width of 24mm. A schematic of the numerical model without control is shown in figure 1.

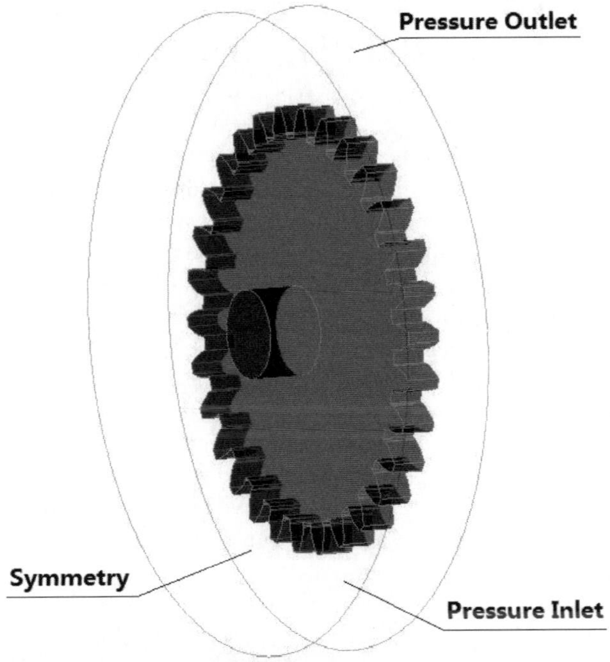

Figure 1. Geometry and boundary conditions of numerical model without control.

The air flow around high speed gear without control has been replicated by using the commercial software Altair HyperWorks AcuSolve. Pressure inlet and pressure outlet have been applied to the inlet and exit of the computational domains. The symmetry boundary condition has been applied to simulate the half computational domain. Unstructured mesh has been used in all regions. The grid number in computational domain is approximately 4 million. The air properties are characterized at the room temperature and atmosphere pressure. As the equations are solved in rotating frame, appropriate boundary conditions have been applied. The rotating velocity of the gear in the rotating frame is set as zero. The rotating velocity of the shrouds in the rest frame is set as zero. The windage power losses at different rotational speeds ranging from 5,000rpm to 7,000rpm have been investigated numerically. Numerical results have been validated by the existing experimental results. The windage power losses at different rotating speeds are shown in figure 2. The numerical results are in excellent agreement with the experimental data.

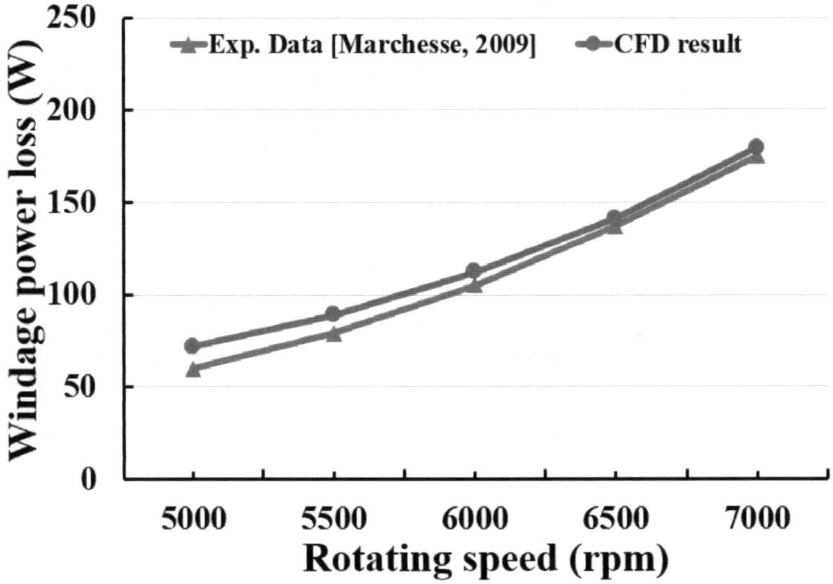

Figure 2. Comparison between numerical model and existing experiment.

3. High speed gear with different shrouds

As we know, the windage power loss of high speed gear is directly proportional to tooth width, pitch diameter and dynamic viscosity of working medium. The windage power loss can be effectively reduced by reducing the above parameters. However, these parameters are determined by the lubrication requirement and difficult to be adjusted.

The windage power loss is also greatly affected by the volumetric flow rate around high speed gear. The power loss can be reduced by decreasing the volumetric flow rate. It is necessary to install the shroud for decreasing the volumetric flow rate around the gear. This study focuses on the influences of smooth and grooved shrouds on windage power loss for rotating speed ranging from 5,000rpm to 7,000rpm. The schematics of the high speed gears with smooth shroud and grooved shroud are shown in figure 3. The terms of c and h are respectively axial and radial distances between shroud and gear. The term c keeps constant at 4mm in this study. The grooved shroud is different than smooth shroud in that grooved shroud has grooves along the circumference. The cross-section of the groove is 0.5mm in width and 0.5mm in height.

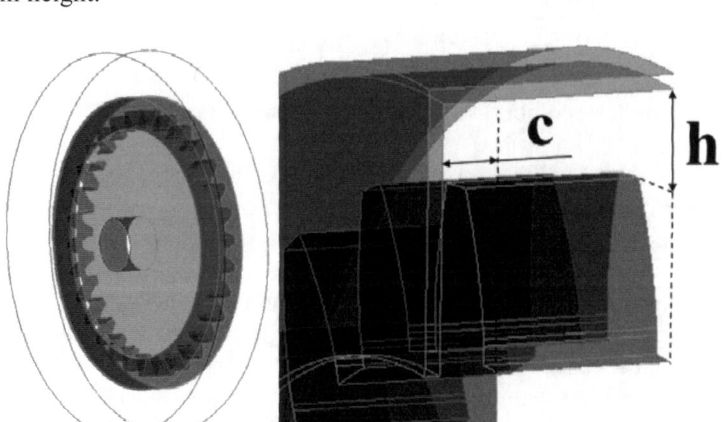

a. smooth shroud

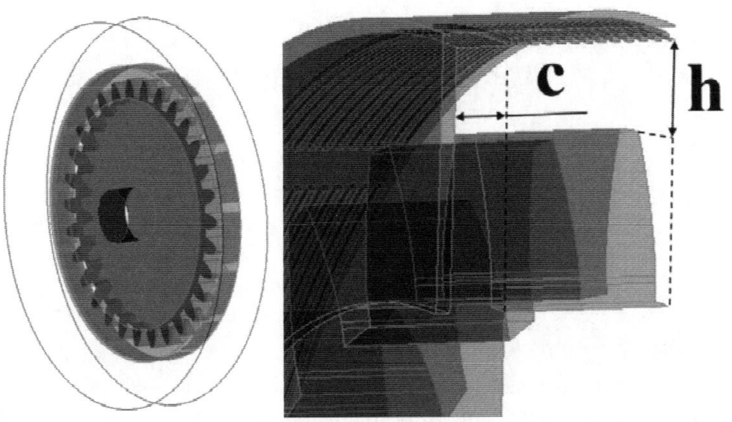

b. grooved shroud

Figure 3. Geometry of computational domains with two kinds of shrouds.

4. Results and discussion

4.1. Effect of smooth shroud on windage power loss

Air flow around the gear teeth has been investigated numerically. The pressure distributions on the tooth surface have been indicated to understand the mechanism of the windage power loss. The air is sucked into the gap between two teeth from both sides of the gear. Then it is thrown out of the gap along the radial direction.

The pressure contours at gear and symmetry surfaces without control are shown in figure 4. It is shown that adjacent tooth surfaces have different pressures. The pressure difference induces the rotating resistance torque of high speed gear. To keep constant rotating speed, the corresponding power has been taken to overcome the resistance torque.

A smooth shroud is applied for enclosing the high speed gear and restricting the pumping action by the gear teeth. The volumetric flow rate around the gear is reduced by enclosing the gear. The pressure difference is much lower with the smooth shroud, as shown in figure 5. Figure 6 indicates the effects of smooth shroud on windage power loss at different rotating speeds. The windage power loss increases with the increase of rotating speed. The smooth shroud is able to reduce the windage power loss by approximately 75% compared to the configuration without control. Shrouding the gear is effective to reduce the windage power loss.

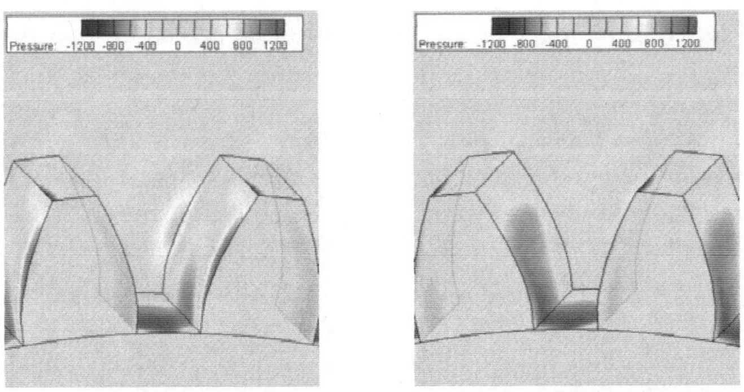

a. pressure surface b. suction surface

Figure 4. Pressure contours without control (ω=7000rpm).

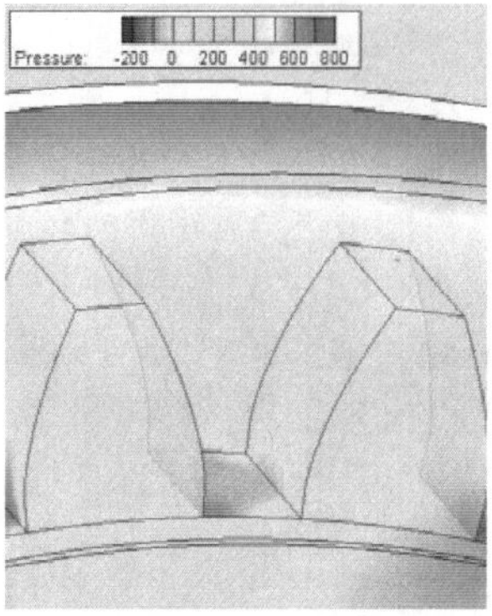

a. pressure surface

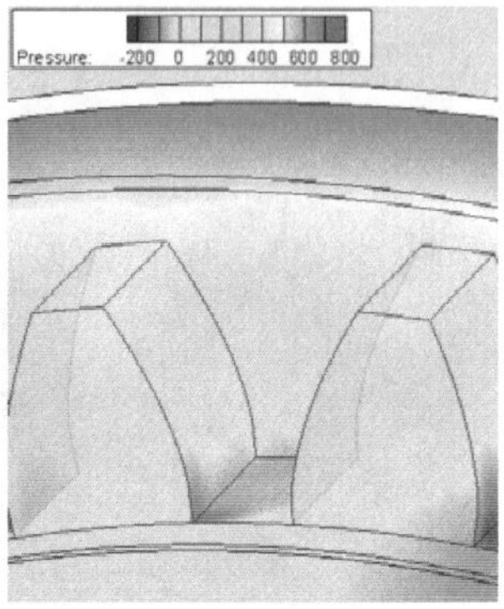

b. suction surface

Figure 5. Pressure contours with smooth shroud (h=7mm, c=4mm, ω=7000rpm).

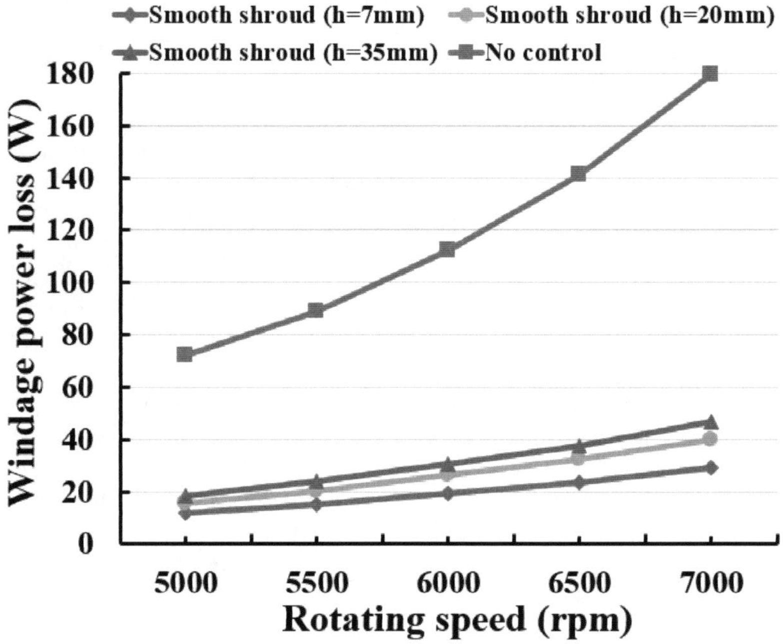

Figure 6. Windage power losses at different rotating speeds.

4.2. Effect of grooved shroud on windage power loss

Due to the space limitation, the size of shroud is sometimes difficult to be adjusted. It is necessary to add the grooved surface on the shroud if higher energy efficiency is needed. Grooved surface is able to further reduce the windage power loss when the space between shroud and gear is fixed. Figure 7 indicates the pressure contours at gear and symmetry surfaces with grooved shroud. It is shown that pressure difference between pressure surface and suction surface is much lower than the condition without control. The grooved shroud is able to restrict the pumping action as well.

In addition, the grooved shroud is able to further reduce the windage power loss by reducing the wall shear stress. The velocity gradient along the radial direction from bottom of the groove is shown in figure 8. The velocity gradient in grooved shroud is much lower than smooth shroud. The surface friction in grooved shroud decreases as the velocity gradient decreases. Windage power losses with smooth and grooved shrouds have been shown in figure 9. It is observed that the grooved shroud provides a considerable increase in energy efficiency at about 10% over the smooth shroud.

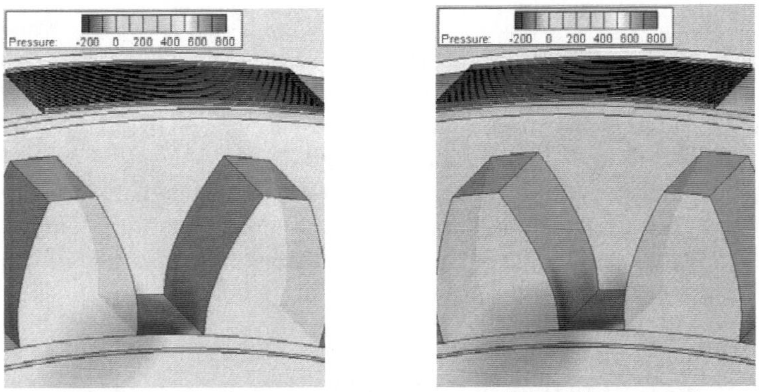

a. pressure surface b. suction surface

Figure 7. Pressure contours with grooved shroud (h=7mm, c=4mm, ω=7000rpm).

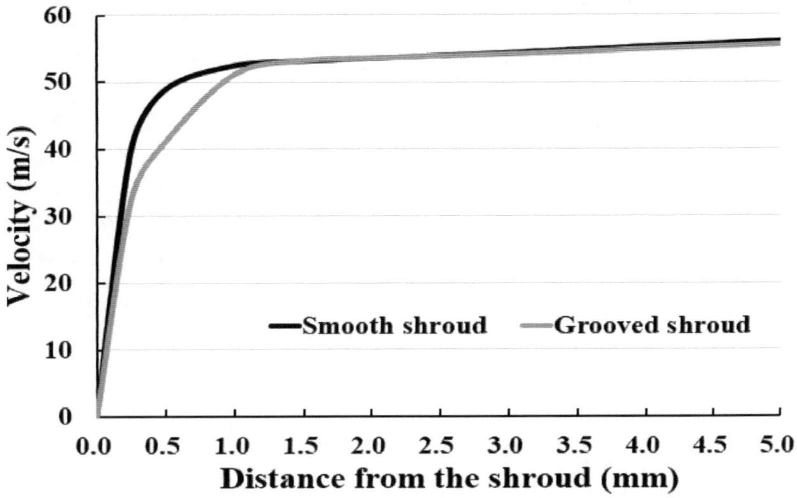

Figure 8. Velocity gradients along the radial direction (h=7mm, c=4mm, ω=7000rpm).

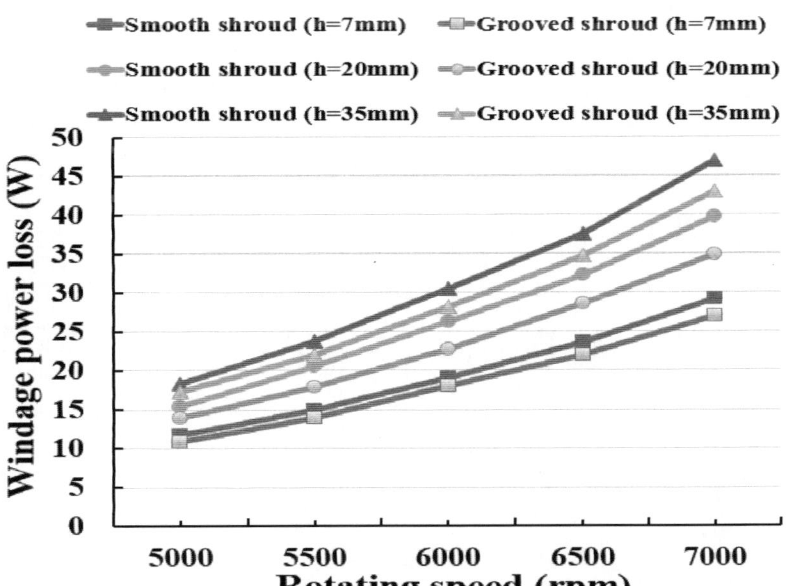

Figure 9. Windage power losses at different rotating speeds.

5. Conclusions

Numerical approach has been taken to understand the physical mechanism of windage power loss. Numerical result has been validated with the existing experiment. It is in good agreement with the experimental result. The effects of smooth and grooved shrouds on windage power loss have been focused in this study. Smooth and grooved shrouds have been applied for enclosing the high speed gear and restricting the pumping action by the gear teeth. The windage power loss has been reduced by shrouding the high speed gear. The comparison between smooth shroud and grooved shroud has been indicated. It is found that the grooved shroud provides a considerable increase in energy efficiency at about 10% over the smooth shroud.

References

[1] Franco C and Carlo G 2016 *Tribol. Int.* **103** 58-68.
[2] Massini D, Fondelli T, Andreini A, Facchini B, Tarchi L and Leonardi F 2017 *Experimental and Numerical Investigation on Windage Power Losses in High Speed Gears.* Proceedings of ASME Turbo Expo 2017: Turbomachinery Technical Conference and Exposition.
[3] Diab Y, Ville F, Velex P and Changenet C 2004 *J. Mech. Des.* **126** 903-8.
[4] Matthew J and Robert F 2012 *A computational Investigation of Gear Windage.* NASA report 2012-217807.
[5] Sylvain P, Yann M, Christophe C, Fabrice V and Philippe V 2013 *Comput. Fluids.* **84** 39-45.
[6] Marchesse Y, Changenet C, Ville F and Velex P 2009 *Investigations on CFD Simulation for Predicting Windage Power Losses in Spur Gears.* 3rd International Conference on Integrity, Reliability and Failure.

Study on the Technique of Beam Coupling for High Power Diode Laser Stack

Yuanyuan Gu[1], Yueming Fu[1], HuiLu, Yan Cui[1]

[1]Tianhua College of Shanghai Normal University, Section of Practice and Training, Shanghai, 201815, China

Abstract. High-power lasers based on GaAs bars meanwhile are well established as reliable and highly efficient laser sources. Continuous improvement of the material itself, mounting and cooling techniques during the last years have led to increased output power and high lifetimes. This paper summarized beam shaping technology of high-power lasers and evaluated these advantages and disadvantages. The beam coupling methods of the high power laser array were expatiated and a new beam coupling structure was present in the paper, also as the results and analysis of polarization coupled experiment.

1. Introduction

In many cases high power lasers are on the way to replace flashtubes for solid state laser pumping. By realizing arrangements of many high-power laser elements, e.g. laser stacks, the overall output power can be increased further up to the kW range. In this power level, we can have many important applications, such as surgery, welding of polymers, soldering, coatings and surface treatment of metals. pumped alkali vapour lasers (DPALs) have attracted increasing attention in the last decade [1]. As sources of such pump-modules Broad-Area (BA) -lasers are normally used because of power scalability [2]. depending on the device geometryThe demand is application driven and has been growing for past few decades thanks to improvements in power rating, efficiency, reliability, and production costs [3]. So all developed countries have attached great importance to high-power laser system and its applications. Germany is developing initiative "BRIOLAS" (brilliant lasers) program, the German Federal Ministry of Education and Research (BMBF) is supporting the BRIOLAS-initiative with about 30 M€. Using coarse wavelength multiplexing, BPP in the range of 20 mm-mrad for kW-class lasers (corresponding to a 400μm core diameter output fiber with 0.1 NA) has been demonstrated [4-6]. However, significant improvement in BPP is still needed to address industrial applications requiring high brightness.

In these programs, it is one of the main goals improving the output power and the brightness. There are two ways to gain high power and high brilliance laser system. One way is taking high efficient beam shaping small spot, the other way is to couple the output beam of several laser arrays to the same optical path, thus, in the case that the beam quality was not changed the output power increases to the original several times; the same brightness at the same time increases the multiplier. So what kind of beam coupling technology is used to achieve high brightness high-quality laser output has become a key issue. Now there are three ways: spacial coupling, wavelength coupling and polarization coupling. Thus, very high power laser systems with stringent demands on the beam quality rely on one additional active laser cavity and, in some cases, additional power amplifying stages to improve the beam quality of the lasers, with fiber lasers and disk lasers being the two most prominent examples [7-

IOP Conf. Series: Materials Science and Engineering **473** (2019) 012045 doi:10.1088/1757-899X/473/1/012045

10]. This paper proposes a new method of beam coupling method, which compared with the traditional coupling method the beam can be not only coupled but also collimation in the slow axis direction. It is propitious to focus into small spot. In addition, we make a series of polarization coupling experiment.

2. Beam coupling method for laser stack

For laser stack, it is generally done in the world by stacking 25 laser strips. This aspect is due to the limitation of heat dissipation. On the other hand, more laminated laser stacks have too much light-emitting area to facilitate focusing. Therefore, in order to further increase the output power and brightness, it is necessary to couple the output beams of the multiple array modules to the same optical path by means of beam coupling. At present, there are three main types of beam coupling methods used abroad: spatial coupling, polarization coupling, and wavelength coupling.

Spatial coupling often requires relatively complex optics, requiring small dimensional tolerances on the prism sheets and high flatness requirements on the end faces to achieve high coupling efficiency; this requires high coating processes. In addition, spatial coupling requires relatively high adjustment of the optical path. Therefore, achieving spatial coupling has certain technical difficulties.

Although the wavelength-coupled device is simple, it couples beams of two different wavelengths. The coupled beam is not a single wavelength, which is not suitable for laser applications requiring a specific wavelength. For the case where the beam of the laser has a certain spectral line width and a wavelength-to-temperature drift, the beam coupling efficiency of the wavelength coupling is high without the polarization coupling.

The polarization coupling can complete the merging of the laser beams of the same wavelength, and can be combined by the reverse direction of the polarization beam splitting prism, and the device structure is simple and the coupling efficiency is high.

So in summary, we first carried out theoretical and experimental research on polarization coupling technology.

3. Thin film interference type polarization beam splitter

The polarizing beam splitter is alternately plated with high and low refractive index films on the slopes of two right angle glass prisms, as shown in figure 1. The natural light is incident on the multilayer film system at an angle of 45 degrees. After the multilayer film is deflected, the linearly polarized light whose two vibration directions are perpendicular to each other is output in the positive direction, and the aperture angle of the device is 5 degrees. When the four-layer film is plated, the splitting ratio is close to one.

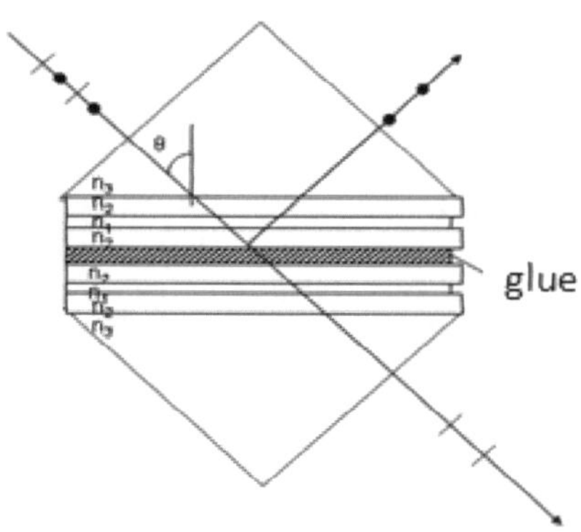

Figure 1. Thin film interference type polarization beam splitter.

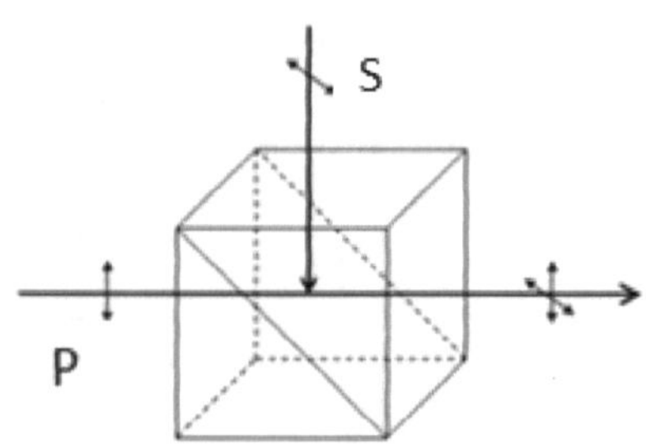

Figure 2. Cubic polarization beam splitting prism.

The relationship between incident angle θ_p and film thickness u of the double-sided coating is calculated by the following formula:

$$\theta_p = \arcsin\left(u^{\frac{1}{2}}\right) \tag{1}$$

Where,

$$au^2 + bu + c = 0 \tag{2}$$

$$a = n_1^8 - n_2^4, b = 2n_1^2 n_2^4 - n_1^8(n_2^2 + 1), c = n_1^4 n_2^2\left(n_1^4 - n_2^2\right) \tag{3}$$

Film thickness d_p is

$$d_p = \frac{1}{4}\left(n_1^4 - u\right)^{-\frac{1}{2}} \tag{4}$$

Ratio of reflectance is

$$R_s = \left[\frac{n_1^4 - n_2^2}{n_1^4 + n_2^2}\right]^2 \tag{5}$$

Splitting angle is

$$\alpha = 180 - 2\theta_p \tag{6}$$

The commercially available thin film interference type polarization beam splitter has an anti-damage threshold of more than 2 kW/cm^{2}for continuous laser and 1 J/cm^2 for a pulsed laser damage threshold of 10 ns, which can meet the requirements of polarization coupling for damage threshold.

For the interference type polarization beam splitter, it is divided into a single wavelength polarization beam splitter and a wide band polarization beam splitter. Although a single-wavelength polarizing beam splitter can achieve higher P-polarized transmittance and S-polarized reflectance, a laser has a certain spectral line width (±3-5 nm), and there is a case where the wavelength drifts with temperature, and the width is wide. The band polarization beam splitter is also very efficient and can meet the needs of polarization coupling, so a wide-band polarization beam splitter should be chosen.

4. Application of a cube polarizing beam splitter

A cube-polarized beam splitting prism (PBS) is used as a polarization-coupled beam-forming device, as shown in figure 2, which is a cube made of two isosceles right-angled triangular prisms coated on a slope of a triangular prism to make P The polarized light is totally transmitted by being incident on the inclined surface at 45°, and the S-polarized light is totally reflected by being incident on the inclined surface at 45°. In this way, the two beams whose polarization directions are perpendicular to each other are incident perpendicularly to the two surfaces of the PBS, and meet on the opposite side, and are merged and inverted to achieve the combination. The angle of incidence of the incident light is 5 degrees, the splitting angle is 90 degrees, and the four surfaces of the light passing through are coated with an anti-reflection film, and the surface reflectance is less than 0.25%.

Figure 3 and figure 4 show the PBS transmittance versus wavelength for the polarization coupling of 808nm and 980nm lasers. It can be seen that the P-polarized light transmittance exceeds 95% and the S-polarized light reflectance exceeds 99%. They are all effective over a wide wavelength range, so in practice we can polarize the coupling of two different wavelength lasers in a certain wavelength range.

IOP Conf. Series: Materials Science and Engineering **473** (2019) 012045 doi:10.1088/1757-899X/473/1/012045

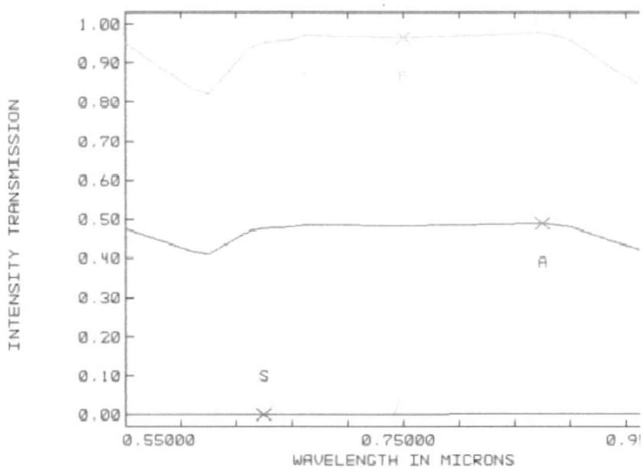

Figure 3. 650-850nm cubic PBS transmittance and wavelength relationship.

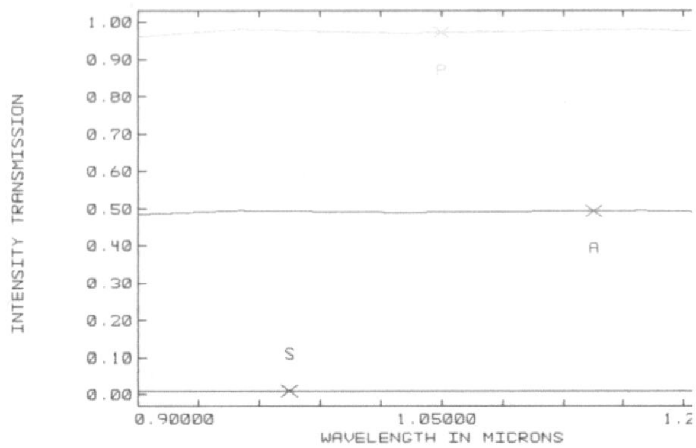

Figure 4. 900-1200nm PBS diagram of the transmittance and wavelength of broadband.

5. Polarization coupling experiment of 808nm seven bar laser array

The laser source parameters used in this experiment are as follows:

- Center wavelength: 808nm
- Output power: 100W/bar (QCW)
- Duty cycle: 20%
- Number of bar: 7
- Array length: 10mm
- Polarization Mode: TM
- Heat sink thickness: 1.8mm
- Beam divergence angle (FWHM): the divergence is 36degree in fast axis, and 10degree in slow axis.

Due to the large number of layers of the laser array, a single cylindrical lens does not achieve a good slow axis collimation effect. As shown in figure 6, after the two beams are coupled in the same optical path, we use a telescope consisting of two cylindrical lenses. The system expands and collimates the slow axis direction of the beam, and finally focuses it with a ball lens to obtain a 4.5*4.5mm2 spot as shown in figure 5 with an efficiency of 67%.

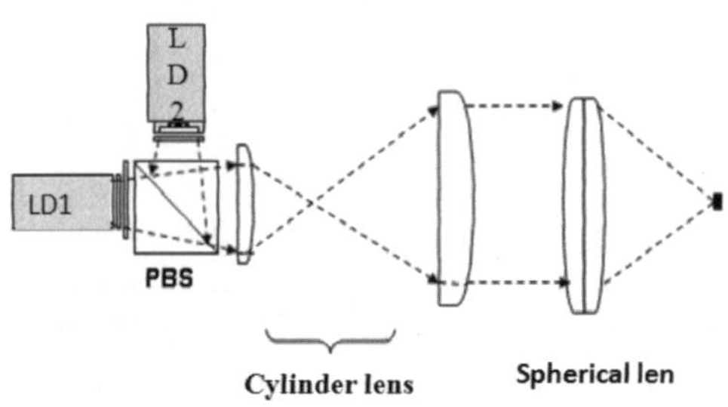

Figure 5. 808nmseven basr Polarization coupling of laser array.

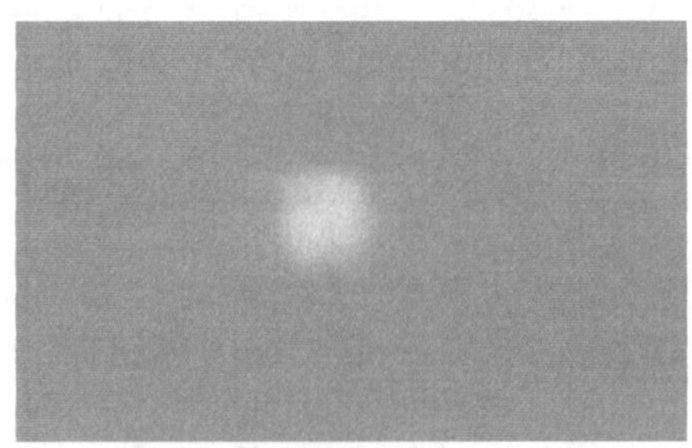

Figure 6. Focused spot after polarization coupling.

6. Analysis of results

The energy loss in the analysis experiment mainly includes the following aspects: the beam of the laser is not the energy loss caused by the complete linear polarization, the degree of polarization is generally in the range of 95%-98%; the incomplete reflection of the polarized light on the PBS And the energy loss caused by transmission; since the beam is not collimated by the slow axis before the polarization coupling, part of the light is not incident on the opposite side of the polarization beam splitting prism at 45O, and the P-polarized transmittance and the S-polarized reflectance are made. Decrease, causing energy loss; the loss of the beam through the optical components during the propagation process, the focusing ball lens we use is not coated with an anti-reflection coating, and its transmission efficiency is less than 90%, while the actual optical device is transparent. The overshoot can reach more than 95%; the deviation of the optics position adjustment also causes the loss of energy. In the experiment, we make the system as compact as possible to reduce the energy loss caused by the beam divergence.

The reason for analysing the large spot size is as follows: First, we use the micro fiber column as the fast axis collimating mirror. Although the divergence angle can be reduced to about 8 mrad, it cannot solve the inherent astigmatism problem of the laser (the fast and slow axis beam is not in the same position) Therefore, it is not easy to focus to obtain a small spot. This problem can be solved by using a graded index lens; the beam quality of the fast and slow axis is not uniform, and the single Bar slow axis optical parametric product (BPP) is several hundred times faster than the fast axis. The solution is to utilize Various beam shaping methods tend to make the fast and slow axes BPP consistent. In addition, the focus lens should be optimized to eliminate aberrations, which is conducive to gathering to obtain small spots.

Acknowledgments
The author would like to thank her collaborators from and support of Tianhua College of Shanghai Normal University, Section of Practice and Training and Lab. of Excited State Processes, Chinese Academy of Sciences, Changchun Institute of Optics, Fine mechanics and Physics, CAS

References

[1] Zhdanov B V, Shaffer M K and Knize R J 2011 *Proc. SPIE* **7915** 791506.
[2] Gilly J, Friedmann P, Kissel H, Biesenbach J and Kelemen M T 2012 *Proc. SPIE* **8241** 82410T1-6.
[3] Berk Y et al. 2011 *Proc. SPIE* **7918** 79180W-1—W12.
[4] Strohmaier S, Tillkorn C, Olschowsky P and Hostetler J 2010 *Optics and Photonics News* **21(10)** 25-29.
[5] Havrilla D, Brockmann R, Strohmaier S and Holzer M 2010 *Proc. SPIE* **7583** 75830B.
[6] Konig H, Gronninger G, Lauer C, Reill, W, Arzberger M, Strauss U, Kissel H, Biesenbach J, Kosters A, Malchus J and Krause V K 2010 *Proc. SPIE* **7583** 75830T.
[7] Hemenway M et al. 2017 *SPIE* **10086** 10086051-10086057.
[8] Chris Ebert* et al. 2017 *Proc. SPIE* **10086** 10086071-10086078.
[9] Huber M et al. 2017 *Proc. SPIE* **10086** 100850B1-100850B8.
[10] Mico Perales et al. 2017 *Proc. SPIE* **10086** 100860G1-100860G18.

Impact of the Method of Analysing Post-Tensioned Flat Slabs on the Amount of Prestressing

Rafał Szydłowski[1], Barbara Łabuzek[1]

[1]Cracow University of Technology, Warszawska 24 Street, Cracow 31-155

Abstract. The post-tensioned slabs have been using for many years. Over the years, many guidelines regarding design methods and calculations have been issued. There are several methods of static analysis and obtained results interpretation. It should be noted that the selection of method has a large impact on the final results in terms of required amount of presstresing steel. It is a great importance for economic aspects of realization of the post-tensioned slabs. In this paper the results of analysis of flat post-tensioned slab are presented. The calculations were carried out using two methods: the equivalent frame method and final elements method. The differences between these two methods and the impact of the selected method on the final usage of the prestressing steel are described in detail.

1. Introduction

The flat slab, i.e. slab without any beams, it is the most commonly used form of slab in concrete structures. A higher slab thickness and higher consumption of reinforcing steel compared to traditional slab-beam systems are compensated by advantages such as lower total height of floor, free space under the slab (lack of beams), no obstacles in running the installation and much easier and faster execution of formwork and reinforcement which translates into lower cost of realization.

The flat post-tensioned slabs are particularly attractive in terms of architecture. Application of prestressed concrete technology allows for enhancement the spacing of columns to 12 meters and in some exceptional cases even to 14-16 meters. The attempts to improve them have been made for a few years at the Cracow University of Technology [1-3].

2. The methods of slabs analysis

Analysis of flat slabs presents many difficulties. It is related to peak value of bending moments over the columns as a results of point supports. This type of structure can be analyzed with Finite Elements Methods or with one of simplified method, as equivalent frame method or grillage method. The value of bending moments near the columns is several times larger than in the center point between the columns. Thus, interpretation of results near the columns is difficult. Further obstacles are encountered in the case of post-tensioned flat slabs. Active reinforcement generates additional loads that have to be added to the numerical model which is time-consuming process. Consequently, the selection of method of analyse is more important than in case of reinforced slabs. The different methods require different work time arrangements, giving different amounts of prestressing steel which assure meeting the conditions of limit states. The two most popular methods and the results obtained with their use are presented in the further part of the work.

2.1. Equivalent frames method

IOP Conf. Series: Materials Science and Engineering **473** (2019) 012046 doi:10.1088/1757-899X/473/1/012046

In the equivalent frame method the slab is represented as a set of two-dimensional frames. Each of the frames consists of the columns located above and below the slab as well as horizontal beam representing the strip of slab between the lines of zero shear (figure 1). The number and arrangement of frames should be taken so that the whole slab is covered with frames. In case of using the equivalent frame method, for estimation of concrete cross-section stresses the bending moments are averaged at the total width of the frame. Therefore, this method does not take into account elastic moment distribution in the transverse direction.

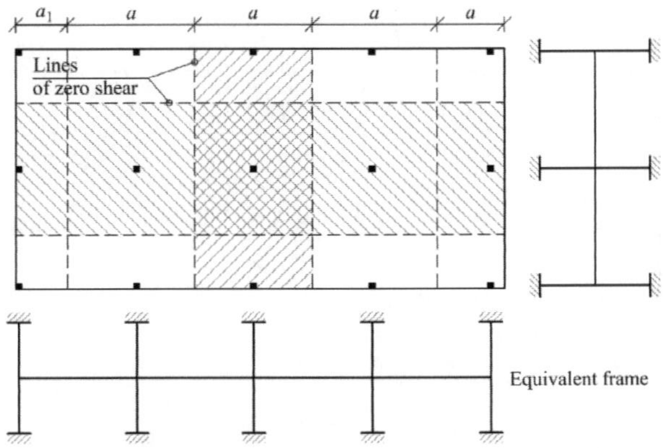

Figure 1. Elastic load distribution effects and slab division on equivalent frames.

2.2. *Finite elements method*

This method allows to analyze the slab as a whole three-dimensional structure. The three-dimensional analysis is possible through the precise modeling with taking into consideration elastic load distribution and actual distribution of prestressing tendons. However, if the calculation are not carrying out with the use of advanced computational program this method is a more time-consuming method than it requires.

2.3. *Limit stresses*

As a result of other distributions of bending moments obtained from both methods the limit stress values are slightly different. Recommendations of *Concrete Society* [4] gives the moments averaging ways for both methods as well as the limit concrete cross-section stress values. For equivalent frame method the bending moment in whole slab width between zero shear lines is considered in all cross-sections. In case of FEM the paper [4] determinates the design strips. These are defined as the entire width of the strips between the lines of zero shear in the span cross-section (*a* in figure 1) and 0.4 of strips width in the column cross-section. Bending moments obtained from FEM analysis are averaged on the design strips. The values of limit stresses for both methods are set in table 1. Presented values are matched to the moments averaging process in each of the methods. Limit stresses for using design strips (FEM) are significantly higher than for full panel width. It results from the other method of averaging moments.

Table 1. Limit stresses in flat slab according to [4].

Location	For full panel width		For using "design strips"	
	In compression	In tension	In compression	In tension
Support	$0.3f_{ck}$	$0.9f_{ctm}$	$0.4\,f_{ck}$	$1.2f_{ctm}$
Span	$0.4f_{ck}$			

3. Slab geometry and characteristic

Analysed slab is presented in figure 2. The columns with cross-section of 0,4×0,4 m are arranged on regular gird 8,5×8,5m. The slab thickness is 250mm. C40/45 class concrete was used. Figure 2 presents the slab geometry as well as its division into equivalent frames with their designation.
The following external loads were adopted in slab calculation: self-weight (25kN/m^3), floor layers (1,80kN/m^2) and service load (3,9kN/m^2).

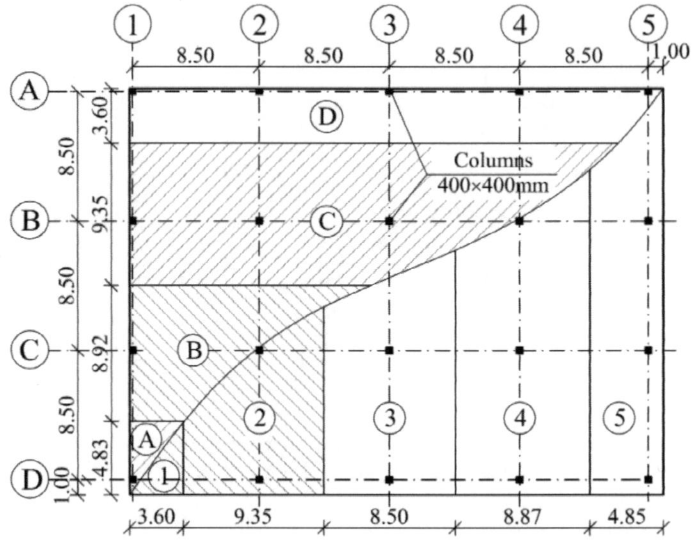

Figure 2. Slab geometry and its division into equivalent frames.

4. The results of analysis

Prestressing with unbonded 15.2mm tendons was assumed. The initial force in tendon is 220kN. Tendon profiles in both directions are presented in figure 3. Effective tendon force $P_{eff} = 0.8P_0 = 176$kN was assumed in analysis. The number of tendons in equivalent frames and tendon arrangement in the slab were adopted so as to meet the limit stresses. The number of tendons in equivalent frames is presented in table 2 and tendon arrangement in the slab (analyzed with FEM) with equivalent load from prestressing in Y direction is shown in figure 4.

Table 3 presents the concrete stresses in two equivalent frames (equivalent frame method) and in two axis for FEM, one in each direction. The axis B and 2 were considered for both of methods. The stress values obtained from the methods are presented in table 3, results of FEM analysis were placed into brackets. Besides the calculated results the table also includes the limit stress values. The limit stress value was used to set the quantity of tendons, which presented in table 2.

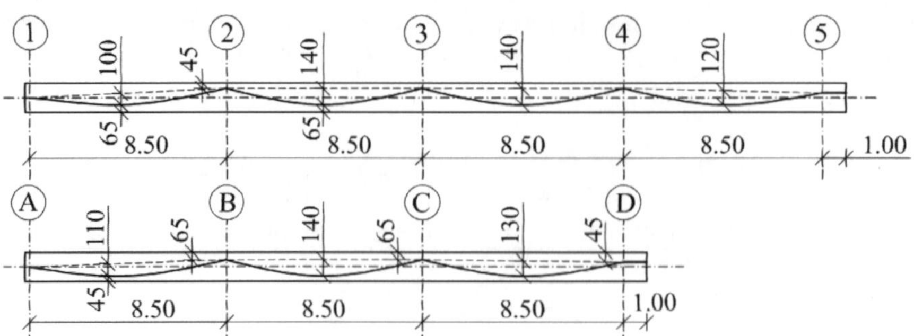

Figure 3. Tendon profiles.

Table 2. Number of prestressing strands in equivalent frames.

Frame	1	2	3	4	5	A	B	C	D
Strand No.	8	22	19	21	16	13	33	32	18

Table 3. Concrete cross-section stresses for equivalent frames method and FEM (in brackets).

Layer	Frame B Location	Dead and service load	Prestress-ing	Final	Frame 2 Location	Dead and service load	Prestress-sing	Finall	Ultimate value
Top	1-2	4.28 (4.95)	0.67 (1.03)	4.95 (5.98)	A-B	3.82 (4.85)	0.22 (0.07)	4.04 (4.92)	14.0 (14.0)
Bottom		-4.28 (-4.95)	3.27 (4.92)	-1.01 (-0.03)		-3.82 (-4.85)	2.79 (4.34)	-1.03 (-0.50)	-2.88 (-3.84)
Top	2	-7.46 (-11.57)	4.60 (9.32)	-2.87 (-2.25)	B	-6.30 (-11.51)	3.54 (8.02)	-2.76 (-3.49)	-2.88 (-3.84)
Bottom		7.46 (11.57)	-0.66 (-3.80)	6.81 (7.77)		6.30 (11.51)	-0.53 (-3.91)	5.77 (7.60)	10.5 (14.0)
Top	2-3	3.06 (3.09)	-0.01 (-0.06)	3.05 (3.03)	B-C	2.36 (2.73)	0.15 (0.34)	2.50 (3.06)	14.0 (14.0)
Bottom		-3.06 (-3.09)	3,95 (5.85)	0,89 (2.76)		-2.36 (-2.73)	2.86 (3.92)	0.50 (1.19)	-2.88 (-3.84)
Top	3	-6,06 (-9.69)	5,24 (11.10)	-0,83 (1.41)	C	-6.22 (-11.27)	3.34 (7.46)	-2.88 (-3.81)	-2.88 (-3.84)
Bottom		6.06 (9.69)	-1,30 (-5.73)	4,76 (3.96)		6.22 (11.27)	-0.33 (-3.28)	5.89 (7.99)	10.5 (14.0)
Top	3-4	3.07 (3.16)	-2.04 (-0,14)	1.03 (3.02)	C-D	3.73 (4.65)	0.43 (0.48)	4.16 (5.13)	14.0 (14.0)
Bottom		-3.07 (-3.16)	4.01 (5.96)	0.94 (2.80)		-3.73 (-4.65)	2.58 (3.98)	-1.15 (-0.68)	-2.88 (-3.84)
Top	4	-7.01 (-11.37)	4.24 (8.88)	-2.77 (-2.49)	D	-2.73 (-3.19)	2.79 (7.87)	0.06 (4.68)	10.5 (14.0)
Bottom		7.01 (11.37)	-0.30 (-3.29)	6.71 (8.08)		2.73 (3.19)	0.22 (0.25)	2.95 (3.43)	-2.88 (-3.84)
Top	4-5	4.18 (4.75)	1,02 (1.63)	5.20 (6.38)					14.0 (14.0)
Bottom		-4.18 (-4.75)	2,92 (4.39)	-1.26 (-0.36)					-2.88 (-3.84)
Top	5	-3.19 (-2.87)	3.65 (9.87)	0.46 (3.49)					-2,88 (-3.84)
Bottom		3.19 (2.87)	0.29 (0.61)	3.47 (7.00)					10.5 (14.0)

The stress values for decisive cross-sections are marked in red. It can be observed, that the cross-section at axis C was crucial in both of methods. Tensile stresses are decisive. It is interesting, that the amount of tendons applied in frame 2 is 22 and in frame B is 33 for equivalent frames method. While the number of tendons for FEM method is 30 and 40 respectively. It means, that the amount of prestressing steel needed for meeting the acceptable concrete stress conditions obtained from FEM analysis is 36% higher than calculated from equivalent frames method.

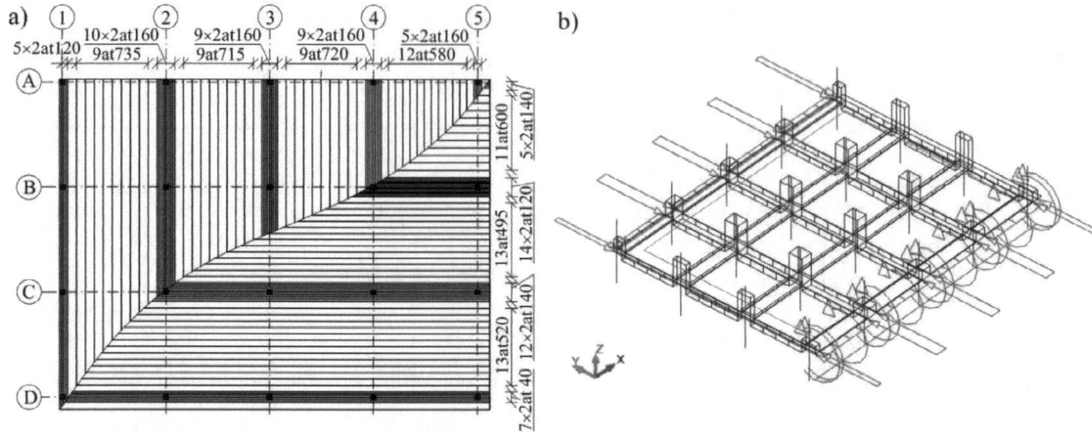

Figure 4. Tendon arrangement (a), equivalent load for tendons in Y direction (b).

5. Conclusions

This paper presents the results of post-tensioned slab analysis based on two methods: FEM and equivalent frames method. FEM method treats the structure more accurately considering the elastic distribution of moments in the transverse direction. It is more time-consuming compared to simplified equivalent frames methods. In spite of these, the simpler method turned out to be more economical in the amount of prestressing needed. It is due to the way of averaging the peak moment above the columns, delivered in recommendations [4]. The range of integration and averaging moments in FEM is larger than one in the equivalent frame method. Thus, unit moments and cross-section stresses obtain in FEM are higher. It is not recompensed by higher values of limit stresses.

The quantitative difference in the values obtained cannot be generalized, it is strongly depended on the prestress arrangement and the way of its modelling. It is indisputable, the simpler and less-time consuming method allows to design more economically. It should be noted, that recommendations used are valid for elastic analysis of the structures. Non-liner analysis taking into consideration crack appearance and bending moments redistribution will give other effects. It will allow to take advantage of FEM analysis and to design more economically.

References

[1] Derkowski W, Skalski P 2017 *Procedia Eng.* **193** 176-83

[2] Szydlowski R, Labuzek B 2017 *Post-Tensioned Concrete Long-Span Slabs in Projects of Modern Building Construction* IOP Conf. Series: Mater. Sci. Eng. **245** 022065.

[3] Mieszczak M, Domagała L 2018 *Mater. Sci. Forum* **926** 140-5.

[4] The Concrete Society 2005 *Post-tensioned concrete floors. Design Handbook* Tech. Report No. 43, Cromwell Press, Wiltshire, UK

IOP Conf. Series: Materials Science and Engineering **473** (2019) 012047 doi:10.1088/1757-899X/473/1/012047

Study on the Matching of Toe-in and Camber of the Double-Front-Axle Steering Automobile

Zhang Peng[1], Wang Hongxin[1]*, Zhang Xiaodong[1]

[1]School of Mechanical and Vehicle Engineering, West Anhui University, Lu'an, China

Abstract. Toe-in and camber of the wheel are important driving performance parameters in the four-wheel positioning parameters of automobile. In this paper, the mathematical geometrical models of toe-in and camber of the first and second bridges of the double-front-axle automobile are established, concluded that the relationship of toe-in and camber between the first bridge and the second bridge. At the same time, based on the side slip mechanism of the tire, the reasonable matching formula for toe-in and camber of the double-front-axle are concluded. Finally, according to the dynamic model established in the Adams/Car, the matching values of toe-in and camber are simulated and verified. The experimental results show that the matching value is reasonable and effective, for the matching of toe-in and camber of double-front-axle steering wheel has a certain guiding role.

1. Introduction

Toe-in and camber of wheels are important front-wheel positioning parameters to the front wheel of automobile. Reasonable matching between toe-in and camber of wheels can reduce the wear condition of tire [9] and improve the handling stability of automobile. At present, there is a lot of literature [1-5] about the matching of toe-in and camber of single-axle steering automobile. Considering the aspect of mechanics in the literature [1, 2], the matching relationship between toe-in and camber is derived by cancelling out the force which generated by toe-in and camber of the tire. In the literature [3, 5], according to the motion trail of wheels, the matching relationship between toe-in and camber can also be deduced. But none of these literatures mentioned above refer to the matching of toe-in and camber of steering wheels in the double-front-axle automobile. In addition, there are few studies on toe-in and camber of the double-front-axle steering wheels at present.

In order to achieve this goal, first, the geometrical relationship between toe-in and camber of the first and second bridges can be derived by establishing the mathematical geometry of toe-in and camber of the double-front-axle steering wheels. Then, the matching relationship between toe-in and camber of double-front-axle steering wheels can also be deduced according to the side slip mechanism of the tire. Finally, the result of simulation experiments can be verified by the dynamic model.

2. Matching study of toe-in and camber of double-front-axle

2.1. The relationship of camber between first and second bridges of double-front-axle

On account of the camber angle of wheels, wheels of automobile will always have the tendency to roll outward during driving. Taking the first bridge camber as an example, the geometry model of rolling wheel camber can be established, as shown in figure 1 [3, 5]:

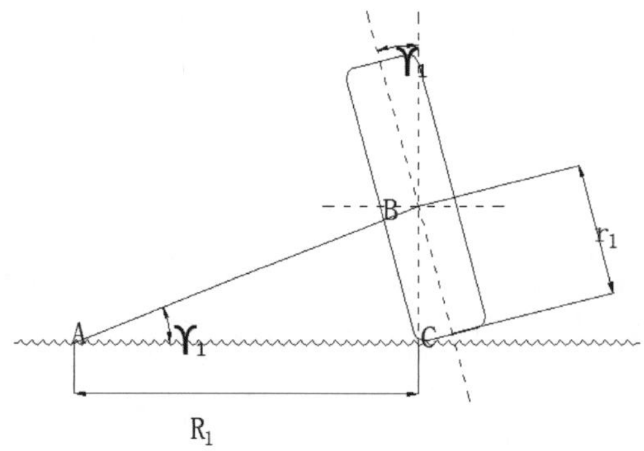

Figure 1. Rolling schematic diagram of the first-bridge camber.

Where, R_1 means the outer rolling radius of the first bridge which are caused by the camber of the tire, r_1 represents the rolling radius under load, and γ_1 indicates the camber of the first bridge of steering wheels.

Therefore, according to the geometric relationship graph of the first bridge, we can get the following relation:

In ΔABC, $R_1 = r_1 / \sin\gamma_1$. Due to the small camber angle, $\sin\gamma 1 \approx \gamma 1$, the rolling radius of the first bridge can be expressed as

$$R_1 = r_1 / \gamma_1 \tag{1}$$

In the same way, the rolling radius of the second bridge can be obtained:

$$R_2 = r_2 / \gamma_2 \tag{2}$$

Where $$r_1 = D/2 - G_1/6K_z \tag{3}$$

$$r_2 = D/2 - G_2/6K_z \quad [6] \tag{4}$$

D is typical of the diameter of wheels under no load, G_1 and G_2 represent the vertical load above the first and second bridges, and K_z is the radial stiffness of the tire.

During steering process of the automobile, the camber angle will changes with the change of the wheel angle, but in the process of this change, we can get such a regularity, that is, the turning radius of each steering wheels (ρ_1, ρ_2) and the rolling cone radius caused by the camber of wheels (R_1, R_2) are as close as possible.

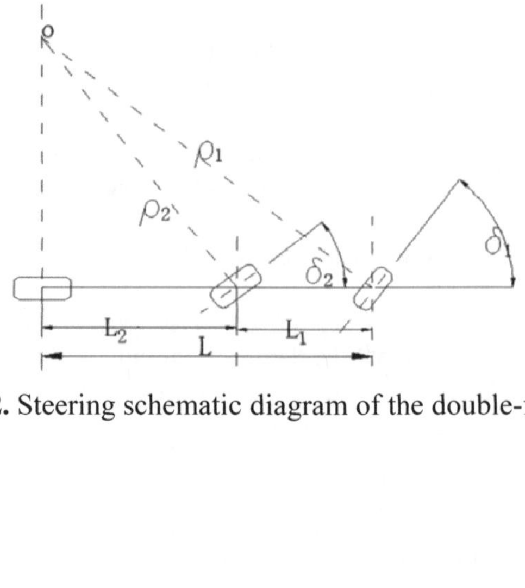

Figure 2. Steering schematic diagram of the double-front-axle.

Where δ_1, δ_2 are the angles of the first and second bridges of steering wheels respectively, ρ_1, ρ_2 indicate the turning radius of the first and second bridges of steering wheels. According to the geometric model of figure 2, we can obtain the following geometric relationship: $\rho_1 = L / \sin\delta_1$, $\rho_2 = L_2 / \sin\delta_2$;

When the conditions $R_1 = \rho_1$, $R_2 = \rho_2$ have been met, the minimum wear of the tire can be guaranteed, we can get the following relationship at present:

$$\gamma_1 = r_1 \sin\delta_1 / L \tag{5}$$

$$\gamma_2 = r_2 \sin\delta_2 / L_2 \tag{6}$$

When steering angles of automobile is small, that is, the condition $\cos\delta_1 / \cos\delta_2 \approx \delta_1 / \delta_2$ can be satisfied. So it can be approximated that the automobile is in straight-line driving state, and the camber angles of the first and second bridges are both initial camber angles.

According to (5) and (6), the relationship between the first bridge and the second bridge of camber can be obtained:

$$\begin{aligned} \gamma_1/\gamma_2 &= (L_2/L)\times(\sin\delta_1/\sin\delta_2)\times(r_1/r_2) \\ &= (L_2/L)\times(\tan\delta_1/\tan\delta_2)\times(\cos\delta_1/\cos\delta_2)\times(r_1/r_2) \\ &= (L_2/L)\times(L/L_2)\times(\cos\delta_1/\cos\delta_2)\times(r_1/r_2) \\ &= (\cos\delta_1/\cos\delta_2)\times(r_1/r_2) \\ &\approx (r_1/r_2) \end{aligned} \tag{7}$$

2.2. The relationship of toe-in between first and second bridges of double-front-axle

Compared with the single-bridge automobile, the double-front-axle automobile has two more steering wheels, so the rolling radius arose from toe-in of the double-front-axle automobile is slightly different from that of the single-bridge automobile. So it can be approximated that the toe-in angle is equivalent to a small corner of the tire. In order to ensure the tire wears minimally during the driving process, the steering wheel should roll around a same point [10].

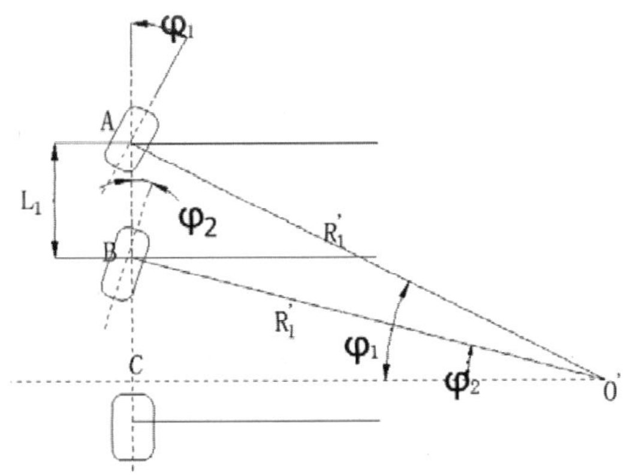

Figure 3. Toe-in rolling schematic diagram of the double-front-axle

Where φ_1, φ_2 indicate toe-in angle of the first and second bridges, R'_1, R'_2 represent the rolling radius of the first and second bridges which caused by the toe-in. φ_1, φ_2 are very small among them, so it can be approximated that $\sin\varphi_1 \approx \varphi_1, \sin\varphi_2 \approx \varphi_2$.

According to the aforementioned toe-in rolling schematic diagram shown in figure 3:

In △OBC:

$$R'_1 \times \sin \varphi_1 \approx R'_1 \times \varphi_1 = L = AC \tag{8}$$

In △OAC:

$$R'_2 \times \sin \varphi_2 \approx R'_2 \times \varphi_2 = L_2 \tag{9}$$

We can conclude that:

$$AC = L_1 + BC \tag{10}$$

by observing the geometric relationship of the graph above.
According to (8), (9), (10) the relationship of toe-in of double-front-axle can be obtained as follow:

$$R'_1 \times \varphi_1 = R'_2 \times \varphi_2 + L_1 \tag{11}$$

In the above-mentioned formula, φ_1, φ_2 can be expressed as

$$\varphi_1 = T_1 / \mathrm{d} \tag{12}$$

$$\varphi_2 = T_2 / \mathrm{d} \tag{13}$$

Where T_1, T_2 represent toe-in of the first and second bridges of steering wheels respectively, and d is the diameter of hub.

In the relation (11), the rolling radius of toe-in of the first and second bridges varies with the change of the toe-in angle of the first and second bridges of steering wheels. The toe-in value of the first and second bridges will be determined there-in-after by matching toe-in with that camber of the first and second bridges.

2.3. Study on the matching relationship between toe-in and camber of the first and second bridges of double-front-axle

In the following, based on the side slip mechanism of the tire, considered the minimum slip of the tire in the double-front-axle automobile during driving, with a view to the parameters of mechanism, combined with the relationship of toe-in and camber of the first and second bridges of double-front-axle above. The reasonable matching formula of toe-in and camber of the first and second bridges of double-front-axle has been determined.

Considering the toe-in and camber synthetically, we can take the first-bridge steering wheels of double-front-axle automobile as an example, the motion curve [3, 5] of tire shown in the following figure is established:

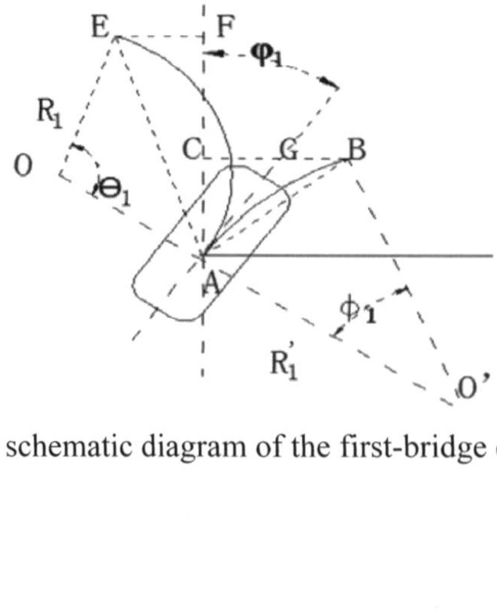

Figure 4. Rolling schematic diagram of the first-bridge camber and toe-in.

In figure 4, $\widehat{AB}$ and $\widehat{AE}$ indicate the grounding mark caused by toe-in and camber of wheels during Δt (time) respectively. φ_1 represents the toe-in angle of the first-bridge steering wheels, ϕ_1, θ_1 respectively express the rolling angle of wheel during Δt(time) under the action of toe-in and camber. R , R'_1 represent the rolling radius of toe-in and camber respectively.

In order to ensure the minimum side slip of tires during driving, EF=BC must be guaranteed.

In ΔAEF:

$$EF = AE \times \sin(\frac{\theta}{2} - \varphi_1),$$

In ΔABC:

$$BC = AB \times \sin(\frac{\phi}{2} + \varphi_1);$$

Available from EF=BC,

$$\theta_1 = \phi_1 + 4\varphi_1 \tag{14}$$

can be obtained.

At the same time, when the rolling radius is large enough, the following formulas are satisfied:

$$\widehat{AB} = AB = \widehat{AE} = AE = l1,$$

where l_1 indicates the actual grounding mark of the first bridge [7].

$$\text{Also, } \widehat{AB} = R_1' \times \phi 1 = l1 \tag{15}$$

$$\widehat{AE} = R1 \times \theta 1 = l1 \tag{16}$$

From the formulas (1), (3), (12), (14), (15), (16), the matching formula of toe-in and camber of the first bridge can be obtained:

$$T_1 = \frac{12K_z L l_1 \gamma_1 d}{(4L + l_1)(3DK_z - G_1)}$$

In the same way, we can also get the matching formula of the second bridge:

$$T_2 = \frac{12K_z L_2 l_2 \gamma_2 d}{(4L_2 + l_2)(3DK_z - G_2)}$$

Combined with the relationship between toe-in and camber of the first and second bridges of the double-front-axle, we can get the matching relationship between the toe-in and the camber of the double-front-axle:

$$T_1 = \frac{12K_z L l_1 \gamma_1 d}{(4L + l_1)(3DK_z - G_1)},$$

$$T_2 = \frac{12K_z L_2 l_2 \gamma_2 d}{(4L_2 + l_2)(3DK_z - G_2)},$$

$$\gamma 1 / \gamma 2 \approx \mathrm{r}1 / r2$$

Where, l_1 , l_2 represent the grounding mark of the first and second bridges of the tires respectively,

$$l_1 = 2\sqrt{(\mathrm{D} - \Delta_1)\Delta_1},$$

$$l_2 = 2\sqrt{(\mathrm{D} - \Delta_2)\Delta_2},$$

$$\Delta_1 = 19.1\,\mathrm{CK}(0.5G_1)^{0.85} / B^{0.7} D^{0.45} P^{0.6},$$

$$\Delta_2 = 19.1 CK(0.5G_2)^{0.85} / B^{0.7} D^{0.45} P^{0.6}.$$

In the formula above, C and K are coefficients, K=0.0015B+0.42, B is the section width of tire, G1, G2 are the loads on the first and second bridges respectively, D represents the diameter of tire, and P is the tire pressure.

3. Practical calculation and simulated analysis

3.1. Practical calculation of toe-in value and camber angle of the steering wheel of the first and second bridges of the double-front-axle prototype vehicle

In order to verify the correctness of the matching formula of toe-in and camber of the double-front-axle mentioned above, the JAC GEFA Y4AP0 double-front-axle heavy truck will be used as the prototype vehicle. According to the data parameters of the prototype vehicle, the toe-in value and the camber angle of the first and second bridges of the double-front-axle will be calculated.

Table 1. Data parameters of the double-front-bridge prototype vehicle.

d/mm	D/mm	G_1/N	G_2/N	B/mm	K_Z/(N/mm)
571.5	1008	37500	19950	300	117.78
L/mm	L_1/mm	L_2/mm	P/(100ka)	C	
4300	1900	2400	7	1.5	

According to the aforementioned vehicle data parameters, combining with the matching formulas of toe-in and camber of the double-front-axle. If a camber angle of the first bridge of a wheel can be given, the toe-in value of the first bridge, the camber angle and toe-in value of the second bridge can all be calculated by matching formula mentioned above.

In the general case, the camber angle of wheels are mostly about 1 °, according to the actual situation of this automobile, the camber angle of the first bridge can be given γ_1=1°, then combining with the parameters and matching formulas of double-front-axle automobile, the toe-in value of the first bridge of wheels are T1=2.47mm, and the toe-in angle is φ_1=0.104°.Similarly,the camber angle of the second bridge of wheels are γ_2=1.05°, the toe-in value of the second bridge of wheels are T2=1.42mm,and the toe-in angle is φ_2=0.059 °.

3.2. Simulation verification of vehicle dynamics model based on Adams/car

According to the vehicle dynamics model established in Adams/car, we conduct the relevant simulation experiments of straight-line driving and steering driving of whole vehicle. In order to verify the rationality of the matching formulas mentioned above, we should compare with the side slip of the four-steering wheels of the first and second bridges, before and after modifying toe-in angles (values) and camber angle of the first and second bridges of double-front-axle.

Simulation of straight-line driving: Straight-line Maintain driving simulation with the dynamic model in Adams/Car are performed, with the duration of 15s, keeping the steering wheel for free.

The following simulation contrast curves were made by the above-mentioned straight-line driving simulation which was carried out separately on the original double-front-axle dynamic model and that after modifying toe-in and camber.

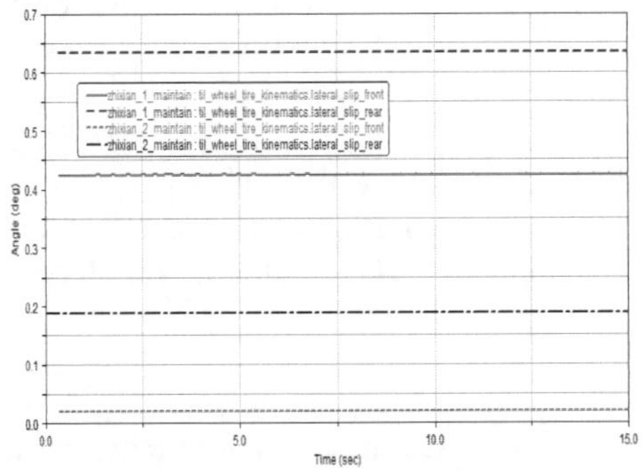

Figure 5. Side-slip comparison of the first and second bridges of tires on the left side of the straight-line driving

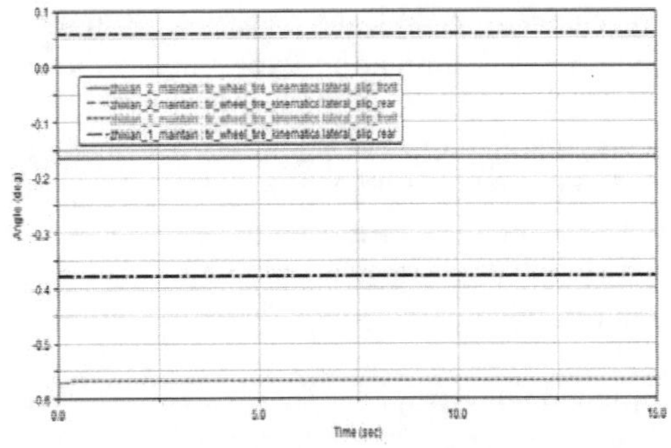

Figure 6. Side-slip comparison of the first and second bridges of tires on the right side of straight-line driving

We can get some information from Figure 5 that the left side slip angle of the first bridge is reduced from 0.43° to about 0.02°,and the left side slip angle of the second bridge is reduced from 0.63° to about 0.18°.Again,as we can see from Figure 6,the right side slip angle of the first bridge is changed from -0.58° to about -0.16°, and the right side slip angle of the second bridge is changed from -0.38° to about 0.06°.Comparing with two simulation lines above, it can be found that the side slip of the tire is obviously improved after modifying the camber and toe-in angle of the first and second bridges of double-front-axle steering wheels.

Simulation of steering driving: Performing the dynamic model of Cornering steering Release driving simulation in Adams/Car, the lateral acceleration and the longitudinal driving speed are set to 0.2g and 28km/m respectively.

The following simulation contrast curves were made by the above-mentioned steering driving simulation which was carried out separately on the original double-front-axle dynamic model and that after modifying the toe-in and camber.

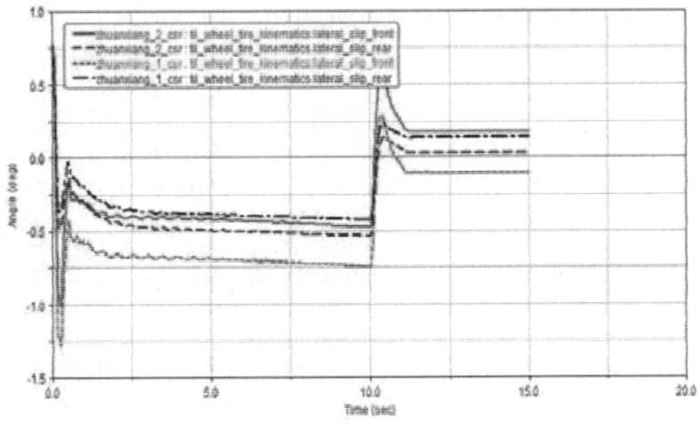

Figure 7. Side-slip comparison of the first and second bridges of tires on the left side of steering driving.

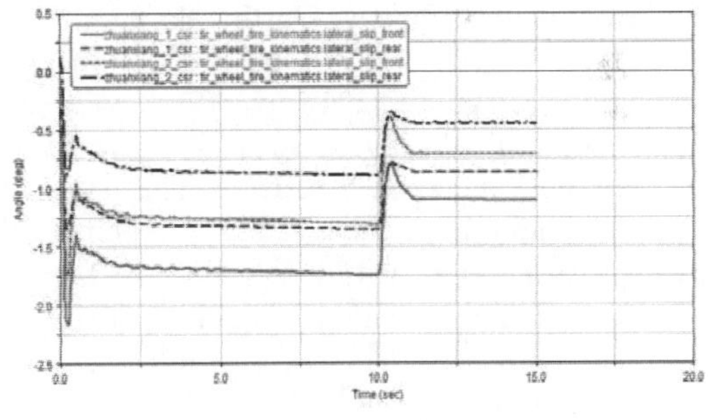

Figure 8. Side-slip comparison of the first and second bridges of tires on the right side of steering driving.

The conclusion about side-slip angle is obtained in figure 7: after modifying toe-in and camber of double-front-axle model, the left side-slip angle curve of the first and second bridges of steering wheels tends to be about 0° compared with the original model. Similar conclusions can also be drawn in Figure 8.

4.Conclusions

• Considering the minimum wear of tires, the geometric relationship between toe-in and camber of the first and second bridges of the double-front-axle is established. At the same time, based on the side slip mechanism of tires, the matching formula of camber and toe-in of the double-front-axle is derived. In addition, the simulation experiment is verified by the dynamic model Y4AP0, which proves the rationality of the matching formula.

• The relationship between the first and second bridges of the double-front-axle was clarified, and exploring the relationship between the first and second bridges of the double-front-axle.

• The matching formula obtained in this paper has a certain theoretical guiding effect on the matching of the four-wheel positioning parameters of steering wheels of the double-front-axle automobile.

Acknowledgement

• Fund project: Key Project of Natural Science Research of Education Department of Anhui province (KJ2016A745); Key Natural Science Projects in Anhui province (KJ2010A329); University-level Natural Science Research Project of West Anhui University (WXZR201810)

• Author: Zhang Peng (1991 --), female, native of qimen, Anhui province, assistant professor, master's degree, school of mechanical and vehicle engineering, West Anhui University, majoring in automobile suspension design and dynamics.

• Author E-mail: 502145956 @qq.com

• Corresponding author: Wang Hongxin (associate professor)

References

[1] Wang B L, Zhou J S, Zhao B et al. 1995 *Trans. Chinese Soc. Agric. Eng.* **11(1)** 89-94.
[2] Pan H D, Su J, Jia Z R et al. 1998 *Chinese J. Highways,* **11(suppl.)** 116-21.
[3] Wei D G, Chen X Q, Hu N J et al. 2003 *Trans. Chinese Soc. Agric. Eng.* **19(6)** 139-42.
[4] Wei D G 2004 *J. Hefei Univ. Technol. (Natural Sci.)* **27(6)** 1594-8.
[5] Ma J, Qian L J 2012 *J. Hefei Univ. Technol. (Natural Sci.)* **35(1)** 25-8.
[6] Dixon J C 1996 *Tire, suspension and Handing 2nd ed.* Warrendal, PA: SAE 81-109.
[7] Zhuang J 1995 *Automotive Tire Science* Beijing: Beijing Inst. Technol. Publishing 127-98.
[8] Xiao H Y 2012 *Study on the influence of four wheel alignment parameters on vehicle frequency characteristics under serpentine conditions* Hefei: Hefei Univ. Technol.
[9] Guo F 2008 *Research and discussion on the relationship between tire wear and wheel alignment* Shenyang: Northeastern Univ.
[10] Zhang Y S, Wen L Z, Liu X F 2009 *Auto Parts,* **4** 74-8.
[11] Sun Y S, Wu Z C 2018 *IOP Conf. Series: Mater. Sci. Eng.*
[12] Ju K, Ye J Wang X H 2017 *Dimensional Variation Simulation Analysis of Front Wheel Camber for Macpherson Suspension* The 19th Asia-Pacific automotive eng. annual conf. and 2017 China automotive eng. soc. annual conf.
[13] Ma W L, Li F G, Liu R X 2006 *Research on Reasonable Matching of Wheel Angle and Front Beam* Tractor and Agric. Trans. Vehicle
[14] Zuo S N, Lu B, *Wheel Angle Accuracy Improvement Scheme Analysis and Application* Automobile Technol. 20

Prestressed Continuous Bridge Evaluation using Structural Health Monitoring System

Woo Hyun Ban[1, 2 a], Jong Wan Hu[1, 2 b] and Mosbeh R. Kaloop[1, 2 c*]

[1]Department of Civil and Environmental Engineering, Incheon National University, Korea
[2]Incheon Disaster Prevention Research Center, Incheon National University, Incheon, Korea

Abstract. Dorim Goh is a prestressed concrete (PSC) continuous bridge in Seuol, Korea. The structural health monitoring is used to assess the behavior of this bridge. 12 strain, 6 displacement and two accelerometer sensors are used to collect the behavior of the bridge under static and dynamic loads. The low-pass filter is used to estimate the static behavior of bridge and extract the accurate values for the bridge performances. Six's cases of static loads and truck's speeds from 10 to 60 Km/h are evaluated. The maximum strain and deflection collect during static loads are 60 με and 1.8 mm, respectively. Moreover, the distribution of stress on the bridge deck is assessed, and we find that the bridge status is safe under static loads. The dynamic factor (DF) is calculated during dynamic test, and the results show that the maximum DF is 0.228. In addition, Fast Fourier Transformation (FFT) is used to estimate the dominant frequency of bridge. The dominant frequency is estimated from acceleration measurements after filtering the collection data; the results show that the dominant frequency of bridge girder is 4.94 Hz. The results are compared with the design simulation model results and we find that the bridge is safe under applied loads.

1. Introduction

The structural health monitoring (SHM) nowadays has been became one of good tools utilized to assess the bridges in Korea. The SHM is applied using different sensors that used to measure the behavior of structures. The one of important sensors that measured the stress, deflection and accelerometer are used in this study. Sohn et al. [1] summarized the SHM methods, sensors and evaluations systems. The semi-static and dynamic of sensors measurements are used to estimate the full behavior of bridges. Therefore, the previous studies are used strain to detect the fatigue behavior of bridges; as well as the acceleration is measured to estimate the frequencies contents and modes of structures; and the deflection of bridge is observed using displacement sensors [2-4].

Many bridges in Korea are constructed in Korea over Han river to connect the two sides of Seoul city [5]. Dorim Goh bridge is one of these bridges that are constructed using prestressed concrete (PSC). A continuous SHM is used to measure the performance of this bridge from it is constructed. In this study the short-period monitoring system is used to measure the performance of bridge. This system is used frequently to assess the bridge while no continuous monitoring system was designed on the bridge.

The time and frequency domains are used to assess the full behavior of bridges [1, 6-8]. The time series under static and dynamic traffic loads are used to assess the behavior of the bridge. Previously, the statistical analysis was used to estimate the behavior of building using acceleration measurements [9]. Moreover, the strain measurements with statistical analyses are used to assess highway steel bridges [1, 10, 11]. In this study the statistical analysis and signal processing is used to evaluate the bridge in time domain; and Fast Fourier Transformation (FFT) is applied to assess the frequency domain. More details for the analyses methods can be found in [1,6].

2. Measurements

Figure 1 show the bridge section, SHM system and case of bridge load. 12 strain, 6 displacement and two accelerometer sensors are used to collect the behavior of the bridge under static and dynamic loads. Static and six's cases of static loads and truck's speeds from 10 to 60 Km/h are evaluated.

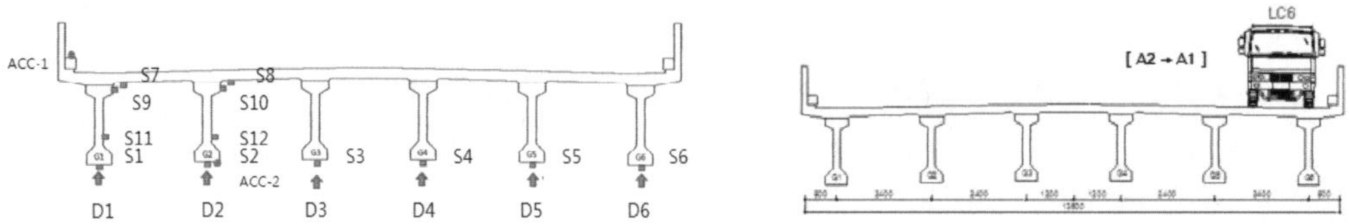

(a) (b)

Figure 1. (a) Bridge section and monitoring system,
(b) case of field-test (D is displacement, S is strain, ACC is accelerometer).

SHM system of the bridge is composed (figure 2). Sensors are connected to a data acquisition device by wire. Measured data are digitized in AD converter and delivered through Bluetooth module and Access Point (AP) by wireless. The collected data are stored in SD memory and PC. A data acquisition device used in this research has one channel; each device was time synchronized by signal sender from PC each time. PC stores data in real time and controls the sensor nodes (data acquisition devices).

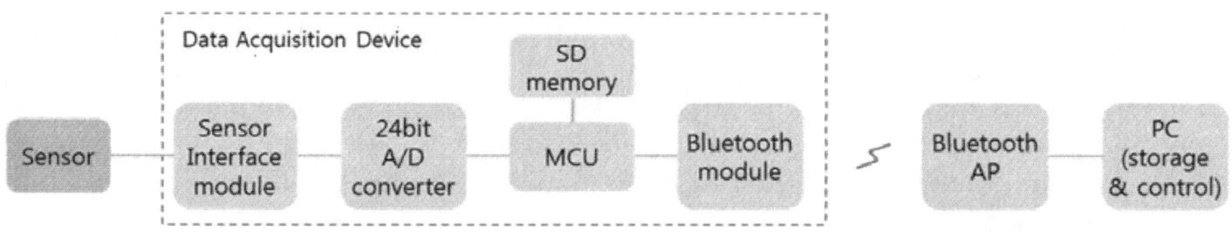

Figure 2. Structural Health Monitoring system composition.

3. Evaluation and Conclusion

The statistical analysis is used to evaluate the static behavior of bridge. The maximum strain and deflection collect during static loads are 60 με and 1.8 mm, respectively. Moreover, the distribution of stress on the bridge deck is assessed, and we find that the bridge status is safe under static loads, as presented in figure 3. From figure 3, it can be observed that the positive stress is 58.76 and 35.47 MPa for the girders G1 and G2, respectively. Moreover, the negative stress is approximately equally. These results mean that the stress is within the design value of stress.

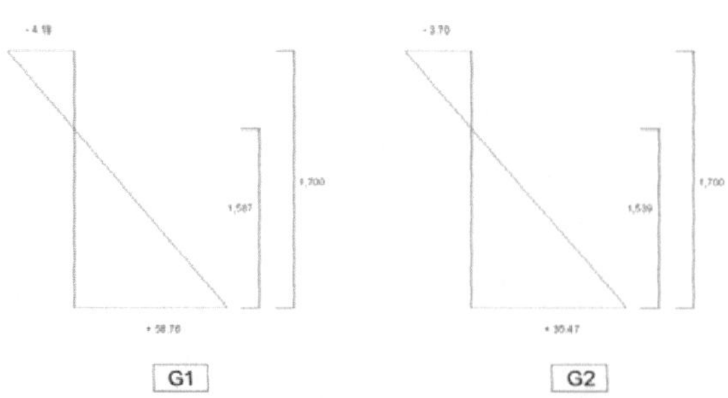

Figure 3. Stress distribution along bridge girder G1 and G2.

In addition, the low-pass filter is used to estimate semi-static performance for the strain measurement and calculate the dynamic factor (DF) for the measurements. The DF can be calculated using the following formula (DF=(Ddyn/Dstat)-1); while, the Ddyn is the maximum strain dynamic behavior observed, and Dstat is the maximum semi-static strain measurements which are smoothed using the filter applied. Figure 4 presents the strain measurements and DF of strain girder. The strain performances exhibit that the range of strain is 60 με (see figure 4.a); this means that the bridge behavior is small under affecting loads. In addition, from figure 4.b, it can be seen that the maximum DF is 0.228, while the design DF value is 0.23; it means that the bridge is safe.

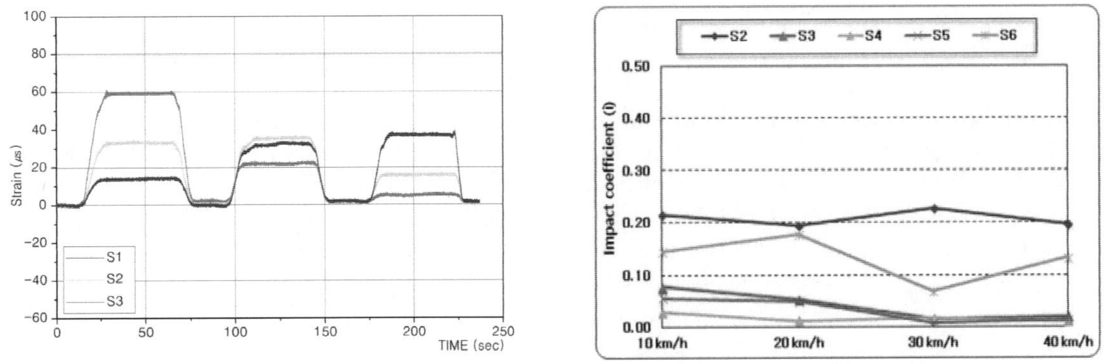

(a) (b)

Figure 4. The (a) strain dynamic measurements and (b) DF of bridge.

Moreover, the FFT is used to determine the bridge dominant frequency. Figure 5 illustrates the acceleration and FFT results. From this analysis, it can be shown that the first mode of frequency is 4.94 Hz. The comparison between the observed and simulated bending frequency (2.34 Hz), it can be summary that the measurement can measure the second frequency mode; and to measure the first mode frequency the traffic load should be redesign with low speed of traffic.

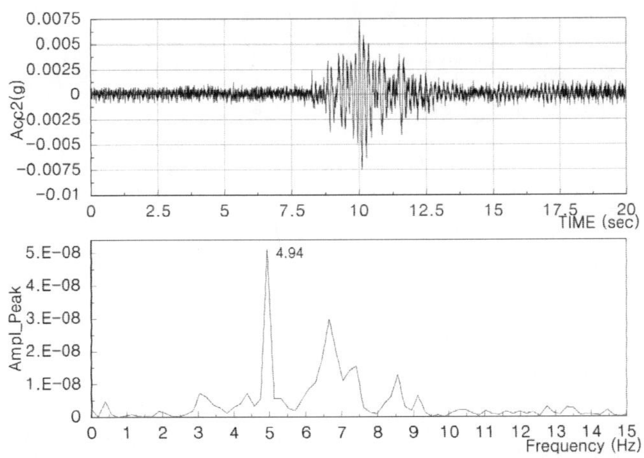

Figure 5. The Acceleration and FFT for the bridge girder.

These results are compared with the design simulation model results and we find that the bridge is safe under applied loads. In conclusion, it can be summarized that the bridge is very safe under affected load.

Acknowledgments
This research was supported by a grant (18TBIP-C144315-01) from Technology Business Innovation Program (TBIP) Program funded by Ministry of Land, Infrastructure and Transport of Korean government.

References

[1] Sohn H, Farrar C R, Hemez F M and Czarnecki J J 2004 *A Review of Structural Health Monitoring Literature: 1996-2001* Los Alamos National Laboratory Report LA-13976-MS.
[2] Song Y, Ding Y, Zhao H, Wang G, Li A and Cao B 2017 *Adv. Mech. Eng.* **9(1)** 168781401668331.
[3] Payo I and Feliu V 2014 *Sens. Actuat. A Phys.* **217** 81–94.
[4] Lü C, Liu W, Zhang Y and Zhao H 2012 *Sens. (Switzerland)* **12(8)** 9987–10000.
[5] Adler M and Ahrend R 2016 *Traffic Safety in South Korea: Understanding the Vulnerability of Elderly Pedestrians.*
[6] Kaloop M, Elbeltagi E, Hu J and Elrefai A 2017 *ISPRS Int. J. Geo-Inform.* **6(12)** 382.
[7] Sousa H, Cavadas F, Henriques A, Bento J and Figueiras J 2013 *Smart Struct. Syst.* **11(4)** 365–86.
[8] Chen C, Wang Z, Wang Y, Wang T and Luo Z 2017 *Sens.* **2017** 1-13.
[9] Kaloop M R, Hu J W, Sayed M A and Seong J 2016 *Shock Vib.* **2016** 1-13.
[10] Zhang P, Chen W and Zhang L2011 *J. Bridg. Eng.* **16(5)** 671–4.
[11] Orton S L, Kwon O S and Hazlett T 2012 *J. Bridg. Eng.* **17(3)** 462–9.

Markov Modeling for the Availability of Firearms

BAI Jun-Min[1, a], YANG Cheng-Wei[1, b] and ZENG Yan[1, c]

[1]Baicheng Ordnance Test Center of China, Baicheng 137000, China

Abstract. In order to give a reasonable forecast of firearms combat readiness, analysis of gun working method is the stochastic process of which time and status are both discrete, to simulate it by Markov process, and to deduce the calculation model of the steady state availability of firearms. Through the calculation model, as long as the known of gun state before start to work, we can just determine the probability of which gun is in the normal state at each firing time, this model can provide an important theoretical basis for maintenance test of the gun.

1. Introduction

The availability of firearm refers to the probability of being useful whenever it is needed, which is the criterion for firearm's performance [1] and it is the comprehensive reflection of reliability, maintainability and supportability for firearm products, meanwhile, it is one of the most important parameters of firearm system effectiveness [2]. For now, the existing model of availability is usually based on the maintenance time which is mainly applied to some large electronic equipment, such as radar and communications satellite. However, the performance for the firearm should be measured by the number of firing ammunition in its life cycle, but not described by working time or other time parameter, which means the model of time based cannot be used. In this article, the time span of firearm is regarded as a stochastic process which switches between normal and failure state, markov process is used to build model of availability, in this way, we can get a reasonable forecast of firearm combat readiness, meanwhile, we can find out the weakness in the process of using and maintenance so that we can improve it in time, the efficiency of firearm will be greatly improved.

2. Availability Analysis of Firearms

The work process of firearms is a discrete random process both in time and position, that means it may be normal or not at any time, in this way, the probability in the normal state is used to describe whether it is available or not. The concept of availability in firearms is the probability that it is normal at any moment during shooting. For any random shooting time t, state of firearms is:

$$X(t) = \begin{cases} 0 & (normal\ at\ time\ t) \\ 1 & (fault\ at\ time\ t) \end{cases} \tag{1}$$

The transient availability at time t is $A(t) = p\{X(t) = 0\}$, which is only related to whether the firearm is normal or not at time t, but not related to whether the fault is happened or repaired before t, higher availability will naturally lead to higher reliability. Even if the reliability is low, but there is something wrong with the quick fix, availability will remain relatively high. Firearms are often in a state of continuous work under the condition of real war, the transient availability $A(t)$ is unreliable, so the steady-state availability is referred [3].

If there is a limit:

$$\lim_{t \to \infty} A(t) = A \tag{2}$$

A is called the steady-state availability ($0 \le A \le 1$) , which means the proportion of firearm is normal during long-time shooting. Firearm belongs to repairable unit [3], for the reasons that it can keep working and continue to work when repaired again and again until it is dead.

3. Gun Availability Model based on Markov Process

Markov process [4] is a random process like this, if state of a process at time t_i is known, then process in the possibility of various state is completely determined at any time t_j after t_i, which will not be affected by state before. Which means the probability of what state the system is after time t can be confirmed by the state before [5]. We can generalize this feature to firearms so that markov process can be used to describe the availability of firearms.

Typical firearms' failure includes bullet/shell stuck or unloaded etc, and these types of failures on the total number of 80%-90% [6], by analysing the principle of fault produced we can know that the restoration belongs to basic repair which means similar failure still happens after being repaired, there is no relationship with when it is in fault last time. The probability of a component fault for firearms is really rare, and the situation of a component broken twice in the same product has never happened, fully repair refers to some components being replaced, but other component still has a continuous failure rate. Model of availability can be built based on assumptions as follows:

1) The life and repair of the unit obeys index distribution, failure rate λ and repair rate μ are both constant;
2) Each time of the maintenance aims at specified unit, other components will not be affected.
3) For the situation of many times failures, failure is not affected by the fault before and it will not affect failures behind, the generation of failure only related to the component's reliability, which means failure is independent in the aspect of time [7].

The working process of firearms includes two kinds of state normal/ fault, it is able to shoot in the state of normal but not in fault. Markov process of firearms can be described as follows:

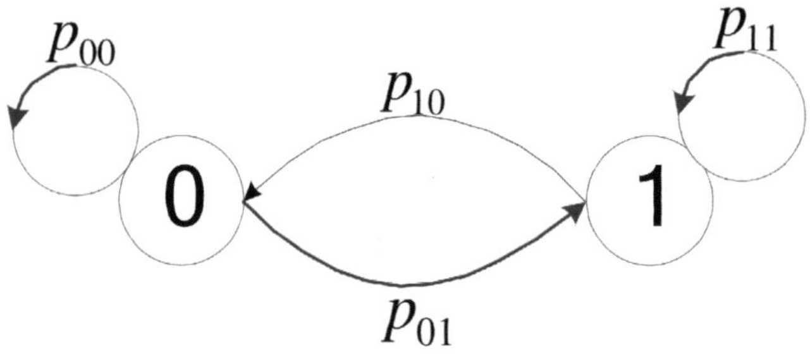

Figure 1. Markov process of firearms.

State "0" and "1" mean normal/fault respectively. $X(t)$ represents the state of firearms at time $t-th$ shooting, it is a homogeneous markov chain and its state space is $E=\{0,1\}$. So that the state of firearms in life test is: $0 \to 0$, $0 \to 1$, $1 \to 0$ and $1 \to 1$ represent C_{00}, C_{01}, C_{10} and C_{11} respectively, state transition probabilities p_{00}, p_{01}, p_{10}, p_{11} can be described as below:

$$\begin{cases} p_{00} = p\{X(t+1)=0 | X(t)=0\} \approx C_{00}/(C_{00}+C_{01}) = 1-\lambda \\ p_{01} = p\{X(t+1)=1 | X(t)=0\} \approx C_{01}/(C_{00}+C_{01}) = \lambda \\ p_{10} = p\{X(t+1)=0 | X(t)=1\} \approx C_{10}/(C_{10}+C_{11}) = \mu \\ p_{11} = p\{X(t+1)=1 | X(t)=1\} \approx C_{11}/(C_{10}+C_{11}) = 1-\mu \end{cases} \tag{3}$$

State transition matrix is:

$$p = \begin{bmatrix} p_{00} & p_{01} \\ p_{10} & p_{11} \end{bmatrix} = \begin{bmatrix} 1-\lambda & \lambda \\ \mu & 1-\mu \end{bmatrix} \tag{4}$$

We can define:

$$p(t) = \begin{bmatrix} p_0(t) \\ p_1(t) \end{bmatrix} = \begin{bmatrix} p\{X(t)=0\} \\ p\{X(t)=1\} \end{bmatrix} \tag{5}$$

The initial work state of firearms is:

$$p(0) = \begin{bmatrix} p_0(0) \\ p_1(0) \end{bmatrix} = \begin{bmatrix} 1 \\ 0 \end{bmatrix} \tag{6}$$

Markov chain's one-dimensional distribution is determined by the initial distribution $p(0)$ and t times of transition probability matrix at any moment $t \in T$ (T is the life of firearms), t times of transition probability matrix is equal to $t-th$ power of one time transition probability matrix.

$$p(t) = p^t p(0) \tag{7}$$

Characteristic equation of matrix p is:

$$|kE-p| = \begin{vmatrix} k-(1-\lambda) & -\lambda \\ -\mu & k-(1-\mu) \end{vmatrix} = [k-(1-\lambda)][k-(1-\mu)] - \lambda\mu = 0 \tag{8}$$

E is second order identity matrix, eigenvalues are $k_1 = 1, k_2 = 1-\lambda-\mu$, eigenvectors can be solved like this.

When $k_1 = 1$, $\varphi_1 = [m_1 \quad n_1]^T$ is defined as eigenvectors, homogeneous linear equations below can be solved.

$$[k_1 E - p]\varphi_1 = \begin{bmatrix} \lambda & -\lambda \\ -\mu & \mu \end{bmatrix} \begin{bmatrix} m_1 \\ n_1 \end{bmatrix} = 0 \tag{9}$$

So that $\varphi_1 = [1 \quad 1]^T$.

When $k_2 = 1-\lambda-\mu$, $\varphi_2 = [m_2 \quad n_2]^T$ is defined as eigenvectors, homogeneous linear equations below can be solved.

$$[k_2 E - p]\varphi_2 = \begin{bmatrix} -\mu & -\lambda \\ -\mu & -\lambda \end{bmatrix} \begin{bmatrix} m_2 \\ n_2 \end{bmatrix} = 0 \tag{10}$$

So that $\varphi_2 = [-\lambda \quad \mu]^T$.

$$\varphi = [\varphi_1, \varphi_2] = \begin{bmatrix} 1 & -\lambda \\ 1 & \mu \end{bmatrix}; \varphi^{-1} = \frac{1}{\lambda+\mu} \begin{bmatrix} \mu & \lambda \\ -1 & 1 \end{bmatrix}; \phi = \begin{bmatrix} 1 & 0 \\ 0 & 1-\lambda-\mu \end{bmatrix} \tag{11}$$

$$p = \varphi\phi\varphi^{-1}; p(t) = p^t = \varphi\phi^t\varphi^{-1} = \frac{1}{\lambda+\mu} \begin{bmatrix} \mu & \lambda \\ \mu & \lambda \end{bmatrix} + \frac{(1-\lambda-\mu)^n}{\lambda+\mu} \begin{bmatrix} \lambda & -\lambda \\ -\mu & \mu \end{bmatrix} \tag{12}$$

We know that $0 < \lambda, \mu < 1$, so $(1-\lambda-\mu)^t \to 0$ when $t \to \infty$.

$$\lim_{t\to\infty} p^t = \frac{1}{\lambda+\mu}\begin{bmatrix} \mu & \lambda \\ \mu & \lambda \end{bmatrix} \tag{13}$$

Matrix limit here means the limit of each matrix element. The result shows that probability of process being state "0" or "1" is $\mu/(\lambda+\mu)$ or $\lambda/(\lambda+\mu)$ after a long time, so we can conclude the model for steady availability is:

$$p(t) = \lim_{t\to T} p^t = \frac{\mu}{\lambda+\mu}(t \in T) \tag{14}$$

4. Instance Analysis

The complete life test data of two 14.5mm anti-aircraft guns is collected in this paper, the cumulative use of bullets 6000×2=12000, and failure statistics is shown in table 1 & 2.

Table1. Failure statistics.

Name / NO.	Ammunition resisted	Unloaded	Ammunition jammed	Ammunition dropped	Stop shooting	Pin breakage	Total
1#	5	2	1	—	1	—	9
2#	6	1	—	2	—	1	10

Table 2. Failure statistics.

Time / NO.	1	2	3	4	5	6	7	8	9	10
1#	1226	1774	2322	2572	2622	2822	4604	5295	5495	—
2#	507	2504	2554	2654	2854	3004	3254	4596	5425	5520

By observing the data above we could know that failures such as ammunition resisted/jammed/ dropped or unloaded account for more than 90% of the total, and relevant repair belongs to basic repair. The main reason is size/strength of components or ammunition belt and the mouth size of bullet box are dimensionally out of tolerance, some of the failures are blamed for inappropriate use and maintenance. Continuous failure rate will not be affected in the case of a normal life consumption, situation of state transition can be arranged in table 3 according to the table 2.

Table 3. State transfer statistics.

State Transfer / NO.	C_{00}	C_{01}	C_{10}	C_{11}
1#	5982	9	9	0
2#	5980	10	10	0

$1-th/t-th$ times of transition probability matrix can be calculated as follows according to formula (4) and table 3.

$$p_1 = \begin{bmatrix} 0.9985 & 0.0015 \\ 1.0000 & 0.0000 \end{bmatrix},\ p_2 = \begin{bmatrix} 0.9983 & 0.0017 \\ 1.0000 & 0.0000 \end{bmatrix}$$

$$p_1^t = \begin{bmatrix} 0.9985 & 0.0015 \\ 0.9985 & 0.0015 \end{bmatrix},\ p_2^t = \begin{bmatrix} 0.9983 & 0.0017 \\ 0.9983 & 0.0017 \end{bmatrix}$$

Steady availability for 1# and 2# are $A_1 = 0.9985$, $A_2 = 0.9983$ respectively, and availability curves are as follows:

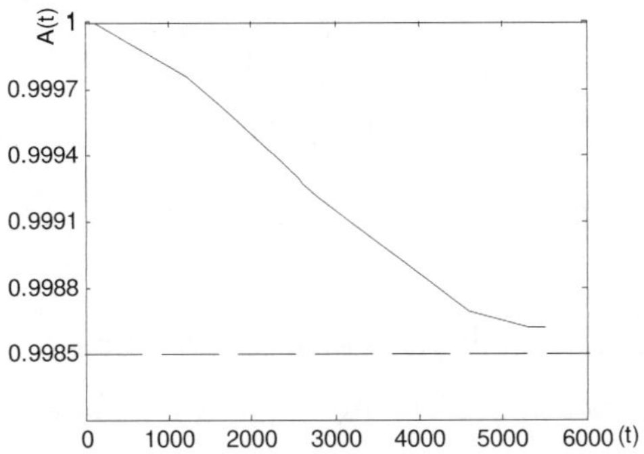

Figure 2. Availability curve for 1#.

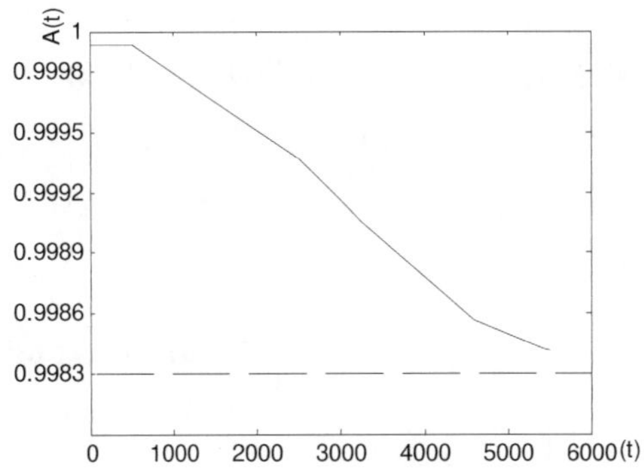

Figure 3. Availability curve for 2#.

5. Conclusion

A conclusion of work time and state are random for firearms is got in this paper, which is based on the mechanism and characteristics of the failure, firearms are defined as products with single unit and repairable; Subsequently, Markov process is used to build the state transition probability matrix for firearms; Model of availability or steady availability is deduced, which is used to calculate firearm's availability in the condition of initial state is known; Finally, the calculation model is verified in the example of 14.5mm anti-aircraft guns. This paper has some reference value for reasonable prediction of weapons and equipment reliability.

References

[1] Gan M Z, Kang J S 2010 Ordnance Gun Repair Engineering. *Beijing: National Defence Industry Press.*

[2] Guo Q S 2005 *Introduction to assess the effectiveness of firearms.* Beijing: National Defence Industry Press.

[3] Xu H. 2008 *Reliability Assessment active duty firearms.* Nanjing: Nanjing University of Science and Technology.

[4] Liu C H 2014 *Stochastic Process and Its Application.* Beijing: Higher education press.

[5] Jin X, Hong Y J 2007 *System reliability and availability analysis method.* Beijing: National Defense Industry Press.

[6] Jia Y F 2004 *Research on reliability of rules of new equipment typical light weapons* Shijiazhuang: Ordnance Engineering College.

[7] Jiang R Y, Zuo M J 1999 *Reliability model and application.* Beijing: Mechanical Industry Press.

The Impact of Information System Implementation to the Integrated System for Increasing the Supply Chain Performance of Manufacturing Companies

Zeplin Jiwa Husada Tarigan[1, a*], Hotlan Siagian[1, b], and Rick Richard Bua[1, c]

[1]Department of Management, Petra Christian University, Surabaya, East Java, Indonesia

Abstract. This study examines the impact of information system management implementation on the company performance with the mediating role of process innovation and process innovation. Data collection is conducted by distributing questionnaires to, and through interviews with the respondents engaged in the department of the information system of 41 companies in Surabaya, Indonesia. The questionnaire is designed using a five-point Likert scale ranging from 1: strongly disagree up to 5: strongly agree. The data analysis uses smart PLS software version 3.0. The study found that the implementation of a management information system directly affects the process of innovation. The implementation of a management information system also directly influences product innovation. The implementation of a management information system does not directly affect the supply chain performance. The process innovation improves product innovation. Process innovation does not provide a direct impact on supply chain performance. Product innovation directly gives an impact on the improvement of the performance. In total, the implementation of management information systems affects supply chain performance through the process innovation and product innovation by the coefficient of 0.526. This study contributes to the current research on supply chain management and paves the way for managers to improve the supply chain performance by implementing the management information system, product innovation, and process innovation.

1. Introduction

Today, the domestic farm industry is one of the sectors which support the Indonesian economy sustainability. This industry contributed 15.38% to the country's domestic product in the industry and trade sector [1]. This means that the farm industry is still growing, and the need for the farm products is consequently increasing as well. The Ministry of Agriculture has published the annual strategic plan with one of the primary targets is to achieve meat self-sufficiency. The fulfillment of meat demand becomes the priority. However, several factors have caused the inability of the farm industry to fulfill the increasing demand for meat and eggs in Indonesia [2]. In line with the current market demands, customer's needs are also rapidly changing. It requires support from all related partner such as suppliers, distributor, and retailer until the final customer [3]. The supply chain management practices could become a solution on this industry regarding how to manage the process of material from the supplier into finished products and then delivery to consumers. The extent to which those companies within a supply chain fulfill the customer demand is defined as the supply chain performance. The supply chain performance is assessed in term of efficiency (cost) and responsiveness of an internal business operation. According to Lai et al., [13] supply chain performance is measured regarding profit, cost, and environmental performance. Cho et al., [14] suggest the supply chain performance measurement is based on the industry sector. Both studies revealed that supply chain management (SCM) is an essential competitive advantage for companies in providing fast service with high product variety and low cost. Al-Sa'di et al., [19] states that process innovation and product innovation have a positive and significant impact on company performance. The implementation of information technology in the farm's industry is still limited to administration only, and its enforcement is still done partially on the company operation. The information technology has not been integrated into all departments, especially in the operational section of the company's farm industry.

Meanwhile, research by Tarafdar and Gordon [4] mentions that information systems and information technology support business processes innovation. Information system and technology support the innovation of the process through the improvement and development of a better company's business processes [5]. Information technology supports the process of innovation in the company during the product design. Information technology supports the innovation of the process as information technology can provide the required data in real time. Research conducted by Ferneley and Bell [6] states that information technology can improve the innovation of corporate processes such as business rationalization and improvement. Another research conducted by Adamides and Karacapilidies [7] state that the information system can result in a process innovation in the form of the improved business process such as easy understanding of the procedure, an operational standard, which enable the transfer of knowledge among individuals. Roper et al., [8] state that the end of a knowledge process is a transformation of innovation to produce new products and processes. Company's innovation achievement is represented by applying industrial process practices in producing new products or processes. The company's product innovation meet the customers need and make the customer satisfied better than competitors, and increases its market share [9].

Basole et al., [10] suggest that the adoption of information technology is vital to process innovation for transforming organizations to adapt to changes. Product innovation in the company is derived from a new idea by doing an innovation process. Performing product innovation and process innovation requires substantial forward investment in many aspects from the development of specialized resources, new equipment, research and development (R&D), new technology and even new business units [11]. The innovation capability in the manufacturing industry is required to develop new technology by conducting R & D to produce a new product required by the customer [12]. The implementation of information technology in the farm's industry is still limited to administration only, and its enforcement is still done partially on the company operation. The information technology has not been integrated into all departments, especially in the operational section of the company's farm industry.

Based on the above description, this study examines the impact of information technology implementation on the company performance with the mediating role of the process innovation and product innovation in the farm industry.

2. Relationship Among Concepts

The adoption of information and communication technology by the SME (small medium enterprise) contributes to the improvement of innovation including product innovation, process innovation and innovation in the overall enterprise system [15]. Process innovation by the company can be realized in the form of changes in work procedures and innovation procedure. Changes made through improvement in manufacturing processes, and using new technology will produce a new product (innovated product) for the company. Explorative learning and improvement creativity gained by the company enhance the ability to innovate [16]. Process innovation reflects the changes how the organizations produce the end products and services of a company. Process innovation is a tool for improving quality as well as cost saving. Product innovation allows the company to introduce a new product or service into the market to meet the customer's needs and market orientation [17]. Lukas & Farel [18] classify the product innovation into three basic types: 1) product line extensions, 2) tailor-made products, and 3) new products. Product line extensions are relatively new products on the market but not new to the company. The tailor-made product is a relatively new product for the company, but it is relatively well known in the market. New to the world products are new products both for the company and for the market. The research conducted by Al-Sa'di et al., [19] states that process innovation and product innovation have a positive and significant impact on the company performance. The process innovation is undertaken at the company related to changes in the blueprint, especially on changes in the business processes, standard operating procedures and work instructions to achieve a shorter and faster process transformation. This concept emphasizes the integration of the flow of information and materials through business networks to support the process of corporate innovation to achieve increased corporate capabilities in meeting the needs and desires of the consumers. The company's built-in integration with partners to build collaboration will provide an effective and efficient process for the flow of products, services, and information [20]. Integration of the company's internal business processes in collaboration with providers, suppliers, customers, and supply chain partners will have an impact on supply chain performance [21]. This process involves the interconnection of various function within the organization to provide fast delivery of products to customers [22]. Implementation of ERP as an integrated information system at the company provides innovation to all business areas through business process re-engineering in improving organizational performance. The implementation of the ERP system as one of the integrated management information systems starting from material planning, material purchase, production planning, and production control improves supply chain performance [21]. Based on the above theoretical review, this study proposes six research hypotheses as follows:

(H1) The information system management implementation influences the process innovation.
(H2) The information system management implementation affects product innovation.
(H3) The process innovation affects product innovation.
(H4) The process innovation influences the supply chain performance.
(H5) The product innovation affects the supply chain performance.
(H6) The information management system implementation affects the supply chain performance.

3. Research Method

The unit of analysis of this study are the firms engaged in the farm industry and located in the region of East Java, Indonesia. Data collection is conducted using a questionnaire designed with a five-point Likert scale. A respondent is a person working for the company for more than two years, using the information system, and understanding the production process. This study used census research which involves all 41 companies consisting of 12 farm breeding companies, 17 poultry feed production, two egg hatching companies, and ten processed meat companies in East Java. Data collection was done in two stages, firstly, data collection through a self-administered survey by distributing the questionnaires to be filled by the respondents, and secondly, the researcher interviews several (by random) respondents to check their face validity [23]. The measurement uses a five-point Likert scale, with 5: strongly agree, 4: agree, 3: neutral, 2: disagree, and 1: strongly disagree.

In this study, the management information system is defined as the extent to which the company has implemented an information system and daily data transaction in decision making [24]. Five indicators are used to measure these variables, i.e., availability of software and hardware, information submitted on time, clear time limits for the information provided, accurate available information, available information according to user requirements and complete presented information. Process innovation is the process of using technology in the added value process of the product. Four indicators used to measure these variables are 1) easier technology upgrades, 2) faster processing time, 3) precise applied task specifications, 4) working mechanisms and simpler used information. Product innovation assesses the extent to which the product performs its function better than the one the competitor offers. Three indicators used to measure these variables are 1) new product development, 2) product variety development, and 3) development of product models and standards. Supply chain performance is the measurement of the extent to which the company adopted supply chain management principles in the pursuit of increased performance. Supply chain performance is a measure of process-oriented company performance, integration of purchasing, production and product delivery to consumers consisting of raw material management, information flow, and finance. This study is adapted to the conditions of SCM performance commonly used in all farm companies including which are on-time product delivery, product quality improvement, raw material availability, and better time flexibility. Data analysis method used in SEM is based on Partial Least Square (PLS).

4. Discussion

Based on the analysis result using the PLS version 3.0 software, it is found that those indicators of the variable are valid in term of convergent and discriminant validity. The recommended minimum value of factor loading for convergent validity is 0.5, and the cross loading of each indicator with other variable is less than its factor loading for discriminant validity. The information system management implementation has six indicators with the factor loadings respectively as follows: software and hardware availability (0.803); information delivered on time (0.839); the information provided has a time limit (0.812); the accuracy of information (0880); the available information is matching the requirements (0.755), and the information presented is complete (0.829).

The process innovation has four indicators with the loading factors respectively as follows: technology is easier to use (0.916); process cycle time is faster (0.753); the standard of the task performed is more precise (0.701); the working mechanism and information used are simpler (0.841). The product innovation is assessed using three indicators with the result of factor loading as follows: new product development (0.928); variation of product developed (0.815); and the model development and standard product size (0.864). The last construct is the supply chain performance which is composed of four indicators namely product delivery (0.855); improvement of product quality with (0.672); availability of raw materials (0.793); and better time flexibility (0.678). This finding implies that those indicators of each variable are valid in term of convergent and discriminant validity.

The result of the analysis with a structural equation modeling (SEM) PLS is detailed in table 1.

Table 1. The Result of Hypotheses Testing.

Hypotheses	Original Sample	Sample Mean	Standard Deviation	T - Statistics
(H1)Information system -> Process Innovation	0.649	0.652	0.044	14.719
(H2)Information system -> Product Innovation	0.174	0.167	0.066	2.642
(H3)Process Innovation -> Product Innovation	0.686	0.697	0.075	9.175
(H4)Process Innovation -> SCP	0.177	0.204	0.132	1.338
(H5)Product Innovation -> SCP	0.665	0.640	0.138	4.825
(H6)Information system-> SCP	-0.021	-0.020	0.093	0.230

Table 1 demonstrated the result of hypotheses testing. It is found that four hypotheses (H1, H2, H3, H5) were supported, and the rest two (H4, H6) were not supported. These findings are discussed in the following section. The first hypothesis (H1) is the influence of information system management implementation on process innovation. A good management information system, when applied to a company, will improve the quality of information that can give a significant impact on process innovation and product innovation. The innovation of these products and processes can be more effective and efficient when applied to information systems and technology, and the value of the company will increase, and at the end, will also improve competitiveness. Results of this study showed the value of the influence of management information system on process innovation is 0.649 with t-value of 15.380. This shows that the implementation of a management information system has a positive effect on process innovation. The result of the research is consistent with H1 stating that there is a definite influence of management information system implementation on process innovation. This shows that the accuracy of information is strongly influenced by the renewal of the technology used. Thus, the users will get the information they need more quickly, precisely, and accurately.

The second hypothesis (H2) is about the influence of information system management implementation on product innovation. Implementing a supply chain management enables the farm industry to meet the needs of most communities engaged in this industry to keep improving its competitiveness. The collaboration in a business network is more emphasized on the integration of information and material flow through the business network to support the company's innovation process which is an important thing to enhance the company's capability in fulfilling the consumer's needs and desires. Providing better value to customers is reflected in the form of dependability, proactivity, flexibility, and delivery. Innovation and speed are central to the company and are the impact of the company's labor and information system. The results of data processing show the value of process innovation effect on supply chain performance of 0.172 with a critical ratio (t-value) of 1.229. The results of this study did not support the H4. There is no influence of process innovation on supply chain performance. In the farm industry itself, technology renewal is done by applying computerization and hi-tech equipment during operational activities. Delivery of the product to the customers is carried out according to their wishes.

The third hypothesis (H3) is the influence of process innovation on product innovation. The main focus of innovation is the creation of new ideas, which will, in turn, be implemented into new products, and new processes as well. Process innovation is a suggestion to improve quality as well as cost saving. This reflects that adoption of an innovated process improves production efficiency and product quality. Results of data analysis show that process innovation affects to product innovation is with the path coefficient of 0.686 and the t-value 10.225. The finding proves that developing a new product require an updated technology in supporting product innovation. Given the increasing need for various animal products requirement, the farm industries in Indonesia are continuing to update their existing technologies and to use them to maximize their services to consumers. The renewal of the technology used is expected to develop the attributes used to support their operational activities such as repairing or replacing old production equipment with the new ones.

The fourth hypothesis (H4), the process innovation did not affect the supply chain performance. The results indicate that the t-value is 1.338, which is not acceptable at a 5% significance level. It seems that implementing a supply chain in the farm industry give no impact on the performance. The process innovation between the supply chain partner is not useful in improving the performance. This finding could mean that the collaboration in the supply chain does not work well because each supply chain partner just considering their benefit. It is essential to achieve the company's capability for improvement in meeting supply chain partners' needs.

The fifth hypothesis (H5) is the influence of product innovation on the supply chain performance. Along with the rapid development of the market, the customers need for animal products will be increasing. Therefore, it is necessary for the suppliers in the management and distribution of products to reach the end customers. This process involves the interconnection of various organizations, such as raw material suppliers, factories, distributors, retailers, and transportation services known as the supply chain. In order to be able to raise the value of a company, it is necessary to support the increased efficiency and effectiveness of internal business operations such as supply chain management (SCM). The results of the data analysis show that the value of product innovation influence to supply chain performance is 0.672 with a critical ratio of 4,801. The results of this study proved that product innovation has a positive effect on supply chain performance.

The sixth hypothesis (H6) states that the information system management implementation affects the supply chain performance. However, this study did not support this hypothesis. There is no direct influence of the information systems management implementation on the supply chain performance. The results of the data analysis show that the path coefficient is - 0.021 with t-value of 0.230. The fact-finding for this result indicated that the use of data provided by the information system is mostly for the administration and financial report purpose. The use of data from the system for integrating the operation is still limited. The companies need to capitalize on the information system in integrating all the system for either operational or nonoperational to improve the performance. The strategic roles of the information system management in the farm industry are to minimize the company weakness, to support business strategy, and to support competitive advantage. The implementation of the information system management will support the strategy pursuit by a company with the goal to enhance the performance of the supply chain, all of which will lead to customer satisfaction. In the farm industry, the accuracy of data related to operational activities such as production, distribution, and sales activities is crucial. It could happen that the quantity of purchased product does not match the quantity of the product already stated in the delivery order due to sudden changes in demand. This event will disturb the sales process as the seller must confirm in advance to the market relating to the changes. However, the accurate data provided by the system could prevent the situation to happen. Based on the result found from table 1 and discussion the research model is presented in figure 1.

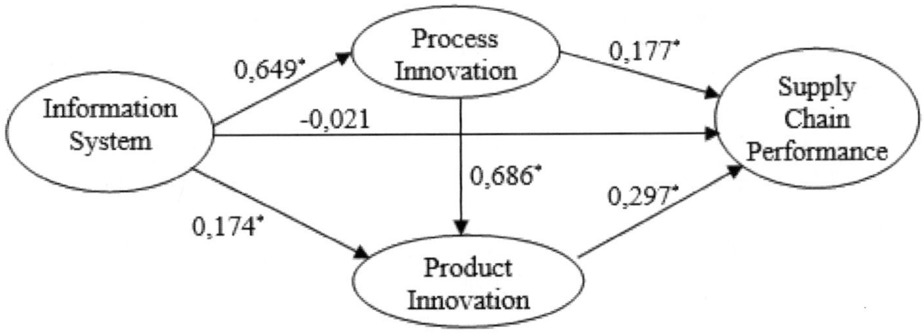

Figure 1. Research Model Result (* Significant at level 0.01).

5. Conclusion

The primary purpose of this study is to examine the influence of the information system management implementation on the supply chain performance with the mediating roles of the process innovation and product innovation. Based on the results of data analysis and discussion above the conclusion are drawn as follow: first, the implementation of an information system management influences the process innovation. Second, the implementation of a management information system influences product innovation. Third, process innovation conducted affects product innovation. Fourth, the process innovations do not affect the supply chain performance. Fifth, the product innovations conducted by the farm industry in East Java give effect to supply chain performance. Sixth, the implementation of a management information system does not give effect to supply chain performance. This study contributes to the existing research in the field of supply chain management. This study could pave the way for the manager, how to improve the supply chain performance through the establishment of the information system management, process innovation and product innovation in the pursuit of the supply chain performance.

References

[1] Information on https://www.bps.go.id/
[2] Information on http://www.pertanian.go.id/file/RENSTRA_2015-2019.pdf
[3] Ferrari M 2011 *European Manage. J.* **29** 181-92.
[4] Tarafdar M and Gordon S R 2007 *J. Strat. Inform. Syst.* **16** 353-92.
[5] Reijers H A and Mansar S L 2005 *Omega* **33** 283-306.
[6] Ferneley E and Bell F 2006 *Technovat.* **26(2)** 232-41.
[7] Adamides E D and Karacapilidis N 2006 *Bus. Proc. Manag. J.* **12(5)** 557-75.
[8] Roper S, Du J and Love J H 2008 *Res. Pol.* **37** 961-77.
[9] Roxas B G 2008 *Asian Acad. Manag. J.* **13(2)** 57-77.
[10] Basole R C, Seuss C D and Rouse W B 2013 *Dec. Sup. Syst.* **54** 1044-54.
[11] Amit R and Zott C 2012 *MIT Sloan Manag. Rev.* **53** 40-9.
[12] Wang C and Kafouros M I 2009 *Int. Bus. Rev.* **18** 606-16.
[13] Lai K H, Wong C W Y and Lam J S L 2015 *Int. J. Prod. Econ.* **164** 445-53.
[14] Cho D W, Lee Y H, Ahn S H and Hwang M K 2012 *Comp. Ind. Eng.* **62** 801-18.
[15] Peñalba J E M, Guzmán G M and de Mojica E G 2015 *J. Bus. Econ. Pol.* **2(2)** 124-31.
[16] Valaei N, Rezaei S and Emami M 2017 *Bus. Proc. Manag. J.* **23(5)** 957-83.
[17] H-Espallardo M, Delgado-Ballester E 2009 *European J. Innovat. Manag.* **12(4)** 470-91.
[18] Lukas B A and Ferrel O C 2000 *J. Acad. Market. Sci.* **28(2)** 239-47.
[19] Al-Sa'di A F, Abdallah A B, Dahiyat S E 2017 *Bus. Proc. Manag. J.* **23(2)** 349-76.

[20] Flynn B B, Huo B and Zhao X 2010 *J. Operat. Manag.* **28** 58-71.
[21] Shatat A S and Udin Z M 2012 *J. Enterprise Inform. Manag.* **25(6)** 576-604.
[22] Hilletofth P and Eriksson D 2011 *Ind. Manag. Data Syst.* **111(2)** 264-81.
[23] Cooper D R and Schindler P S 2008 *Business Research Method Tenth Ed.*, McGraw Hill.
[24] Suprpato W, Tarigan Z J H and Basana S R 2017 *The influence of ERP system to the company performance seen through the innovation process, information quality, and information sharing as the intervening variables* **ICEMT' 17 Proc.** 87-91.

The Impact of Organizational Commitment on Upgrading ERP for Maintaining the Quality of Information and the ERP Performance

Zeplin Jiwa Husada Tarigan [1, a*], Lianto[1, b] and Sautma Ronni Basana[1, c]

[1]Lecturer Management, Petra Christian University, Surabaya, Indonesia

Abstract. This study examines the impact of organizational commitment on upgrading ERP for maintaining the quality of information and the performance. Many manufacturing companies in East Java have been implementing ERP in managing their business. Although the ERP post-implementation has been continuously maintained by these companies, the inevitable changes and development in the current business practices have forced them to upgrade their ERP infrastructure. All ERP system upgrades require a commitment from both management and employees, which is also known as the organizational commitment. This research used the population of 110 manufacturing companies in the region of East Java which have been implementing ERP for more than two years. The response rate is quite good, as many as 90 questionnaires out of 110 distributed questionnaires are completed by the respondents. Data analysis has been performed using partial least square (PLS) technique. The result demonstrated that the organizational commitment brings a significant impact to the ERP infrastructure upgrade. While the organization commitment shows a significant impact on the information quality. The organization commitment does not show a direct impact towards the ERP performance. Similarly, the ERP infrastructure upgrade also shows a significant impact on the information quality. The ERP infrastructure upgrade does not show a direct impact on the ERP performance. The information quality contributes a positive impact on the ERP performance. In total the organizational commitment affects directly and indirectly the ERP performance. The management commitment empowers the key user to be aware of the ERP system upgrading all the time. By keeping the ERP to the most up-to-date condition will contribute to the high quality of the information. Furtherly, the best information quality provides the appropriate information in term of the right time and the right information in the decision making process. The appropriate information enables the top management to decide in the pursuit of improving the company performance.

1. Introduction

The intensity of business competition has forced manufacturing companies to build the business system to survive the competition at the global level. Dealing with global competition, management should be building an integrative information technology (IT) system to raise the synergy between departments which will produce a more effective and efficient process. On the industry level, the implementation of IT will create a better competitiveness by making it easier to acquire accurate data, and improving the information infrastructure for companies [1]. The IT implementation in the form of ERP technology needs to be continuously maintained by the company. The previous research has been discussing the critical success factors in ERP implementation, such as the organizational improvement as being on time, and on a budget [2]. Matende and Ogao [3] in their research stated that users' involvement in providing ERP system and running it will contribute to the successful ERP implementation. A research by Nah et al., [4] suggests that the ERP implementation in the stage of monitoring and evaluation of performance, as well as in the stage of maintenance will be affected by some factors as business process re-engineering and minimum customization, software development, testing and troubleshooting, monitoring and evaluation of performance.

Engineering systems require an automation system to run production machines that is directly integrated with the company's computer system. This integration process is carried out with business process reengineering automation systems in producing operational and process improvements in manufacturing companies. The automation process requires updated software and hardware to fit. The current process of automation is widely used by companies in running production machines that are automatically connected to the company's administrative system.

Currently, ERP has become an integrated system in the implementing companies, as it is used by all departments and decision makers in these companies. The ERP post-implementation needs to be continuously adjusted to match the companies' functions and the ever-changing situation in the companies. The changes will result in a cultural shift in the company to pay closer attention to the information quality, information technology infrastructure, and top management's commitment to ensuring ERP existence in order to benefit the company. Previous researchers have been discussing the ERP post-implementation, such as one conducted by Ajer and Hustad [5] which postulates several determining factors of ERP successful post-implementation on franchise companies. ERP post-implementation success factors will be determined by the technology aspects, which consist of the project management and system configuration, the environmental aspects, and the organization aspects, which consist of the leadership involvement and organization fit [6]. A research by Yu [7] suggests many variables contributing to the successful implementation of ERP in different stages, starting with pre-implementation, during implementation and post-implementation. Determining factors in the post-implementation stage, such as Degree of data accuracy, Degree of system stability, Degree of user-friendliness, Size of the induced ERP system, and Integration of the ERP system implementation. Nicolaou and Bhattachary [8] suggest that the post-implementation will be determined by add-ons or ERP upgrade which has been adjusted to the organization's needs and the company's financial performance. The impact of ERP system in post-implementation from the user's perspective is being discussed in Hsu et al. [9], which state the significant positive impact of the quality system towards service quality and information quality in the post-implementation stage towards the increase of employee extended use. This is normally described as the post-implementation stage. In this stage of post-implementation, the company will continuously apply renewal of ERP system in order to be able to maintain connection and integration with the rest of the system run by the company. This would create a fully integrated system which is commonly known as the extended ERP.

The application of extended ERP will enable the company to obtain a high quality information. This high quality of information will benefit the company in improving the company's performance due to the shorter response time to the departments' needs [10]. Conceptually, the information quality dimensions are determined by the data accuracy, data quantity, completeness, objectiveness, reliability, and verifiability [11]. The information quality in the context of manufacturing companies will require data completeness which can be accessed by other departments and connected to the ERP system to produce a good collaboration. The existence of good collaborate on in the company will create integration in sustaining ERP system. However, the fact indicates that many companies do not pay attention to upgrade their ERP software to the latest version. This unawareness of the company makes the ERP system, i.e., hardware and software, only function as an administrative tool only. The consequence of this situation is resulting in incapability of the management in accessing appropriate information required in decision making. Only very few studies focus on the direct influence of ERP upgrading to the company performance.

2. Organization commitment

ERP system has become a common system that is used by almost all employees and departments in many companies. The benefits from the usage of the ERP system will only be gained from the constant updates of the system, which allow the company to cope with the changes in the operation. Any changes in the company process require the management commitment in the form of funding, assigning new responsibility to key users, and providing proper trainings for the ERP users and developers. The system will continuously evolve until it is considered as sufficient by the top management. The management commitment will become the prerequisites to synchronize the company's internal situation with the changes in the companies' external environment. This process of comparing company's internal and external conditions will help the company comprehend better its strengths and weaknesses, which will help the company to come up with a strategy to modify its ERP system. A research by Zhu et al., [6] suggests that the management's role in the ERP post-implementation will be in the form of organizational readiness. The commitment built by the organization will be in the form of how the organization successfully convinces its employees to be voluntarily and loyally involved in the organizational. Organization management needs to fully build a good relationship between each department and each employee within the organization to achieve the targets that are set by the company.

Building organization commitment will preserve the process stability and productivity in their efforts to reach the goals. This can be identified in the following three components, affective, continuance, and normative commitment. The affective commitment is described as an employee's emotional involvement towards his/her organization, and manifested in the feeling of love to the organization. Continuance commitment is described as employee's perception of costs and benefits of leaving the organization. Normative commitment is described as a moral dimension based on obligatory feelings and a sense of responsibility towards the organization. The measurement in this research is adopting Caki et al., [13] with three components, consists of affective, continuance, and normative commitment. The development applied and being used as the indicators are delightful in working, delightful in the involvement of completing the jobs, the absence of intention to leave the organization, doing the job well, and responsibility towards the job.

3. Information quality

ERP implementation in the company will result in the interconnection between modules in the company's business system. Order acceptance in manufacturing company will provide data entry in sales and distribution module. This entry in the sales and distribution module will be automatically connected to modules in production planning, modules in materials management, and other modules in the company. Interconnection will ensure the validity of the data and process. This will produce the desired report which matches the needs of the users in the departments and the output of the ERP system. The fast transformational process of the data entered into an accurate report will become a necessity for the company's decision making. This output of the ERP system in manufacturing companies will be known as information quality.

Information quality is very important for companies as it will provide the right information for decision-makers. Information quality users include customers, employees, and the company's management. This is a multi-dimensional concept. According to Marinagi et al., [10] information quality dimensions are believability, interpretability, value-add, reputation, completeness, objectiveness, reliability, timeliness and response time, price, accuracy, availability, latency and response time. The information quality would be influenced by accuracy data, amount of data, completeness, objectiveness, reliability, and verifiability. However, it is not influenced by availability, believability, concise representation, consistent representation, relevancy, timeliness, and understandability in the case of information security practitioners in Malaysia [11]. This research will use indicators which have been adjusted to manufacturing companies' condition such as data accuracy, completeness, objectiveness, reliability, timeliness and relevancy data.

4. Update ERP in Automtic System

Ensuring the ERP system running smoothly during its implementation can be considered as suitable for the company's goals. Building IT structure in the company can be done simultaneously with the day-to-day operation of the company. Manufacturing companies must be able to provide IT infrastructure which is met by the basic requirements of the ERP system. The ERP implementation requires two infrastructures regarding ERP softwares and ERP hardwares. ERP software can be fulfilled by either buying available existing ERP packages, as well as building and developing company's personalized softwares. Buying an existing ERP package means that the company buying a standardized system which requires further customization to match with the company's operational system. The success of ERP implementation will heavily depend upon the quality of company IT infrastructure. The already installed ERP system is able to run the automatic system of the machines that exist in the company so that there is a synergy between the running production system and the company's computer system. The availability of a high quality IT infrastructure will produce a high-quality system with good data integration in all the departments in the company, especially in the effectiveness and efficiency of back office and front office [14]. Companies which implement ERP will most likely have software and hardware for their daily operation. Nevertheless, the company needs to update this software and hardware to match the requirement of the more advanced system. This continuous update is a necessity for the implementation of a sustainable ERP which knows as the extended ERP or ERP post-implementation. Adjusting the IT development for ERP system will provide a good administration system, making it easier for the user in applying the ERP system, and would enable information sharing across departments to support fast and accurate decision making [15]. ERP in the post-implementation stage is not only relying on the infrastructure as being used in company operation, but it requires a continuous update to increase its capabilities in dealing with job activities. Post-implementation can be acknowledged when the companies have been using ERP system for at least two years and continuously conducting the necessary upgrades infrastructures to fulfill the companies' demands [8]. The indicators for measuring ERP infrastructures: data integration, system stability, and system upgrade and technology utilization.

5. ERP performance

ERP technology provides a positive contribution to the companies which have been implementing it. The ERP technology implementation will improve the company's operational capabilities and provide both tangible and intangible benefits [8]. The ERP application can also reduce the number of stocks or company inventory because decision-makers can quickly follow the inventory level in the fields through the ERP report. Under this circumstances, they can manage the purchasing activity, and control the output and input in the company's warehouse. Applying ERP can also help to control the inventory turnover which will increase the company's profitability. Business performance can be reflected by the effective and efficient usage of the company resources, productivity, quality improvement and product/service innovation [16]. Another research suggests that company performance is measured using five indicators, which are the improvement in job performance, the pace in completing the individual task, the productivity of job, making the job more manageable, and overall satisfaction towards ERP system as a whole [17]. The indicators being used to measure the performance are easiness in decision making, shorter business process, easiness in resource management, decreased inventory level and easiness in analyzing company's condition.

6. Conceptual the research

Preserving ERP implementation stability for the companies in the post-implementation will be defined as improving or at least preserving the quality of information. A research by Zhou et al., [18] postulates that overall business is determined by the information quality. Similarly, the ERP implementation, which has been useful in helping companies, needs to be sustained by customizing the system through upgrading the ERP system. This process must be supported by the organizational commitment. The infrastructure upgrade requires a substantial cost and extended time in customization to match with the company situations. Customization which implemented in redesigning company's processes needs to assisted by an expert consultant or experienced capable key users [19]. The upgrade process is considered completed when the system is running smoothly and can be used by the key users in the day-to-day operation. There are six hypothesis constructed. The first hypothesis is that the organizational commitment contributes significantly to updating ERP infrastructures. The second hypothesis is that the organizational commitment has a significant direct impact to the ERP performance. The third hypothesis is that the organizational commitment contributes significantly towards the information quality. The fourth hypothesis is that the updated ERP infrastructure improves the information quality in maintaining ERP. The fifth hypothesis is that the updated ERP infrastructure improves the companies ERP performance. The sixth hypothesis is that the well-maintained information quality brings improvement to the ERP performance.

7. Research method

This research rook data from companies which implemented ERP in East Java, Indonesia. Based on the data from the government service on Trades and Industries, there were 234 manufacturing companies in the province of East Java. The sample was determined using purposive sampling to fulfill the sets of criteria made by the researchers. The questionnaire was distributed to 110 manufacturing companies which had been implementing and upgrading the ERP system. The distribution and visitation by ten surveyors resulting in 90 valid questionnaires could be processed further. Eight questionnaires were not submitted on time, 12 questionnaires were not fully filled, and two questionnaires had some partial lost data.

Based on the ERP software used by the companies, the respondents' profile was as following: 26 companies were using SAP (29%), seven companies were using Oracle (8 %), three companies were using MFG Pro (3 %), one company was using JD Edwards (1%), six companies were using People soft (7 %), eight companies were using Dynamics Microsoft NAV (9 %) and other 39 companies were using their self-developed software (43 %) by hiring local consultants or the companies' internal programmers. Based on their level in management of the companies, the respondents could be classified as following: manager 37 respondents (41 %), IT coordinator 2 respondents (2 %), owner/CEO 9 respondents (8 %), superintendent 17 respondents (18 %), sssistant manager 11 respondents (13 %), supervisor 12 respondents (13 %) and financial controller 2 respondents (2 %).

The research was using PLS as its analyzing tools by computing the model fit of the tools in measuring outer model and inner model. The measurement model used was following the variables represented in the measurement indicators. The outer model was evaluated for ensuring the validity and reliability of the construct. Measuring reliability in PLS was done using composite reliability. The reliability value had to be higher than 0.7 to be considered as reliable. The data processed resulting 0.871 for organizational commitment, 0.850 for upgrade ERP, 0.846 for information quality and 0.899 for ERP performance. The third measurement model was computed the value of influence for each variable (table 1).

Table 1. Path Coefficient.

Effect of Variable	Original sample estimate	Mean of subsamples	Standard deviation	T-Statistic
Org. Comm -> Upgrade ERP	0.537	0.601	0.092	5.828
Org. Comm -> Inf. Quality	0.408	0.380	0.113	3.605
Upgrade ERP -> Inf. Quality	0.408	0.426	0.081	5.009
Org. Comm -> ERP Perform	0.274	0.293	0.158	1.730
Upgrade ERP -> ERP Perform	0.054	0.061	0.165	0.329
Inf. Quality -> ERP Perform	0.389	0.354	0.177	2.199

Coefficient path (table 1) resulted in the influence of organizational commitment towards upgrade ERP infrastructure is 0.537 with t-statistic of 5.828 > 1.96. This result indicates that there is a significant positive influence on organizational commitment to upgrade ERP infrastructure. This can be interpreted as the increase in organizational commitment significantly increases the upgraded ERP infrastructure. Therefore the first hypothesis can be accepted. The coefficient path for organization commitment towards information quality is 0.408 with a t-statistic of 3.605, which is higher than 1.96. This can be concluded that there is significant positive influence between the organization commitments towards the information quality. This means that an increase in the organization commitment will result in a significant increase in the information quality. Thus the second hypothesis can be accepted. As for the organization commitment towards the ERP performance, the coefficient path is 0.274 with t-statistic of

1.730 < 1.96. This indicates that there is no significance between the organizational commitment and the ERP performance. Therefore, the third hypothesis is not accepted. The coefficient path for the influence of the upgraded ERP infrastructure towards the information quality is 0.408 with t-statistic of 5.009 > 1.96. This suggests that the increase in upgraded ERP infrastructure variable will result in the significant increase of the information quality. Thus, the fourth hypothesis can be accepted. The impact of the upgraded ERP infrastructure to the ERP performance is 0.054 with t-statistic of 0.329 < 1.96. This suggests that the increase in the upgraded ERP infrastructure will not be able to bring any significant increase in the ERP performance. In the case, the fifth hypothesis is rejected. The path coefficient for the information quality to the ERP performance is 0.389 with t-statistic of 2.199 > 1.96. This result concludes that there is a significant positive influence of the information quality to the ERP performance in manufacturing companies. An increased information quality will bring a significant increase in the ERP performance, which leads to the acceptance of the sixth hypothesis.

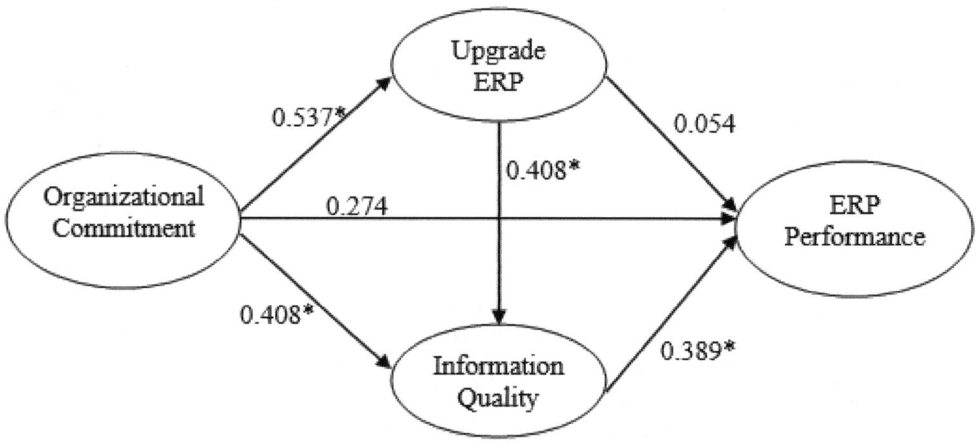

Figure 1. Full Model Effect Organizational Commitment to ERP Performance.

This result indicates that the total influence of the organizational commitment on the ERP performance with information quality and upgrade ERP are intervening variabel high as shown by the path coefficient of 0.518.

8. Discussion and conclusion

The ERP post-implementation requires active efforts to maintain its system stability to be able to produce improvement in the companies' performance. The research hypothesis tests are resulting in four accepted hypothesis and two rejected hypothesis. The first accepted hypothesis is the organizational commitment having a significant positive influence to the upgraded ERP infrastructure in East Java manufacturing companies. This influence indicates that organizational commitment in the East Java manufacturing companies, in the form of employees active involvement in the task completion and the absence of intention to leave the companies in the foreseeable future, will fasten the upgrade of the ERP system. This will improve the data integration and help the companies in making a more accurate decision. The second accepted hypothesis is the significant positive influence of the organizational commitment towards the information quality. This suggests that the organization commitment, in the form of active involvement and lack of desires to leave the company, will shape the data completeness on the ERP system and the good reliability of the system, will be preserved, which will positively contribute to a better proper decision. The third accepted hypothesis is the fast upgraded ERP infrastructure and better data integration will complete the data in the ERP system. This will result in the well-maintained reliance on the ERP system. This research is aligned with a research by Hsu et al., [9] which suggest a stable ERP system supporting a good information quality. The fourth accepted hypothesis is suggesting that the information quality which depends on a reliable well-maintained system will result in a lower level of inventory. The rejected hypothesis suggests that the organizational commitment would not directly impact to the ERP performance. This is probably caused by the fact that organizational commitment is merely a statement which would not produce a real result in terms of companies' operation without a tangible action in this case by having upgraded ERP infrastructure and through good information quality measurement would affect the sustainability of ERP system and would result in performance improvement [10]. This research could be further developed by focusing more on the level of management commitment in upgrading the ERP infrastructure.

References

[1] Mandal P and Bagchi K 2016 *Ind. Manag. Data Syst.* **116(6)** 1259-78.
[2] Bradley J 2008 *Int. J. Account. Inform. Syst.* **9(3)** 175-200.
[3] Matende S and Ogao P 2013 *Procedia Tech.* **9** 518-26.
[4] Nah F F H, Lau J L S and Kuang J 2001 *Bus. Proc. Manag. J.* **7(3)** 285-97.
[5] Ajer A K and Hustad E 2015 *Procedia Comp. Sci.* **64** 948–56.
[6] Zhu Y, Li Y, Wang W and Chen J 2010 *Int. J. Inf. Manag.* **30** 265–76.
[7] Yu C S 2005 *Ind. Manag. Data Syst.* **105(1)** 115-32.
[8] Nicolaou A I and Bhattacharya S 2006 *Int. J. Account. Inform. Syst.* **7** 18-35.
[9] Hsu P F, Yen H R and Chung J C 2015 *Inf. Manag.* **52(8)** 925-42.
[10] Marinagi C, Trivellas P and Reklitis P 2015 *Procedia Soc. Behav. Sci.* **175** 473-9.
[11] Shamala P, Ahmad R, Zolait A and Sedek M 2017 *J. Inf. Sec. Appl.* **36** 1-10.
[12] Ifinedo P, Rapp B, Ifinedo A, Sundberg 2010 *Comput. Human Behav.* **26(5)** 1136-48.
[13] Caki N, Asfuroglu L and Erbas O 2015 *Procedia Econ. Fin.* **26** 1007-13.
[14] Candra S 2012 *Procedia Soc. Behav. Sci.* **65** 141-9.
[15] Nahar N et al 2006 *Inf. Manag.* **43(5)** 663-77.
[16] Suprapto W, Tarigan Z J H and Basana S R 2017 *The influence of ERP system to the company performance seen through innovation process, information quality, and information sharing as the intervening variables* **ICEMT' 17 Proc.** 87-91.
[17] Park J H, Suh H J and Yang H D 2007 *Inf. Manag.* **44** 300-12.
[18] Zhou H, Shou Y, Zhai X, Li L, Wood C and Wu X 2014 *Int. J. Prod. Econ.* **147** 624–33.
[19] Wu J H and Wang Y M 2007 *Comput. Human Behav.* **23(3)** 1582-96.

The Effect of Procedure Change, TQM and ERP Implementation to Company Performance on Manufacturing Industries

Zeplin Jiwa Husada Tarigan[1, a*], Widjojo Suprapto[2, b*], Sautma Ronni Basana[3, c]

[1]Lecturer Master Management, Petra Christian University, Surabaya, Indonesia
[2]Lecturer Business Management, Petra Christian University, Surabaya, Indonesia
[3]Lecturer Finance Management, Petra Christian University, Surabaya, Indonesia

Abstract. The initial changes in standard procedures made by manufacturing companies when implementing ISO 9000: 2008 system (SNI/ National Indonesia Standard) are establishing procedures related to the system in each department of the companies. These changes are related to the standard operating procedures, work instructions, and forms. The second stage of changes happens during the implementation of the Enterprise Resource Planning (ERP) in which the companies have to integrate various departments into one integrated data system. These processes of changes in the companies are called business process redesign or re-engineering (BPR). The changes are conducted by making adjustments to all procedures in the companies so that they can fit into applying ERP modules. Proper adjustments will increase the employment performance for manufacturing companies. The data are collected from 100 questionnaires distributed to the manufacturing companies in East Java, and can further analyze from 77 questionnaires with a 77% response rate. Data processing is performed by using structural equation modeling of PLS and finds that BPR does not provide a direct improvement to the implementation of ERP in the companies. The changes to the standard procedure of the company, or BPR, give an impact of 0.422 to the implementation of ISO 9001. The implementation of ISO 9001 brings an impact of 0.763 on the implementation of ERP. The implementation of ISO 9001 does not affect the company performance, but the implementation of ISO gives a significant impact to the company's performance of 0.664. The changes in standard procedures performed in the company, often called BPR, can improve the company's performance by 0.2138, obtained through the implementation of ISO 9001 and the implementation of ERP.

1. Introduction

Business competitions among manufacturing industries are extreme, especially in producing competitive goods. The crucial point now for many companies is how they can build their competitive advantage by increasing their target markets and how they can implement changes to develop the business. One change management that is often done by those companies is by applying some business process mapping using the latest information technology. Restructuring business process systems are conducted by doing some rationalization for all related departments to interconnect data and integrate business processes. The radically redesigned business process is conducted to create a fast recovery, to increase business performance, to reduce the cost of production, to stabilize the product quality, to response the customers properly, and the increasing customer satisfaction through customer service. The total change to the whole system done by many companies brings extra values to the company's customers internally and externally. The stiff competition among global companies forces them to build effective and efficient business systems as their competitive advantages. In Indonesia, many companies install Total Quality management (TQM) systems to improve their standardized processes to guarantee the produced products [1]. The implementation of ISO can also improve the company's performance better and provide the strength for the companies to compete because the implementation of ISO can improve the credibility of the companies in their consumers.

One of the essential business process reengineering is redesigning business process and technology improvement. The implementation of the redesigned business process in the Indonesian manufacturing industries is carried out through the application of TQM (Total Quality Management). This is used to make improvements to product quality and company productivity. TQM is a comprehensive and cross functional system among departments within a company that is continuously controlled in making continuous improvements. TQM integrated with the Enterprise Resource Planning (ERP) system becomes an accelerated process for companies to make rapid system engineering changes. ERP is an integrated information technology system that is able to connect data of all departments in the company in the real time. Both TQM and ERP systems must be designed specifically to match with the appropriate standard operating procedures (SOPs) in order to support each other and to improve company's performance.

The implementation of ISO 9000: 2008 is one form of TQM concept execution in a management system. This management system requires that the organizational management must have the standard (clauses) as required by this ISO standard. These management system clauses consist of management responsibility clauses, resource management clauses, product realization clauses, measurement and evaluation clauses, and documentation system clauses. TQM system is the basis for the management in implementing the ISO 9001: 2008 Quality Management System. The implementation of TQM provides some improvements in the company performance, such as timely company product delivery, improvement in corporate innovation, decreased company overhead costs, and increased sales in the company [2]. Research conducted by Brah et al. [3] is comparing some companies in Singapore that are implementing ISO with the help of consultants and without using any consultants, and the result suggests that companies with ISO consultants perform better. While conducted by Joiner [4] states that consultants used by companies in Australia have a positive impact on the implementation of ISO. The use of tax consultants in companies will provide the accelerating process of financial reporting because the consultants have adequate experiences and understand the taxation rules well [5].

Nowadays, there is a system called TQM (Total Quality Management) with the SNI (Standard Nasional Indonesia) modification ISO 9000:2008 in Indonesia [6]. Often called as the integrated quality management system, this SNI or ISO 9000:2008 is the result of adopting and modifying the previous ISO, the standard ISO 9000:2008. This standard is used by many companies to measure the capability of an organization or manufacturing company to fulfill the customer requirements, the policies, the regulations, and the law in producing and marketing their product in Indonesia. This standard suggests adopting the process approach when planning, implementing and evaluating the effectiveness of the quality management system to meet customer satisfaction by fulfilling the order. The system in manufacturing companies in Indonesia is required by the government of Indonesia to adopt ISO so that the products can comply with the specifications that are set by the government through the Indonesian National Standard (SNI). Manufacturing products and imported goods must be declared to have ISOs so they can be marketed in Indonesia.

Total quality management (TQM) can be considered as the best and most effective way to reach the goals and objectives of educational institutions, as TQM contains the philosophy of continuous improvements which can provide practical tools for the educational institutions to fulfill the needs, wants, and hopes of their customers [7]. TQM is also one of the best ways to improve the competence of human resources and the quality of an organization so that it increases the product quality (both goods and services) that are done by all organization members who do the task [8]. Based on the latter statement, the product quality must be done by the whole organization or institution members, because the TQM cannot be implemented without cooperation with all organizational members.

A proper TQM implementation in manufacturing companies manages to organize interaction processes in the whole company through working procedures, job instructions, and work result forms. Several changes manufacturing companies have to make on their processing procedures, especially in undertaking their business activities processes. First of all, the marketing department has to make some adjustments to the procedures related to the standardized acceptance of customer order sales, standardized demand schedule management, standardized delivery order, standardized customer satisfaction. Second, the production and planning inventory control (PPIC) departments have to standardize their activities related to the production schedule procedure, inventory control procedure, production reporting procedures, and others. Third, the warehouse departments have to alter their procedures, primarily related to the material retrieval procedures, material return procedures, inventory checking procedures, material storage procedures, standardized inventory level procedures, and others. Fourth, the purchasing departments have to make some adjustment related to the standardized purchase procedure of material, the standardized procedure of material acceptance, the standardized procedure of making the material name. The alterations are also for other departments that are related to the implementation of ISO 9000.

Besides implementing ISO, most companies also implement ERP technology to integrate all data from various departments so that the top management can make an accurate decision. To implement the ERP, a company has to redesign all business processes to match with the software used by the company [9]. The alterations made by the company are tailored to the ERP modules which are used in each department. The intake of data has been done manually, and the entry to the computer software system is still departmental. There are still many manufacturing companies that each department uses information technology systems that cannot be integrated with other departments. The implementation of integrated ERP will result in the adjusted data in each department so that other departments can use the data, and all departments must agree on the data integration so that the implementation can be appropriately conducted. The companies have to ensure that all departments have agreed upon the standard procedures for all processes so that the disputed opinion among departments can be avoided, and this whole process is called business process redesign or re-engineering. ERP is software to integrate all data that are related to production processes in the company [10]. With the implementation of ERP, the top management can build the operating system to monitor the work procedures and can also get the latest integrated data to make accurate and fast decisions. With the right decision making and well-maintained procedures, the top management can boost the company performance. The practical importance of this research is to implement ERP in manufacturing companies. In implementing the ERP, many companies create new problems which are caused by the different ISO system and the abundant numbers of personnel. The personnel pay more attention to the ERP system but neglect the ISO system. Therefore, many standard operating procedures are ignored to implement the ERP. With the two systems are implemented in one company, the synchronization of two systems is needed, especially in documenting systems. The previous researches do not pay much attention to the successful implementations of two systems, ERP and ISO, which will bring impacts to the company's performance.

2. Research hypothesis

To increasing the competitive advantage by improving the company performance can be accomplished by implementing enterprise resources planning (ERP) technology and total quality management (TQM). The ERP implementation is aimed to create efficiency by making decisions accurately and fast, while the TQM implementation is to create standardized and consistent processes. Implementing these two systems are very complex because a company needs to spend plenty of time and money starting from pre to post implementing activities. In preparing to implement both TQM and ERP, the company needs to conduct an appropriate redesign business process or BPR so the operational process can be synchronized with the software in the ERP system. Enterprise Resource Planning (ERP) is one way to manage company's resources using the information technology. The ERP technology can coordinate and integrate data from different business divisions in the company so the given information can be shared entirely to all departments, especially the top management in making appropriate business decisions.

The ERP implementation requires the related documents, primarily the corporate business blueprints, and the quality manual. The ISO quality manual is one document which contains the implementation guidelines of a company and becomes the primary requirement for the ISO implementation. Conducting all activities are described thoroughly in standard operating procedures (SOP), job instructions, and various forms to record quality in the company. Best practices in the companies can be developed from the lessons learned in implementing ISO 9000, which are documented using ERP system so that they can contribute a strong relationship with the indicators to measure the financial performance of the companies.

In preparing to implement both TQM and ERP, the company needs to conduct an appropriate redesign business process or BPR so the operational process can be synchronized with the software in the ERP system. The initial change that the companies have to do is to customize the business practice so that the software system complies with the ERP system. The success of implementing ERP by doing some changes to the adjusted business process system is called BPR [11]. The integrated information technology (ERP) implementation needs alterations to the business process to increase performance [12]. Research conducted by Lee et al. [13] state that BPR brings a positive impact on the implementation of technology in creating business innovations. The BPR (business process re-engineering) in a company will create customized product innovations that possess new standard operating procedures. The changes of the business process are usually conducted by altering procedures, work instructions, and forms to obtain new ISO 9000:2008. SNI is often considered as the Total Quality Management system in manufacturing industries in Indonesia. Research by Costa et al. [14] state that one benefit for a company, that is doing some alterations on the procedures and documents, and is creating new standard operating procedures, is the positive impacts on the changing organizational culture. BPR (Business process re-engineering) is used to redesign the continuous improvement process in companies [15]. A standardized system in a company will ease the end users of ERP to enter the data into the software, so all data are also standardized for the needs of the company. The decision to use the standardized process is used to implement ERP for better company performance and cutting costs [16]. Research by Tsai et al. [17] mentions that the ERP system needs a precise procedure and standardized forms as the integrated data solution. The standardized process with real-time data in an organization can speed up the data integration in a company, so the ERP implementation is faster and smoother [18]. A company must produce standardized goods that comply with the requirement set by the customers. Those products must not violate the law within the jurisdiction areas or regions where the company is located. One research states that the standardized materials ordered by the purchasing department must follow the needs of the production department and bring positive impact to the business performance of the company. ERP is an information technology that can integrate all departments in a company. Kang et al., [19] reveal that the ERP implementation gives a positive impact on the company performance. Zhang et al., [20] mention that the ERP function with the help of application software can aid organizations to control the business because it can reduce inventory, reduce orders, increase work productivity, maximize communication level, and increase profits. Five hypothesis for this research:

H_1: The business process re-engineering influences on the ERP implementation.
H_2: The business process re-engineering influences the ISO 9000:2008 (SNI).
H_3: The SNI ISO 9000:2008 has an impact on the ERP implementation.
H_4: The SNI (ISO 9000:2008) has an impact on the enterprise performance.
H_5: The implementation of ERP brings impacts to the enterprise performance.

3. Research method

The best practice used by manufacturing companies in Indonesia is obtained from the implementation of SNI and ERP. The manufacturing companies concentrate on the quality management system by focusing on product quality. This is causal research that investigates the relationship among BPR, ISO 9000:2008 (SNI), ERP, and the performance in manufacturing companies in East Java. This research is using the quantitative method. The quantitative research is a research that uses empirical approaches to collect, analyzes, and presents the numerical data. According to Cooper and Schindler [21], quantitative research is a research that tries to measure a phenomenon accurately.

To measure accurately, this research is using some analytical tools; among others are the analytical indexing tool and Structural Equation Modeling (SEM) Partial Least Square (PLS). The analytical indexing tool is used to measure every indicator in each variable. Meanwhile, SEM is used to analyze the causal structural relationship with the help of Smart PLS. The questionnaires are directly distributed and collected by the researcher from 100 companies in East Java. The returned questionnaires are 80, but three questionnaires are not valid because they are not entirely marked. Then, the data are collected and further processed from 77 questionnaires, with a 77% response rate.

4. Finding and discussion

Based on the collected questionnaires, it is obtained that the full structural model is reflected inner model in table 1. After doing the whole bootstrapping data, to test the hypothesis comes as follows:

Table 1. The Result of the Inner Model and the Hypothesis Testing.

Effect of Variable	Original sample estimate	Mean of subsamples	Standard deviation	T-Statistic
BPR -> ERP	0.108	0.079	0.094	1.144
SNI -> ERP	0.763	0.783	0.061	12.610
BPR -> SNI	0.422	0.501	0.071	5.983
ERP -> Performance	0.664	0.649	0.180	3.689
SNI -> Performance	0.051	0.112	0.190	0.270

From the results in table 1, the first hypothesis (H1) is rejected because the t-statistic value (1.144) is less than 1.96; therefore, there is no influence between Business Process Re-engineering and Enterprise Resources Planning in manufacturing industries in East Java. This is caused by some certain circumstances happening in the business environment. However, the changes are not adjusted to the ERP system, such as the requests from vendor software are not fully justified, instead the software is merely installed to integrate the system. As the result, the vendor software cannot conduct many analyses that are happening in the company. This research is different from the result of a research conducted by Schniederjans and Kim [22] stating that BPR is essential in implementing ERP because it is useful to integrate all departments to produce an effective and efficient process. This research is also different from the results stating that BPR can have a positive impact on the implementation of information technology [23, 24]. Meanwhile, the second hypothesis (H2) is accepted with the t-statistic value 5.983 greater than 1.96. Therefore Business Process Re-engineering influences ISO 9000:2008 (SNI). This result shows that the working alteration is only changing the working system through altering work procedures, work instructions, and work forms. This condition is caused by the implementation of BPR that has to comply with the minimum requirements of SNI ISO 9000:2008. This minimum requirement of SNI is also acting by the Indonesian government regulation in producing manufacturing goods. This research supports the result of research by Li et al., [25] describing that BPR is central to changes and adjustments to the TQM implementation. BPR is one of the three principles of TQM. The reengineering process is a reflection of an improvement on the company's business process and company procedures to support the company's operations.

The third hypothesis (H3) is accepted with the t-statistic value of 12.610, which is greater than 1.96, so ISO 9000:2008 (SNI) has an influence on Enterprise Resources Planning in manufacturing industries, however, the established standard cannot increase the company performance directly. This condition happens because the business process, standard operating procedures, and various working forms needed by the ERP are available and complete when implementing ISO 9000:2008 (SNI). In building the ERP system, the key users have already understood the system as they are accustomed to construct it in implementing SNI. This research also supports the result of Loukis et al., [24, 25] which states that the implementation of TQM can have a positive impact on the implementation of information and communication technology (ICT), and can increase the added value of the company.

This is confirmed by the rejection of the fourth hypothesis. The fourth hypothesis (H4) is rejected because the t-statistic value is .270, which is less than 1.96. SNI ISO 9000:2008 does not influence enterprise performance in manufacturing industries. This happens because the changing and adjusted business process are merely bringing some paper works which surrender orderliness in the production reports to the top managers. The orderliness of production reports does not surrender better productivity as the production processes in manufacturing industries still rely heavily on manual processes or human resources. The result also supports Flynn and Flynn [28] statements that the implementation of TQM is not capable of impacting the company's performance because its system has not been able to run properly by employees.

The fifth hypothesis (H5) is accepted with the t-statistic value of 3.689, which is higher than 1.96. Therefore, it can be said that the implementation of ERP in manufacturing companies in East Java brings a positive and significant impact to the company performance. The implementation of ERP for manufacturing companies brings many benefits, especially in the quick data analysis so that it speeds up the decision making by the top management. This research is in accordance to some previous researches that the implementation of ERP in manufacturing companies can bring positive impacts to the company performance [18- 20, 22]. Based on the result of hypothesis testing (H1 to H5), it can be summarized in figure 1. That depicts the total research results after conducting all hypothesis tests.

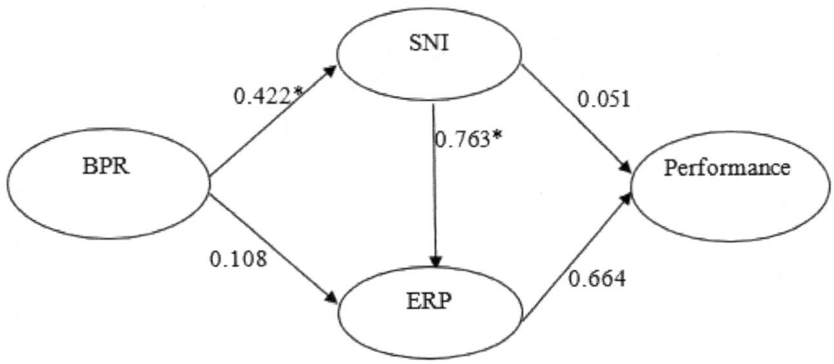

Figure 1. The effect of BPR, SNI, and ERP on Company Performance.

Based on figure 1, it can be concluded that there is no significant influence between BPR and ERP in a company, but there is an influence between BPR and TQM, in this case, is represented by the implementation of ISO 9000:2008 (SNI), which is caused by the alteration of work interaction among departments. TQM through the restructuring work standard, work procedures, work instructions, and work forms can bring a positive and significant impact on the implementation of ERP, but cannot improve the company performance. The companies must be able to restructure and redesign the SNI system so that the documentation systems needed by the ERP implementation are well managed and controlled. The well-managed ERP system can increase the information quality improvements on all business functions in the companies, such as the functions of finance, operations and productions, inventory, marketing, and human resources. Such a condition will create good business networks, increase company efficiency, and business target achievement. Business process re-engineering which is conducted through SNI and BPR will synchronize the data system and will bring the best practices for the companies. Finally, the implementation ERP gives impacts to the improvement of the manufacturing company performance.

5. Conclusion

From the data analysis, it can be concluded that there is no significant influence between BPR and ERP in a company, but there is an influence between BPR and TQM, represented by the implementation of SNI ISO 9000:2008, which is caused by the alteration of work interaction among departments. TQM through the restructuring of work standards, work procedures, work instructions, and work forms can bring a significant positive impact on the implementation of ERP, but cannot improve the company performance. Finally, ERP gives impacts to the improvement of the company performance. The next step for this research will add another variable, that is the organizational culture variable, to investigate the role of the culture in driving the implementations. The implementation of TQM and ERP together will have an impact on the system changes in the company so that the product data obtained by the marketing department can be understood by the engineering design department and the production department as well as in the industrial engineering section. Orders from the customer requests can be granted by the company and can comply to the product specifications and product design requested by the SNI.

References

[1] Syduzzaman R, Islam M, Habib M and Habib S 2014 *European Sci. J.* **10(34)** 341-58.

[2] Munizu M 2013 *IOSR J. Bus. Manag.* **9(2)** 55-61.

[3] Brah S A, Lee S L and Rao B M 2002 *Int. J. Quality Reliabil. Manag.* **19(4)** 356-79.

[4] Joiner T A 2007 *Int. J. Quality Reliabil. Manag.* **24(6)** 617-27.

[5] Rotimi O, Udu U S A and Abdul-Azeez A A 2013 *European J. Bus. Soc. Sci.* **1(10)** 25-35.

[6] Badan Standarisasi Nasional Indonesia (BSNI): *Sistem Manajemen Mutu Persyaratan, SNI ISO 9001:2008*, bsn. go. id/index. php?/sni_main/sni/detail_sni/9511.

[7] Zabadi M A 2013 *J. Fin. Econ.* **1(1)** 42-60.

[8] Zehir C, Ertosun O G, Zehir S and Muceldilli B 2012 *Proc. Soc. Behav. Sci.* **41** 273-80.

[9] Greasley A and Wang Y 2017 *Bus. Proc. Manag. J.* **23(1)** 2-15.

[10] Garg P and Garg A 2013 *Bus. Proc. Manag. J.* **19(3)** 496-514.

[11] Huq Z and Martin T N 2006 *Bus. Proc. Manag. J.* **12(5)** 576-87.

[12] Shin I 2006 *Small Bus. Econ.* **26(3)** 241-56.

[13] Lee Y C, Chu P Y and Tseng H L 2009 *J. Manag. Organ.* **15** 170-85.

[14] Costa M M, Lorente M R A and Choi Y T 2008 *Int. J. Prod. Econ.* **113** 23-8.

[15] Jain R, Chandrasekaran A and Gunasekaran A 2010 *Benchmarking: an Int. J.* **17(1)** 77-94.

[16] Fulford R 2013 *J. Syst. Inform. Tech.* **15(1)** 117–32.

[17] Tsai W H, Shaw M J, Fan Y W, Liu J Y, Lee K C and Chen H C 2001 *Dec. Sup. Syst.* **50(2)** 480-90.

[18] Nikookar G, Safavi S Y, Hakim A and Homayoun A 2010 *Inf. Syst.* **35** 271-7.

[19] Kang S, Park J H and Yang H D 2008 *J. Comp. Inf. Syst.* **48(4)** 25-38.

[20] Zang Z, Lee M K O, Huang P, Zhang L and Huang X 2005 *Int. J. Prod. Econ.* **98** 56-80.

[21] Cooper D R and Schindler P S 2008 *McGraw Hill* **10 ed**

[22] Schniederjans M J and Kim G C 2003 *Int. J. Operat. Prod. Manag.* **23(4)** 418-29.

[23] Loukis E, Pazalos K and St. Georgiou 2009 *J. Enterprise Inform. Manag.* **22(5)** 564-86.

[24] Ramirez R, Melville N and Lawler E 2010 *Dec. Sup. Syst.* **49(4)** 417-29.

[25] Li L, Markowski C, Xu L and Markowski E 2008 *Int. J. Prod. Econ.* **115(2)** 569-80.

[26] Flynn B and Flynn E 2005 *Int. J. of Prod. Res.* **43(16)** 3421-36.

The Cause Analysis of Collapse of a Granary

Guoliang Zhu[1]

[1]GuiZhou Construction Science Research & Design Institute Co., Ltd, GuiZhou, China

zhuguoliang2371@sina.com

Abstract. The direct cause of the collapse of granary is irregular design, old breakage, lack of routine maintenance, and improper operation in the process of staff operation. The indirect reason for the collapse of granary is the customary operation of staff. If the grain height exceeds the grain line, it goes beyond the capacity allowed by the warehouse design, overloading the grain, and its side pressure exceeds the strength that the wall can bear, resulting in cracks and even collapse in the wall of the bungalow, which eventually leads to the occurrence of a buried accident. A granary was built in the 80s of last century, and it collapsed in the process of stacking grain. This paper analyzes the causes of collapse from the aspect of stress, according to the exploration situation, the actual calculation model is determined and the force calculation is carried out. The calculation parameters should be collected in force analysis, including: section size, strength, load and structural layout.

1. Introduction

The direct cause of the collapse of granary is irregular design, old breakage, lack of routine maintenance, and improper operation in the process of staff operation. Many accidents in the warehouse are caused by improper grain removal methods and the workers are buried by grain. Most of the hidden warehouses are hidden in the construction. Most of the warehousing enterprises where the grain is buried in the warehouse are mostly due to the use of the grain plate instead of the fixed type of grain retaining door, the grain plate needs to be manually disassembled, and the design of the grain retaining plate is unreasonable and there is no fixed safety rope (belt) on the top of the warehouse. The safety device is obviously insufficient. It promotes the occurrence of the accident; if the grain height exceeds the grain line, it exceeds the capacity allowed by the warehouse design, overloading the grain, and its side pressure exceeds the strength that the wall can bear, resulting in cracks and even collapse in the wall of the bungalow, which eventually leads to the occurrence of a buried accident, and the hidden danger of the silo, grain. In the process of feeding out the storehouse, once the cone hopper is not fixed well, the collapse of the steel plate or the falling of the cone can lead to the collapse and burial of the grain pile, and the hidden danger of the shallow round barn. In the process of grain production, an inverted conical funnel will be formed on the surface of the grain pile, and the "active funnel" on the surface of the grain pile can drag into any object in the grain pile and cause a major safety hazard to the production operators.

The indirect reason for the collapse of granary is the customary operation of staff. In recent years, accidents have found that most of the casualties are foreign workers, and most of them are outsourced workers, foreign construction workers and foreign grain sales personnel. Because of the relatively tired work of the grain depot, the outsourced personnel engaged in work are generally older and less educated, and the safety production training organized by the safety production supervision and management department is difficult to accept for them. The accidents are difficult to avoid when the foreign workers who are almost safely producing "zero training" do their homework. At the same time, many warehousing enterprises often do not operate according to rules and regulations, such as unguardianship on the job site, dismantling the food and other operations without safety rope (belt) and so on, which will cause the occurrence of accidents of grain pile buries.

The granary building was built in the 80s of last century. Its structural form is single brick wood structure, and its walls are fired brick walls. There are no structural measures such as gird. The triangular wooden frame, the purling of the gables, is directly shelved on the mountain wall by hard mountain purling. The thickness of the wall below the grain line is 370mm, and the thickness of the wall above the grain line is 240mm. The collapse of the barn is located at the end of the building, mainly for gables and roof spanning. The collapse of the gable is 12.63m and there are no girds in the middle. There are only three wooden columns with diameters of 180 to 200mm.

2. Analysis of force mechanism

The wall is the important bearing and maintenance component of the barn of the flat house, bearing the load of wind load and grain side pressure in the grain height range. The traditional cottage granary buildings are mostly built with brick concrete structure. Due to the special advantages of the grain storage function of the granary building and the unique advantages of the clay brick in moisture proof, moisture absorption and heat preservation. The granary is mostly fired clay brick, and its construction depends largely on the height of the pile. When the height of stacked grain is high, the cross section of the wall is thicker, and the vertical load of the upper structure is transferred to the foundation. The lateral pressure of grain is transferred to the silo wall by masonry, coupling beams, structural columns and foundation. In the vertical direction, the friction force is generated when the structure is self weight and the grain is transported, so the wall is a compression bending member. Because the grain friction is small and beneficial to the force of the silo wall, it can be calculated according to the bending member.

The middle of the caving wall of the collapsed granary is not set in the middle of the wall, and only three logs with a diameter of 180 to 200mm are set in the middle, so the central wall can be analyzed by the cantilever member embedded in the lower part and does not consider the pulling effect of the purling on the wall. The two ends of the gable are subjected to force analysis according to the lower part and one side articulated member. The most unfavorable part of the gable is the middle part, and the unit width is adopted to analyze the force.

The strength of the wall material is measured by the actual strength standard. The strength of masonry mortar is extracted by the penetration method and the strength of the masonry mortar is detected by the penetration method, and the strength of the block is by the rebound method [1]. Before the collapse, barns were stacked with sorghum, and the standard combination was applied to the wall.

3. Calculation of the actual bearing capacity of the wall

The actual strength of masonry mortar is 0.4Mpa and the actual strength grade of brick is MU7.5.

The calculation of bearing capacity of non-reinforced masonry member in bending is based on the following formula [2]:

$$M \leq fW \tag{1}$$

The shear capacity of non-reinforced masonry member in bending is calculated according to the following formula [2]

$$V \leq f_v bz \tag{2}$$

$$f = k_4 \sqrt{f_2} \tag{3}$$

$$z = \frac{I}{S} \tag{4}$$

$$f_v = k_5\sqrt{f_2} \tag{5}$$

Flexural capacity of 370mm wall:

$$M = fW = 2.08kN.m \tag{6}$$

Shear capacity of 370mm wall:

$$V \leq f_v bz = 22.52kN \tag{7}$$

In the last-written formulas, M is the bending moment. M is the shearing force. b is the cross section width. h is the cross section height. W is the cross section resistance moment, for rectangular cross section, $W=bh^2/6$.

f is the flexural tensile strength of masonry is taken as the tensile strength of the longitudinal joint.

f_v is the shear strength of masonry.

z is the internal force arm, for rectangular cross section, z=2h/3.

I is the section inertia moment.

S is the Section area moment.

4. Loading calculation of sorghum on wall

As the granary is a masonry barn, the roof is a wooden house cover and a non reinforced concrete granary, the simplified calculation method is used for the mining of the barn, which does not consider the continuity of the storehouse along the height direction, the upper part of the silo wall, the wall section of a silo that does not bear the horizontal pressure stress in the lower part of the barn wall, and the constraints of the simplified roof and the floor [3]. The horizontal pressure standard value and vertical friction standard value of grain acting on the unit area of the warehouse wall at S depth is calculated according to the following formula [4]:

$$P_h = k\gamma s \tag{8}$$

$$P_t = k\gamma s \tan\delta \tag{9}$$

$$k = \frac{\cos^2\varphi}{\cos\delta\left[1+\sqrt{\frac{\sin(\varphi+\delta)\sin\varphi}{\cos\delta}}\right]^2} \tag{10}$$

$$k = \frac{\cos^2\varphi}{(1+\sin\varphi)^2} \tag{11}$$

In the last-written formulas, P_h is the standard value of horizontal pressure acting on the unit area of the warehouse wall.

P_t is the standard value of vertical friction acting on the unit area of the warehouse wall.

k is the side pressure coefficient of grain. When calculating horizontal pressure, use equation 10 and rest equation 11.

s is the distance from the top of the grain to the calculated section, unit is meters.

γ is the grain gravity density, unit is kN/m^3.

φ is the internal friction angle of grain.

δ is the external friction angle of the grain to the wall brush surface.

The stowage in the granary is sorghum, physical parameters [5] is:

$\varphi = 34^0$

$\delta = 27^0$

$\gamma = 7.74kN/m^3$

The bending moment of sorghum on the calculation depth of the cross section of the silo wall is:

$$M' = \frac{1}{6}P_h s^2 - 0.7\int_0^s p_t \frac{0.37}{2}dx \tag{12}$$

The shearing force of sorghum on the calculation depth of the cross section of the silo wall is:

$$V' = \frac{1}{2}P_h s \tag{13}$$

5. Allowable heap height calculation

According to flexural capacity:

$$\begin{aligned} M = 2.08kN.m \geq M' &= \frac{1}{6}P_h s^2 - 0.7\int_0^s p_t \frac{0.37}{2}dx = \frac{1}{6}k\gamma s^3 - 0.7\int_0^s k\gamma s \tan\delta \frac{0.37}{2}dx \\ &= \frac{1}{6}k\gamma s^3 - 0.7\int_0^s k\gamma s \tan\delta \frac{0.37}{2}dx \end{aligned} \tag{14}$$

Figure out s ≤ 1.85m

According to shear capacity:

$$V = 22.52kN \geq V' = \frac{1}{2}P_h s = \frac{1}{2}k\gamma s^2 \tag{15}$$

Figure out s ≤ 4.06m

Taking the smaller values of the two, the actual allowable heap of Sorghum s=1.85m, see table 1 for details.

Table 1. Allow high calculation results of heap height.

Computational state	Allowable heap height (m)	Final value (m)
According to flexural capacity	1.85	1.85
According to shear capacity	4.06	

6. Cause analysis

If the grain height exceeds the grain line, it will exceed the capacity allowed by the warehouse design, and its side pressure exceeds the strength that the wall can bear, resulting in cracks and even collapse in the wall of the bungalow, which eventually leads to the occurrence of a buried accident.

According to the calculation of warehouse receipts, 200741 kg of sorghum was stacked before the barn collapsed, and the average grain height was 2.82m.

Before the collapse of the barn, the height of the sorghum pile is 2.82m, which exceeds the permitted height of 1.85m, resulting in the collapse of the gable. In addition, no structural measures such as gird and structural column can be set up in granary, which can not provide beneficial effects for gables and delay its destruction and collapse.

Summary

1. After calculation, the height of the sorghum before the collapse of the granary was 2.82 m, which exceeded the allowable height of 1.85 m, resulting in the out-of-plane collapse of the gables, which was the direct cause of the collapse of the granary.

2. The structure measures such as ring beams and structural pillars are not set up in the granary, which can not provide a beneficial effect for the gables and can not delay their destruction and collapse.

3. This analysis shows that the calculation parameters should be collected first, including the section size of the component, the strength of the material, the load action and the arrangement of the structure, and then the calculation model is determined according to the actual structural arrangement. The calculation model can be determined according to the code or structural mechanics. The calculation results should be analyzed to see if they are consistent with the actual situation, and if they are not, the correctness of the parameters should be checked.

References

[1] He X H, Di X T and Gao X W 2004 *Technical standard for inspection of building structure*.**2004**:28-33.

[2] Gao L Y, Xu J and Wan Z F 2011 *Code for design of masonry structure*. **2011**:35-38.

[3] Cai J Y, Jiang C R, Hou F Z and Qiu H T 1999 *Internal force analysis of lateral pressure of reinforced concrete silo*, Proceedings of the Eighth National Conference on Structural Engineering (first volume):132-136.

[4] Wang Z Q and Yang M Y 2014 *Code for design of grain storehouses*. **2014**:18-21.

[5] Jin X Y, Wang J and Wang G Y 2012 *Load code for the design of building structures*. **2012**:79-82.

Institute of Physics
Dirac House, Temple Back
Bristol BS1 6BE UK

ISSN: 1757-8981
ISBN 978-1-5108-8219-5

CURRAN ASSOCIATES INC.
proceedings.com